# Genetic Disorders, Syndromology and Prenatal Diagnosis

**Advances in the Study of Birth Defects**

VOLUME 5

# Genetic Disorders, Syndromology and Prenatal Diagnosis

EDITED BY

**T. V. N. Persaud**

**MTP** **PRESS LIMITED**
*International Medical Publishers*

Published by
MTP Press Limited
Falcon House
Lancaster, England

First published 1982

**British Library Cataloguing in Publication Data**
Genetic disorders, syndromology and prenatal
    diagnosis.—(Advances in the study of
    birth defects; v. 5)
    1. Abnormalities, Human
    I. Persaud, T.V.N.        II. Series
    616'.043        RG626

ISBN-13: 978-94-011-6671-3    e-ISBN-13: 978-94-011-6669-0
        DOI: 10.1007/978-94-011-6669-0

Typeset by Macmillan India Ltd., Bangalore and
printed by Robert MacLehose & Co. Ltd., Renfrew, Scotland

# Contents

# List of Contributors

**E. T. BERSU**
Department of Anatomy
University of Wisconsin
Madison, Wisconsin 53706, USA

**M. M. COHEN Jr.**
School of Dentistry
Health Sciences Building, SB-24
University of Washington
Seattle, Washington 98195, USA

**L. D. EDMONDS**
Birth Defects Branch
Bureau of Epidemiology
Center for Disease Control
Atlanta, Georgia 30333, USA

**JANE A. EVANS**
Department of Pediatrics (Genetics)
University of Manitoba
Faculty of Medicine
Winnipeg, Manitoba, Canada R3E OW3

**P. K. GHOSH**
Cytogenetics Laboratory
Department of Anthropology
University of Delhi
Delhi-110007, India

**KATERINA HAKA-IKSE**
Department of Pediatrics (Neurology
Division)
The Hospital for Sick Children
555 University Avenue
Toronto, Ontario, Canada M5G 1X8

**I. HALBRECHT**
B. Gattegno Research Institute of Human
Reproduction and Fetal Development
Hasharon Hospital
Petah-Tiqva, Israel

**I. LANGE**
The Division of Maternal Fetal Medicine
Department of Obstetrics and Gynaecology
Women's Centre, Health Sciences Centre
The University of Manitoba
Winnipeg, Manitoba, Canada

**P. M. LAYDE**
Birth Defects Branch
Bureau of Epidemiology
Center for Disease Control
Atlanta, Georgia 30333, USA

**LENORE S. LEVINE**
Department of Pediatric/Pediatric
Endocrinology
The New York Hospital
Cornell University Medical College
New York, New York 10021, USA

**F. A. MANNING**
The Division of Maternal Fetal Medicine
Department of Obstetrics and
Gynaecology
Women's Centre, Health Sciences Centre
The University of Manitoba
Winnipeg, Manitoba, Canada

**M. S. MATTEVI**
Departamento de Genética
Instituto de Biociências
Universidade Federal do Rio Grande do Sul
Caixa Postal 1953
90000 Porto Alegre
RS, Brazil

**D. T. MININBERG**
Department of Surgery/Urology
The New York Hospital
Cornell University Medical College
New York, New York 10021, USA

**REITA NAND**
Cytogenetics Laboratory
Department of Anthropology
University of Delhi
Delhi-110007, India

**MARIA I. NEW**
Department of Pediatrics/Pediatric
Endocrinology
The New York Hospital
Cornell University Medical College
New York, New York 10021, USA

**J. C. PETTERSEN**
Department of Anatomy
University of Wisconsin
Madison, Wisconsin 53706, USA

**BETTY J. POLAND**
Department of Obstetrics & Gynecology
Faculty of Medicine
The University of British Columbia
Vancouver General Hospital
Vancouver, B.C., Canada V5Z 1M9

**M. RAY**
Division of Genetics
Health Sciences Centre
700 William Avenue
Winnipeg, Manitoba, Canada R3E OZ3

**F. M. SALZANO**
Departamento de Genética
Instituto de Biociências
Universidad Federal do Rio Grande do Sul
Caixa Postal 1953
90000 Porto Alegre
RS, Brazil

**F. S. SHABTAI**
B. Gattegno Research Institute of Human
Reproduction and Fetal Development
Hasharon Hospital,
Petah-Tiqva, Israel

**A. SHALEV**
Department of Genetics
The Hebrew University of Jerusalem
Jerusalem, Israel

**NANCY E. SIMPSON**
Department of Pediatrics
Queen's University
20 Barrie Street
Kingston, Ontario, Canada K7L 3N6

# Preface

Birth defects have assumed an importance even greater now than in the past because infant mortality rates attributed to congenital anomalies have declined far less than those for other causes of death, such as infectious and nutritional diseases. As many as 50% of all pregnancies terminate as miscarriages, and in the majority of cases this is the result of faulty intrauterine development. Major congenital malformations are present in at least 2% of all liveborn infants, and 22% of all stillbirths and infant deaths are associated with severe congenital anomalies. Not surprisingly, there has been a great proliferation of research into the problems of developmental abnormalities over the past few decades.

This series, *Advances in the Study of Birth Defects*, was conceived in order to provide a comprehensive focal source of up-to-date information for physicians concerned with the health of the unborn child and for research workers in the fields of fetal medicine and birth defects. The first four volumes featured recent experimental work on selected areas of high priority and intensive investigation, including mechanisms of teratogenesis, teratological evaluation, molecular and cellular aspects of abnormal development, and neural and behavioural teratology. It seems logical and timely that the clinical aspects should now be presented. Accordingly, leading experts were invited to review a broad range of common problems from the standpoint of embryology, aetiology, clinical manifestations, diagnosis and management. This volume deals with genetic disorders and prenatal diagnosis.

I am greatly indebted to the distinguished panel of contributors. Their enthusiasm, many useful suggestions and cooperation have made this volume a reality. My sincere thanks go to the publishers, especially Mr D. G. T. Bloomer, Managing Director, MTP Press Limited, for their encouragement and for extending to me every kindness. Once again, I owe much to my secretary, Mrs Barbara Clune, for her invaluable and unstinting help. I should also like to thank Mr Roy Simpson, medical photographer, for his assistance with several of the illustrations.

This work is affectionately dedicated to Indrani, Sunita, and Ren.

Winnipeg, Canada
May, 1981

T. V. N. Persaud

# 1
# Chromosomal abnormalities in single gene disorders

## M. RAY

The cytogenetics of several recessive inheritance disease states in human will be discussed. These diseases belong to the categories of (1) autosomal recessive and (2) X-linked recessive disorders.

## AUTOSOMAL RECESSIVE DISORDERS

A trait transmitted as an autosomal recessive is expressed only in those individuals who have received the recessive gene, one from each parent, thus becoming homozygous for that gene. Typically the trait appears in some of the sibs of the propositus but not in relatives outside the sibship and the parents are not affected. Nearly 1100 clinical disorders with autosomal recessive inheritance have been recognized[1].

### Chromosomal breakage syndromes

A number of rare inherited disease states in man have been found to have markedly increased frequencies of chromosome aberrations in peripheral blood lymphocytes and cultured fibroblasts. These disorders (autosomal recessive) are known as 'chromosomal breakage syndromes'. The term was coined by Dr Victor A. McKusick[2]. The best known of these syndromes are (1) ataxia telangiectasia (Louis–Bar syndrome)[3,4], (2) Bloom's syndrome[5], (3) Fanconi's anaemia[6,7] and (4) xeroderma pigmentosum[8,9].

*Clinical symptoms*
*Ataxia telangiectasia syndrome*
This disorder is characterized by an overall hypoplasia of the individual and immunological deficiency with abiotrophic features of deterioration in the central nervous system, skin and respiratory tract.
*Bloom's syndrome*
Facial telangiectatic erythema which involves the butterfly area and is exacerbated by exposure to sunlight usually develops in the first year of life.

Café-au-lait spots, variable microcephaly, malar-hypoplasia and prenatal onset of shortness of stature are other identifying symptoms.

### Fanconi's anaemia

This disease is marked by pigmentation of skin, short stature, small cranium, mental retardation, strabismus, abnormal ears, abnormal thumbs, renal anomaly and hypoplasia of marrow with time.

It is of interest to note that certain common features exist between Fanconi's anaemia, Bloom's syndrome and ataxia telangiectasia. In each of these disorders there occurs a generalized growth retardation, skin disorder and propensity to develop lymphoreticular malignancy, with a high frequency of chromosomal breakage in cultured leukocytes and fibroblast cells.

### Xeroderma pigmentosa syndrome

This syndrome is characterized by undue sunlight sensitivity, atropic and pigmentary skin changes, and actinic skin tumours.

Xeroderma pigmentosum is caused by several non-allelic autosomal recessive genes, each of which results in defective repairs of u.v.-light induced damage to DNA when present in the homozygous state[8,9]. This syndrome is associated with defects in the repair of DNA after exposure to ultraviolet light. The natural history in this case is progressive disfigurement, tumours often by 3–4 years of age and most die of malignancy before 20 years.


### Conventional chromosome analysis

The vast majority of cells tested had a modal number of 46 chromosomes with a normal karyotype. The structural rearrangements were seen in some cells (Figure 1.1). The majority of the aberrations (gaps and breaks) consist of chromatid and chromosome types (Figure 1.2).

Lymphocytes from ataxia telangiectasia patients showed approximately 50 % fewer chromosome damage (0.18 breaks/cell) than those patients with Bloom's syndrome (0.34 breaks/cell) or Fanconi's anaemia (0.37 breaks/cell)[10].

Elevated levels of chromosome damage were found in fibroblasts of individuals with Bloom's syndrome and Fanconi's anaemia. Studies of lymphocytes from ataxia telangiectasia propositi showed a 50 % reduction in chromosome damage when compared with the other syndromes, whereas fibroblast studies on similar patients showed chromosome damage in excess. The average rate of chromosomal damage in fibroblast cells of the ataxia telangiectasia propositi was 0.88 breaks/cell, while in Bloom's and Fanconi's syndromes it was 0.26 and 0.60 breaks/cell, respectively[10]. The comparison of these two tissues from patients with Bloom's syndrome and Fanconi's anaemia failed to show any dramatic increase in chromosome damage in fibroblast cells. It is more desirable in ataxia telangiectasia patients to perform chromosome analysis from fibroblast cells in order to distinguish them from Bloom's syndrome and Fanconi's anaemia. Another characterizing factor for ataxia telangiectasia may be the frequent presence of a large D group chromosome (14q +). There is not much to summarize concerning the cytogenetics of xeroderma pigmentosum. Although an increased tendency for chromosomal disruption and rearrangements has not been observed directly,

2

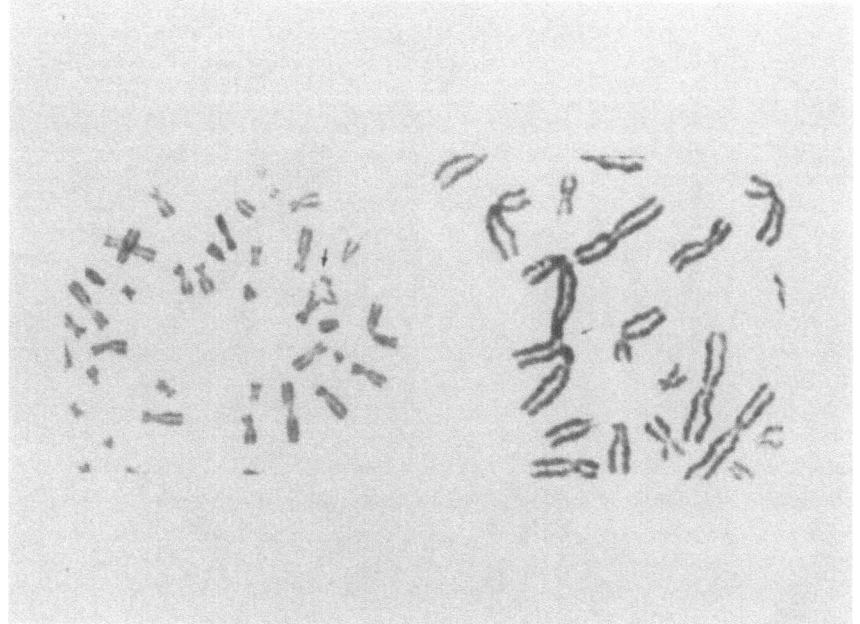

**Figure 1.1** Chromosome structural rearrangements in Fanconi's anaemia, showing quadriradial exchanges between non-homologous chromosomes

clones of skin fibroblasts with abnormal complements have been detected growing amid normal diploid fibroblast cells[11].

It is of interest that of the structural rearrangements observed in ataxia telangiectasia, dicentrics appeared most frequently[10,12]. The quadriradials (QR) are also observed in ataxia telangiectasia; they are very different from those found in Bloom's syndrome. In both ataxia telangiectasia and Fanconi's anaemia symmetrical exchanges between apparent homologues rarely occur, and the vast majority are of the non-homologous asymmetrical type (Figure 1.1)[10,13]. A highly characteristic feature of the chromosome instability in Bloom's syndrome is the tendency for exchanges to occur between chromatids of homologous chromosomes at homologous sites (Figure 1.3) (symmetrical type)[5].

Quadriradial exchanges between homologous and non-homologous chromosomes in somatic tissues could result in the formation of recombinant chromosomes. The mechanism of somatic recombination has not yet been elucidated. The recombinant chromosomes will result and cause reshuffling of genes located on the corresponding chromatids. This may be related to the predisposition of developing cancer in tissues with high mitotic indices (bone marrow, lymphoid tissue, mucosal cells of the gastrointestinal tract), a risk that apparently increases with age[14].

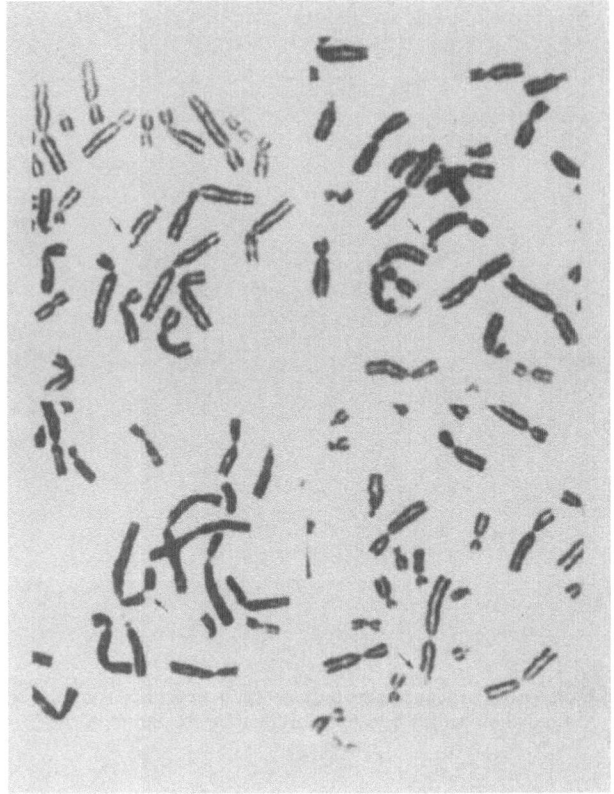

**Figure 1.2** Chromosome aberrations in Fanconi's anaemia, showing gaps and breaks

### Sister chromatid exchanges (SCEs)

Recent techniques to detect sister chromatid exchanges (SCEs) in somatic metaphase chromosomes[15-20] afford an opportunity to examine chromosomal instability occurring in cultured cells of some hereditary disorders. The results of chromatid exchanges can be observed as differences in the uptake of certain stains by the two chromatids of each metaphase chromosome after the cell has replicated twice in the presence of 5-bromodeoxyuridine (BudR) prior to harvest. This thymidine analogue, when incorporated into DNA, reduces the ultraviolet light fluorescence of fluorochromes such as Hoechst 33258[17] or acridine orange[16,18]. Permanently Giemsa-stained preparations can be obtained following exposure to fluorescent dyes[18,20]. Two cell cycles in the dark (to prevent artificial induction of SCEs by photolysis of BudR containing DNA) are required to observe a sharp demarcation between one brightly fluorescent or deeply Giemsa-stained chromatid, and one that is not (Figure 1.4). The former contains DNA with only one strand, the latter with both strands, substituted with the base analogue[20].

4

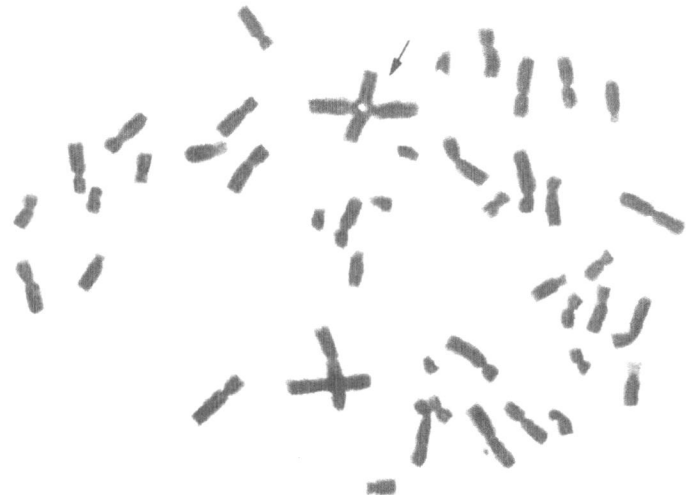

**Figure 1.3** Metaphase cell from an individual with Bloom's syndrome, showing quadriradial exchanges between homologous chromosomes. (Photomicrograph courtesy of Darrell Tomkins, Children's Hospital of Eastern Ontario, Ottawa)

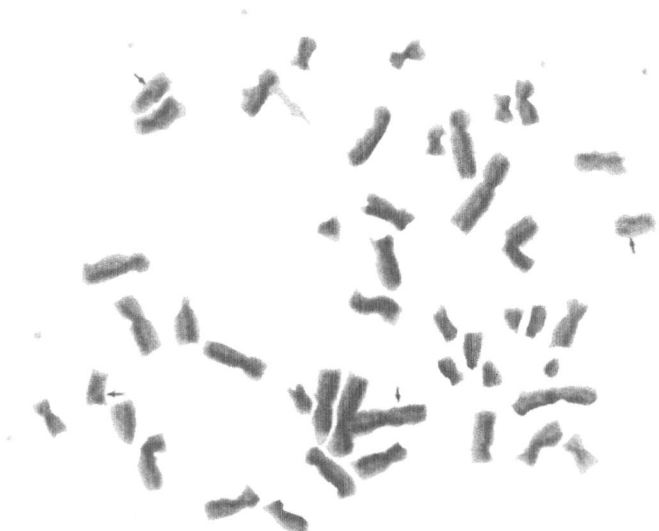

**Figure 1.4** Metaphase cell from a normal individual, showing few sister chromatid exchanges

Using these methods, the occurrence of sister chromatid exchanges has been studied in cultured lymphocytes from patients with ataxia telangiectasia, Bloom's syndrome, Fanconi's anaemia and xeroderma pigmentosum. A ten- to twelvefold increase in frequency of spontaneous SCE has been reported in cultured lymphocytes and fibroblast cells from Bloom's syndrome patients (Figure 1.5), whereas it was normal in others[21-24]. The frequency of SCEs in Bloom's syndrome was $90.6 \pm 12.7$/cell, whereas no increase was found in Bloom's heterozygotes (parents) $(10.5 \pm 3.5$/cell)[21], who had results similar to control $(10.9 \pm 3.6$/cell). The frequency of SCEs was not increased in the lymphocytes of patients with ataxia telangiectasia, Fanconi's anaemia or xeroderma pigmentosum.

In xeroderma pigmentosum, an increased frequency of exchanges, as compared with controls, could be induced by exposure of 30 h culture to ultraviolet light[21]. This cytogenetics response could be used as a test for xeroderma pigmentosum, if controlled and standardized exposure levels of u.v. light are first established. The induction of SCE by u.v. in xeroderma pigmentosum lymphocytes can be readily explained by the defective excision of u.v. induced pyrimidine dimers from DNA[9,25]. Schönwald and Passarge[26] observed increased SCEs response to u.v. light in six of nine patients. The reasons for no response in three cases out of nine are not clear. With some limitations, this test could serve as a rapid diagnostic tool for this particular disorder and as well it could be applicable in prenatal diagnosis of this disorder. Kaback et al.[27] described a technique for prenatal diagnosis in xeroderma pigmentosum using DNA sedimentation method. Ramsay et al.[28] used autoradiographic technique for measuring the DNA repair synthesis in fibroblast cells from xeroderma pigmentosum in order to establish a reliable prenatal diagnosis for this syndrome.

The spontaneous, about ten- to twelvefold, increase in SCE frequency in the

**Figure 1.5** Metaphase cell from an individual with Bloom's syndrome, showing several sister chromatid exchanges (Photomicrograph courtesy of Darrell Tomkins, Children's Hospital of Eastern Ontario, Ottawa)

Bloom's syndrome appears to be very specific for this disorder. It has not been found in other hereditary diseases which fall under the category of chromosomal breakage syndromes. There is no overlap in the range of SCEs in Bloom cells and normal controls. Theoretically, the frequency of SCEs in a single cell would suffice for the diagnosis. The rate of SCEs in lymphocytes metaphase cells of the heterozygous parents was in the range of normal controls (about ten per cell). Thus the specific increase in SCE levels in Bloom's syndrome, its manifestation in fibroblasts, together with the feasibility of discriminating between the homozygous and the heterozygous condition with respect to SCE frequencies are the prerequisites for application of this cytological phenomenon to prenatal diagnosis.

The high frequency of chromosomal abnormalities (Figures 1.1 and 1.2), predominantly breaks, suggests that there might be an error in DNA repair in Fanconi's anaemia. It was observed that the baseline frequency of sister chromatid exchanges in lymphocytes from four males with Fanconi's anaemia differed little from that of normal lymphocytes. However, addition of the bifunctional alkylating agent mitomycin C (0.01 or 0.03 $\mu$g/ml) to the Fanconi's anaemia cells in culture, induced less than half of the increase in exchanges found in identically treated normal lymphocytes[29]. This reduced increment in exchanges is accompanied by a partial suppression of mitosis and a marked increase in chromatid breaks and rearrangements. The reduced inducibility of sister chromatid exchanges observed in Fanconi's anaemia lymphocytes may ultimately be of diagnostic use.

## Roberts syndrome and SC phocomelia

Roberts syndrome and SC phocomelia are autosomal recessive conditions with prenatal and postnatal growth retardation, symmetrical limb reduction, and craniofacial abnormalities. The relationship between the SC phocomelia and Roberts syndromes remain unclear. Herrmann et al.[30] were the first to describe the former condition and to raise the question about the difference between it and the disorder described by Appelt et al.[31], which is now considered identical with the Roberts[32] syndrome. The question is still to be settled whether the two conditions are due to different recessive genes, different alleles or the same recessive gene. All the malformations associated with Roberts syndrome could have resulted from a disturbance in development before the seventh week of gestation[33].

A distinction has been made between the two syndromes on the basis of relative severity of these manifestations. Where chromosome studies have been carried out, most have been reported as normal. However, there have been two reports of consistent centromere abnormalities: one in a patient with SC phocomelia (pseudothalidomide syndrome), the other in a patient with Roberts syndrome. Four patients with similar phenotypic manifestations have been shown to have the same centromeric puffing or splitting[34]. Recently the chromosomes from a male patient with SC phocomelia have been studied and the similar centromeric puffing or splitting was observed[35] (Figure 1.6a). The C-banding studies of these chromosomes indicate paracentromeric heterochromatin puffing or splitting (Figure 1.6b). The Y chromosome

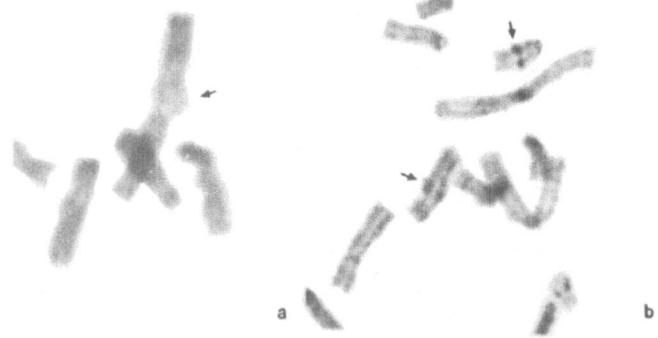

**Figure 1.6** Partial metaphase cells from a patient with SC phocomelia. a, Giemsa stain, showing puffing of heterochromatin region of chromosome 1; b, C-banding, showing distinct separation of C-positive regions

(Figure 1.7) shows splitting in the heterochromatic region rather than in the paracentromeric areas as in other chromosomes. These results appear to indicate that heterochromatin regions (paracentromeric or non-paracentromeric) are involved in the abnormality. The puffing or splitting is clear in all the acrocentric chromosomes, especially in heterochromatic regions of 1, 9, 16 and Y chromosomes. It might be postulated that there might exist a regulatory gene controlling the pairing affinity of these heterochromatic regions in human chromosomes. Sister chromatid exchange analysis was carried out on fibroblast cells, and the average number of exchanges was within the normal range (Figure 1.8). The cytogenetic findings in Roberts syndrome and SC phocomelia may be useful for prenatal diagnosis when used in conjunction with other diagnostic tools such as ultrasound[36].

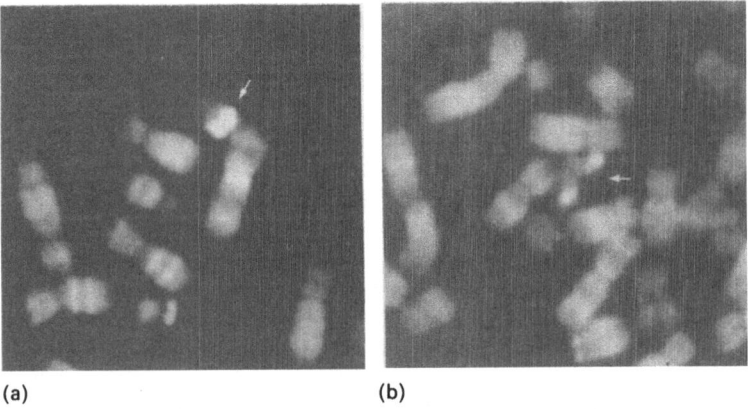

(a)                    (b)

**Figure 1.7** Partial Q-banding metaphase cells with Y chromosomes. a, Y chromosome from a normal individual, showing closeness of brightly fluorescent heterochromatin regions; b, Y chromosome from a patient with SC phocomelia, showing separation of brightly fluorescent heterochromatin regions

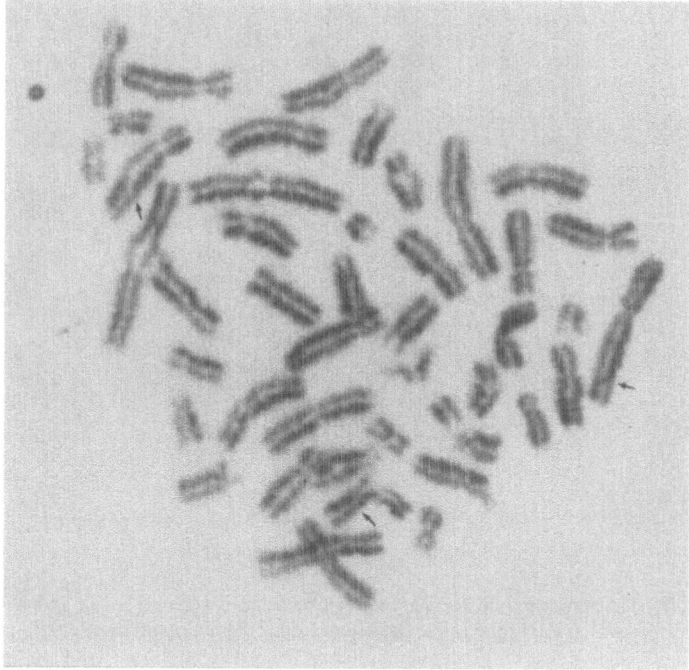

**Figure 1.8**  Sister chromatid exchanges in fibroblast cell from SC phocomelia patient. Arrows indicate the exchange sites

## X-LINKED RECESSIVE DISORDER

### X-linked mental retardation

The inheritance of recessive genes on the X chromosomes follows a well-defined pattern. A trait inherited as an X-linked recessive is expressed by all males who carry the gene, but the females are affected only if they are homozygous. The X-linked recessive disorders are restricted to males and the females are carriers for that gene in heterozygous condition.

Mental retardation with macro-orchidism in males, with probably X-linked recessive inheritance, has been described by Turner et al.[37] and Cantú et al.[38,39]. It is now established that the fragile site in the distal part of Xq 27 or 28 (Figure 1.9) is associated with, and may even be the cause of this form of X-linked mental retardation with macro-orchidism[40-44]. Nevertheless, demonstrations of these fragile sites are not easy and are subject to much controversy[43-47]; in particular the determination of carrier status in females remains difficult and unreliable[41]. The carrier detection may be fairly reliable in the age group below 20–25 years. However, the cytogenetic carrier's detection in obligate mothers above this age is difficult. This may be due to the decreased manifestation of the fragile site in human chromosomes, with

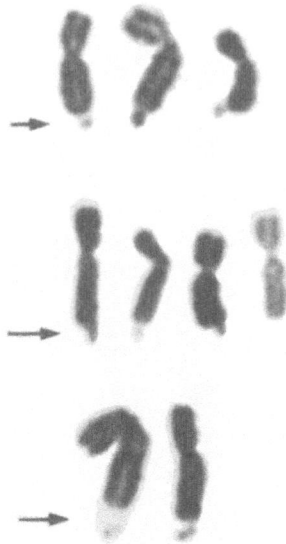

**Figure 1.9** The appearance of the marker X chromosomes, showing fragile sites at the distal ends. The chromosomes are taken from different metaphase cells

increasing age. Sutherland[41] has reported that the fragile site on Xq is not detectable in male skin fibroblast or lymphoblastoid cell cultures with this fragile site. Jacky and Dill[48] demonstrated in female carrier fibroblast cells that an increase in the frequency of expression of the marker chromosome occurred when cells were examined after exposure to BudR during their last 5–6 h of incubation in MEM-FA medium plus 5 % fetal calf serum (FCS).

The diagnosis of X-linked mental retardation with macro-orchidism remains difficult. Not all the retarded males with the fragile site have macro-orchidism[41]. Furthermore, the fragile site can often be demonstrated in only a small proportion of metaphases from some retarded males even when the diagnosis is virtually certain on clinical grounds, and from a study of the family history and other affected relatives. Despite the methods developed for manipulation of the composition of the culture medium to increase the frequency of fragile site expression in males[41], this method remains totally inadequate for females (carriers) and less than satisfactory for males. We have studied the expression frequency of fragile site in X from a male in different composition of the medium: (1) 199 medium, HEPES buffered and 5 % FCS, (2) 199 medium and 5 % FCS, (3) HB 597 medium and 5 % FCS and (4) McCoy's 5A medium and 20 % FCS. The expression of X marker was highest in HB 597 medium with 5 % FCS (36 %)[49]. Further experiments are required to enhance expression of the fragile sites in a higher proportion of lymphocytes, particularly from females, and in other cell types. The further development of this technique may provide a more reliable prenatal diagnostic procedure for this disorder.

The nature of the association between the fragile site and the mental retardation remains obscure. Not all families with X-linked mental retardation show the fragile site. It may be that demonstration of these sites in some families is more difficult than in others. Even in those retarded males studied, the proportion of metaphase cells in which the sites are expressed ranged from less than 5% up to more than 30%[41]. There is, however, no reason why X-linked mental retardation could not be a group of different conditions, only one of which is associated with the fragile site. There is another uncertainty in this disorder. The carrier females with X-linked mental retardation are generally normal. Deroover et al.[50] and Howard-Peebles and Stoddard[51] reported families having both normal carrier females and mildly retarded carrier females. There is no doubt that more families have to be investigated, because of heterogeneity of this syndrome.

## References

1 McKusick, V. A. (1978). *Mendelian Inheritance in Man. Catalogs of Autosomal Dominant, Autosomal Recessive and X-linked Phenotypes*. 5th Edn. (Baltimore: Johns Hopkins UP)
2 McKusick, V. (1969). *Birth Defects: Orig. Art. Ser.*, **5**(5), 126. (New York: National Foundation)
3 Gropp, A. and Flatz, G. (1967). Chromosome breakage and blastic transformation of lymphocytes in ataxia telangiectasia. *Humangenetik*, **5**, 77
4 Hecht, F., Koler, R. D., Rigas, D. A., Dahnke, G. S., Case, M. P., Tisdale, V. and Miller, R. W. (1966). Leukaemia and lymphocytes in ataxia telangiectasia. *Lancet*, **2**, 1193
5 German, J. (1969). Bloom's syndrome. I. Genetical and clinical observations in the first twenty-seven patients. *Am. J. Hum. Genet.*, **21**, 196
6 German, J. and Crippa, L. P. (1966). Chromosomal breakage in diploid cell lines from Bloom's syndrome and Fanconi's anemia. *Ann. Genet.*, **9**, 143
7 Schroeder, T. M., Auschwitz, F. and Knopp, A. (1964). Chromosomen-aberrationen bei familiarer Panmyelopathie. *Humangenetik.*, **1**, 194
8 Cleaver, J. E. and Bootsma, D. (1975). Xeroderma pigmentosum: Biochemical and genetic characteristics. *Ann. Rev. Genet.*, **9**, 19
9 Cleaver, J. E., Bootsma, D. and Frieberg, E. (1975). Human diseases with genetically altered DNA repair process. *Genetics*, **79**, 215
10 Cohen, M. M., Shaham, M., Dagan, J., Shmueli, E. and Kohn, G. (1975). Cytogenetics investigations in families with ataxia telangiectasia. *Cytogenet. Cell Genet.*, **15**, 328
11 German, J. (1972). Genes which increase chromosomal instability in somatic cells and predispose to cancer. In Steinberg, A. G. and Bearn, A. G. (eds.) *Progress in Medical Genetics*, Vol VIII, pp. 61–101. (New York: Grune and Stratton)
12 Pfeiffer, R. A. (1970). Chromosomal abnormalities in ataxia telangiectasia (Louis Bar's Syndrome). *Humangenetik*, **8**, 302
13 Shaw, M. W. and Cohen, M. M. (1965). Chromosomal exchanges in human leukocytes induced by mitomycin C. *Genetics*, **51**, 181
14 German, J. (1974). Bloom's syndrome. II. The prototype of human genetic disorders predisposing to chromosome instability and cancer. In German, J. (ed.) *Chromosome and Cancer*, pp. 601–617. (New York: Wiley)
15 Dutrillaux, B., Fosse, A. M., Prieur, M. and Lejeune, J. (1974). Analyse des échanges de chromatides dans les cellules somatiques humaines. *Chromosoma*, **48**, 327
16 Kato, H. (1974). Spontaneous sister chromatid exchanges detected by a BudR labelling method. *Nature*, **251**, 70
17 Latt, S. A. (1973). Microfluorometric detection of deoxyribonucleic acid replication in human metaphase chromosomes. *Proc. Natl. Acad. Sci. USA*, **70**, 3395
18 Perry, P. and Wolff, S. (1974). New Giemsa method for the differential staining of sister chromatids. *Nature (London)*, **251**, 156
19 Perry, P. and Evans, H. J. (1975). Cytological detection of mutagen-carcinogen exposure by sister chromatid exchange. *Nature (London)*, **258**, 121

20 Wolff, S. and Perry, P. (1974). Differential Giemsa staining of sister chromatids and the study of sister chromatid exchanges without autoradiography. *Chromosoma*, **48**, 341

21 Bartram, C. R., Koske-Westphal, T. and Passarge, E. (1976). Chromatid exchanges in ataxia telangiectasia, Bloom syndrome, Werner syndrome and Xeroderma pigmentosum. *Ann. Hum. Genet.*, **40**, 79

22 Chaganti, R. S. K., Schonberg, S. and German, J. (1974). A manyfold increase in sister chromatid exchanges in Bloom's syndrome lymphocytes. *Proc. Natl. Acad. Sci. USA*, **71**, 4508

23 Hayashi, K. and Schmid, W. (1975). The rate of sister chromatid exchanges parallel to spontaneous chromosome breakage in Fanconi anemia and to Trenimon-induced aberration in human lymphocytes and fibroblasts. *Humangenetik*, **29**, 201

24 Sperling, K., Wegner, R. D., Riehm, H. and Obe, G. (1975). Frequency and distribution of sister chromatid exchanges in a case of Fanconi's anemia. *Humangenetik*, **27**, 227

25 Kraemer, K. H., Coon, H. G., Petinga, R. A., Barret, S. F., Rahe, A. E. and Robbins, J. H. (1975). Genetic heterogeneity in Xeroderma pigmentosum. Complementation groups and their relationship to DNA repair rates. *Proc. Natl. Acad. Sci. USA*, **72**, 59

26 Schönwald, A. D. and Passarge, E. (1977). U.V. light induced sister chromatid exchanges in Xeroderma pigmentosum lymphocytes. *Hum. Genet.*, **36**, 213

27 Kaback, M. M., Howell, R. R., Klein, E. and Burgess, G. (1971). Xeroderma pigmentosum: a rapid sensitive method for prenatal diagnosis. *Science*, **174**, 147

28 Ramsay, C. A., Coltart, T. M. and Giannelli, F. (1974). Prenatal diagnosis of Xeroderma pigmentosum. *Lancet*, **2**, 1109

29 Latt, S. A., Stetten, G., Juergens, L. A., Buchanan, G. R. and Gerald, P. S. (1975). Induction by alkylating agents of sister chromatid exchanges and chromatid breaks in Fanconi's anemia. *Proc. Natl. Acad. Sci. USA*, **72**, 4066

30 Herrmann, J., Feingold, M., Tuffli, G. A. and Opitz, J. M. (1969). A familial dysmorphogenetic syndrome of limb deformities, characteristic facial appearance and associated anomalies: The 'pseudothalidomide' or 'SC-syndrome'. In Bergsma, D. (ed.) *Limb Malformations, Part III*. (Baltimore: Williams & Wilkins) (Published for National Foundation-March of Dimes. BD:OAS V/3, 81)

31 Appelt, H., Gerken, H. and Lenz, W. (1966). Tetraphokomelie mit Lippen-Kiefer-Gaumenspalte und Clitorishypertrophie – ein Syndrome. *Paediatr. Paedol.*, **2**, 119

32 Roberts, J. B. (1919). A child with double cleft of lip and palate, protrusion of the intermaxillary portion of the upper jaw and an imperfect development of the bones of the four extremities. *Ann. Surg.* **70**, 252

33 Smith, D. W. (1976). *Recognizable Patterns of Human Malformations*, p. 164. (Philadelphia: Saunders)

34 Tomkins, D., Hunter, A. and Roberts, M. (1979). Cytogenetic findings in Roberts – SC phocomelia syndrome(s). *Am. J. Med. Genet.*, **4**, 17

35 Ray, M., Greenberg, C. and Davies, D. (1980). A case of SC phocomelia. (In preparation)

36 Kaffe, S., Rose, J. S., Godmilow, L., Walker, B. A., Kerenyi, T., Beratis, N., Reyes, P. and Hirschhorn, K. (1977). Prenatal diagnosis of renal anomalies. *Am. J. Med. Genet.*, **1**, 241

37 Turner, G., Eastman, C., Casey, J., McLeay, A., Procopis, P. and Turner, B. (1975). X-linked mental retardation associated with macro-orchidism. *J. Med. Genet.*, **12**, 367

38 Cantú, J.-M., Scaglia, H. E., Medina, M., González-Diddi, M., Morato, T., Moreno, M. E. and Pérez-Palacios, G. (1976). Inherited congenital normofunctional testicular hyperplasia and mental deficiency. *Hum. Genet.*, **33**, 23

39 Cantú, J.-M., Scaglia, H. E., González-Diddi, M., Hernández-Jáuregui, P., Morato, T., Moreno, M. E., Giner, J., Alcántar, A., Herrera, D. and Pérez-Palacios, G. (1978). Inherited congenital normofunctional testicular hyperplasia and mental deficiency. *Hum. Genet.*, **41**, 331

40 Harvey, J., Judge, C. and Wiener, S. (1977). Familial X-linked mental retardation with an X chromosome abnormality. *J. Med. Genet.*, **14**, 46

41 Sutherland, G. R. (1979). Heritable fragile sites on human chromosomes. III. Detection of fra (X) (q27) in males with X-linked mental retardation and in their female relatives. *Hum. Genet.*, **53**, 23

42 Sutherland, G. R. and Ashforth, P. L. C. (1979). X-linked mental retardation with macro-orchidism and the fragile site of Xq27 or 28. *Hum. Genet.*, **48**, 117

43 Turner, G., Gill, R. and Daniel, A. (1978). Marker X chromosomes, mental retardation and macro-orchidism. *N. Engl. J. Med.*, **229**, 1472

44 Turner, G., Gill, R. and Daniel, A. (1978). Carrier detection in X-linked mental retardation. *Med. J. Aust.*, **2**, 624

45 Mulcahy, M. T. (1978). Carrier detection in X-linked mental retardation. *Med. J. Aust.*, **2**, 489

46 Sutherland, G. R. (1978). Carrier detection in X-linked mental retardation. *Med. J. Aust.*, **2**, 624

47 Wiener, S. and Judge, C. (1979). Carrier detection in X-linked mental retardation. *Med. J. Aust.*, **1**, 128

48 Jacky, P. B. and Dill, F. J. (1980). Expression in fibroblast culture of the satellited-X chromosome associated with familial sex-linked mental retardation. *Hum. Genet.*, **53**, 267

49 Ray, M., Schulz, D. and Josifek, K. (1980). Expression of X marker in different composition of medium. (In preparation)

50 Deroover, J., Fryns, J. P., Parloir, C. and Van den Berghe, H. (1977). X-linked recessively inherited nonspecific mental retardation: report of a large family. *Ann. Genet.*, **20**, 263

51 Howard-Peebles, P. N. and Stoddard, G. R. (1979). X-linked mental retardation with macro-orchidism and marker X-chromosomes. *Hum. Genet.*, **50**, 247

# 2
# The role of chromosome heteromorphism in developmental anomalies

P. K. GHOSH AND REITA NAND

## INTRODUCTION

Morphological variability of the human chromosomes has been known to occur since the 1960s. These variations are predominantly localized in the heterochromatic regions, especially the short arm and satellite regions of the acrocentric chromosomes and the long arm of the Y chromosome. The advent of new staining techniques has helped in the identification of sites of morphological variation in human chromosomes. The variable segments correspond to the regions which are genetically inert and late replicating and contain a high quantity of repetitive DNA. These variations can be detected partly by conventional staining but mostly with C- and Q-banding techniques. Chromosome variant patterns seem to be unique for each individual. The term 'polymorphism' has frequently been used to describe the variation in human karyotype. Since these variations are not discrete but continuous, the term 'variant' has been recommended (Paris Conference, 1971)[1]. The supplement to the Paris Conference (1975)[2] subsequently advocated the term 'heteromorphism' to be used in describing situations where deviations from the norm of chromosome morphology are observed. Besides conventional staining, the most frequently encountered heteromorphisms are visualized by CBG (C-bands by barium hydroxide using Giemsa) and QFQ (Q-bands by fluorescence using quinacrine mustard) techniques. Less frequent heteromorphisms are revealed by RFA (R-bands by fluorescence using acridine orange) technique, silver staining method for detecting heteromorphism of nucleolar organizing regions (NOR) and the BrdU/Hoechst technique for studying the variation in lateral asymmetry of constitutive heterochromatin.

The chromosomal heteromorphisms are stable units in the sense that they are present in all the cells of an individual and are transmitted from parent to offspring in the regular Mendelian fashion but are variable from person to person. The heteromorphic sites contain different amounts of various classes of highly redundant DNA. Classifications of heteromorphisms are based on size, position and staining intensity of the variable bands rather than simple measurements. Consequently different scoring methods have been employed in different laboratories. As a result, comparison of the existing work on heteromorphisms must be carried out with great caution. Since there is no relationship between a heteromorphism identified by one technique and that identified by another, they are described separately.

## CHROMOSOME HETEROMORPHISM IN NORMAL INDIVIDUALS

### CBG heteromorphism

The C-banding technique developed by Arrighi and Hsu[3] and later modified by Sumner[3a] into a relatively simpler method has been widely used for staining constitutive heterochromatin. This heterochromatin is localized in the centromere of all chromosomes, the satellites of the acrocentric chromosomes and the secondary constriction (h) regions of chromosomes 1, 9 and 16 (Figure 2.1). Variation in the length of the C-band is quite common and can be

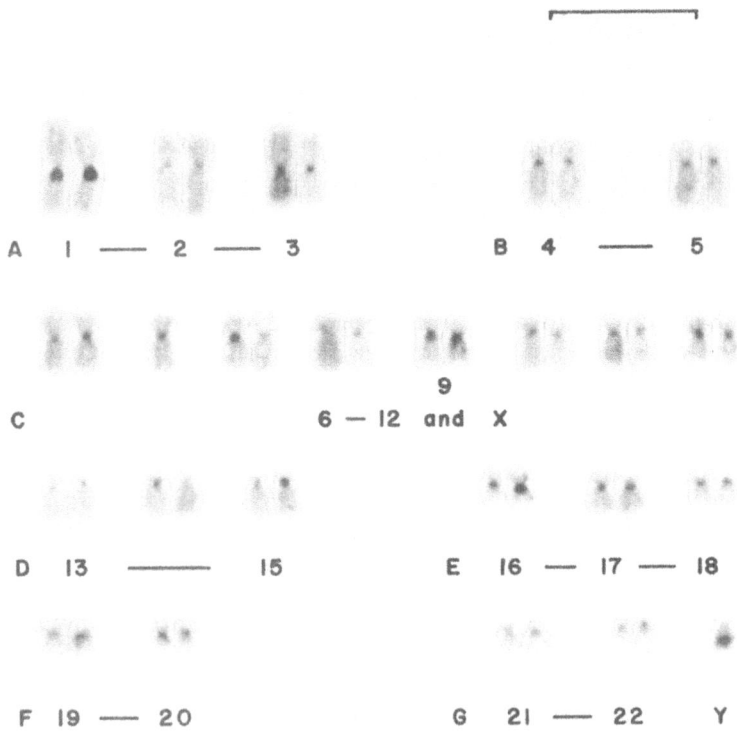

**Figure 2.1**  A karyotype showing C-bands revealed by CBG technique of Sumner[3a]. Bar denotes 10 μm

classified into normal, large or small. Large and small categories can be further classified into very small or small and large or very large. Besides these size variations, position variations of centromeric heterochromatin are also noticed. Normally the C-band is located on the long arm, but occasionally it is shifted to the short arm, and this is termed as pericentric inversion. Pericentric inversion may be complete or partial depending on the amount of C-band material inverted. The frequency and distribution of the C-band variants from different surveys on normal individuals are recorded in Table 2.1. It can be observed from this table that large C-bands are predominantly located in chromosome 1 and 9 whereas the small C-bands are observed at a high frequency in chromosome 16. Inversion of the C-band segment is much more common in chromosome 9 as compared to any other chromosome. There are some other striking heteromorphisms of the C-band which are not restricted to secondary constrictions. These are heteromorphism of chromosome 6 and 12[4] and chromosome 19[5].

## QFQ heteromorphism

The fluorescence pattern of certain bands revealed by QFQ technique shows heteromorphism of fluorescence intensity. This heteromorphism is restricted to the centromere of chromosomes 3 and 4 and the short arm and satellite regions of the acrocentric chromosomes (Figure 2.2). The intensity of

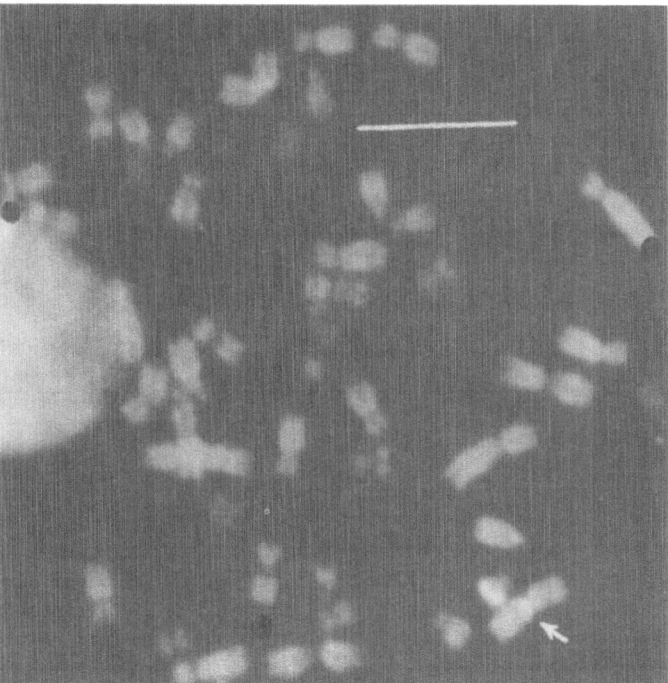

**Figure 2.2** A metaphase showing brilliantly fluorescing centromere of chromosome 3. Bar denotes 10 μm

Table 2.1 Incidence of CBG heteromorphisms in normal individuals

| Population | No. of individuals studied | No. of variants observed | C-band heteromorphism in chromosome | | | | Authors |
|---|---|---|---|---|---|---|---|
| | | | 1 | 9 | 16 | others | |
| American whites | 5 | 7 | 1 | 1 | 1 | 4 | Craig-Holmes and Shaw[99] |
| American whites | 20 | 31 | 5 | 6 | 7 | 13 | Craig-Holmes et al.[100] |
| Japanese | 19 | 20 | 9 | 10 | 1 | — | Iinuma et al.[101] |
| Orientals | 49 | 81 | 20 | 26 | 24 | 11 | Park and Antley[102] |
| English adults | 367 | 116 | 38 | 30 | 16 | 32 | Ferguson-Smith[103] |
| American newborns | 77 | 166 | 13 | 24 | 26 | 103 | McKenzie and Lubs[104] |
| American whites | 33 | 99 | 16 | 13 | 33 | 66 | Craig-Holmes et al.[105] |
| American newborns | 339 | 527 | 185 | 166 | 166 | — | Mueller et al.[106] |
| American whites | 95 | 150 | 13 | 12 | 35 | 90 | Lubs et al.[65] |
| American blacks | 97 | 203 | 17 | 13 | 42 | 131 | Lubs et al.[65] |
| American whites | 182 | 90 | 24 | 36 | 30 | — | Tharapel and Summitt[75] |
| American blacks | 12 | 3 | 2 | 0 | 1 | — | Tharapel and Summitt[75] |
| Jats | 250 | 352 | 62 | 59 | 64 | 167 | Nand[107] |

**Table 2.2  Criteria for QFQ intensity level***

| Code | Description | Comparison |
|---|---|---|
| 1 | Negative | No fluorescence |
| 2 | Pale | As in distal 1p |
| 3 | Medium | As in major bands 9q |
| 4 | Intense | As in major distal 13q bands |
| 5 | Brilliant | As in distal Y |

* Paris Conference (1971)[1]

**Table 2.3  Percentage of QFQ heteromorphisms in normal individuals**

| Population | No. of individuals studied | Q-band heteromorphism in chromosome | | | | | | | Authors |
|---|---|---|---|---|---|---|---|---|---|
| | | 3 | 4 | 13 | 14 | 15 | 21 | 22 | |
| Normal individuals (Holland) | 221 | 48.4 | 2.7 | 5.0 | 14.3 | 21.5 | 24.4 | 21.9 | Geraedts and Pearson[108] |
| American newborns | 77 | 41.0 | 41.0 | 46.7 | 4.5 | 1.3 | 2.6 | 9.0 | McKenzie and Lubs[104] |
| Estonian adults | 208 | 65.0 | 27.8 | 88.5 | 9.8 | 6.2 | 8 | 44.0 | Mikelsaar et al.[39] |
| American newborns | 376 | 55.3 | 13.1 | 81.8 | 15.9 | 13.6 | 19.6 | 62.3 | Mueller et al.[106] |
| Scottish newborns | 482 | 64.9 | 48.3 | 46.8 | 10.3 | 12.5 | 10.3 | 18.0 | Buckton et al.[109] |
| Scottish 74 yr old | 109 | 68.4 | 33.5 | 37.6 | 13.3 | 10.5 | 16.5 | 15.1 | Buckton et al.[109] |
| Canadian newborns | 930 | 55.5 | 14.1 | 33.0 | 1.0 | 1.0 | 1.2 | 0.6 | Lin et al.[110] |
| American adult | 100 | 62.0 | 14.0 | 56.5 | 10.0 | 10.0 | 15.5 | 10.0 | Verma et al.[111], Verma and Dosik[112] |
| American whites | 205 | 86.0 | 20.0 | 57.0 | 11.0 | 11.0 | 8.5 | 8.0 | Lubs et al.[65] |
| Americans blacks | 210 | 99.0 | 9.0 | 97.0 | 13.0 | 13.0 | 15.0 | 18.0 | Lubs et al.[65] |

fluorescence of these chromosomal regions is classified according to the criteria laid down by Paris Conference (1971)[1], which are described in Table 2.2.

The variation of Q-band intensity as reported from different studies is listed in Table 2.3. It can be observed from this table that the intensity heteromorphisms are quite common in the centromere region of chromosome 3 (41 – 68.4 per cent) whereas in chromosome 4 it ranges from 2.7 to 48.3 %. In the acrocentric chromosomes, no. 13 was found to be the most heteromorphic whereas chromosome 21 was the least heteromorphic.

## CHROMOSOME HETEROMORPHISM IN CLINICALLY ABNORMAL INDIVIDUALS

The chromosome heteromorphisms were believed to be without any clinical significance. However, a lot of data have accumulated which show that these heteromorphisms are not meaningless variations but do affect normal development. This is based on the observation of a higher incidence of these heteromorphisms in clinically abnormal individuals as compared to normal controls. These studies are briefly described, as follows.

### Y chromosome heteromorphism

Considerable variations in the length of the Y chromosome, ranging from a small fragment to an unusually long element, have been known to occur in both normal and phenotypically abnormal individuals. Clinical cases associated with deleted or short Y chromosome have been reported by Conen et al.[6], Muldal and Ockey[7], Vaharu et al.[8], Van Wijck et al.[9], Nakagome et al.[10], Nuzzo et al.[11], Surana[12] and Meisner and Inhorn[13]. Clinically abnormal individuals with long Y chromosomes have been reported by Bender and Gooch[14], Bishop et al.[15], Kallen and Levan[16], Makino et al.[17], Tonomura and Ono[18], Makino and Muramoto[19], Sugahara and Sakurai[20], Nielsen[21] and Jeske and Huebner[22]. The mean length of the Y chromosome has been found to be significantly higher in criminals as compared to normal controls by Kahn et al.[23], Huebner[24], Nielsen and Friedrich[25], Soudek and Larya[26] (1974), Nielsen and Norland[27], Martin-Lucas and Abrisqueta[28], Tajmirova and Ondrejcak[29] and Funderburk et al.[30]. Contrary to these observations, Benezech et al.[31], Schwinger and Wild[32], Urdal and Brogger[33], Benezech et al.[34,35], Genest and Dumas[36] and Akesson and Wahlstrom[37] found no difference in the length of Y chromosome in prisoners as compared to normal controls. Long Y chromosome has also been associated with increased activity and psychiatric disorders in boys[27,38,39]. Increased rates of abortions have been observed in families carrying long Y chromosomes[40,41]. An increased frequency of short Y chromosome has been reported in fathers of Down's syndrome children[42,43]. Small or deleted Y chromosome has also been related to severe oligospermia or azoospermia[44]. However, no difference in the mean length of Y chromosome has been observed in fertile and infertile men by Kadotani et al.[45].

## Heteromorphism of acrocentric chromosomes

There is suggestive evidence indicating that short arm and satellite variants may have developmental effects. In Down's syndrome patients, an elevated frequency of acrocentric chromosome with increased short arm has been observed by Dekaban et al.[46], Subrt[47], Starkman and Shaw[48] and Hamerton[49]. Bott et al.[50] found a high frequency of heteromorphism of the acrocentric chromosome in parents of Down's syndrome patients as compared to normal controls. However, Giraud et al.[43] failed to notice increased frequency of aberrant short arms in parents of Down's syndrome patients. Nielsen et al.[51] also did not find any harmful effect of deleted short arm of acrocentric chromosomes. A survey conducted on newborn babies by Lubs and Ruddle[52] showed an association between malformations and G-group satellites. Christensen and Nielsen[38] observed a high frequency of G-group satellites in a population of child psychiatric patients. Funderburk et al.[53] failed to observe any IQ–satellite size correlation in child patients with mental retardation or psychiatric disorders. Higher frequency of satellite variants in adults with reproductive disorders has been observed by Kulazhenko et al.[54], Kaosaar and Mikelsaar[55], Papp et al.[56] and Tsenghi et al.[57].

## Secondary constriction heteromorphism of chromosomes 1, 9 and 16

Many studies have reported significant association between the heteromorphism of the secondary constriction regions of chromosomes 1, 9 and 16 and clinical conditions. Developmental defects similar to Patau's and Meckel's syndrome have been assigned to the pathogenic effect of 1qh+ variant by Gardner et al.[58]. Families with non-specific malformations have also shown the presence of 1qh+ variant[59]. Evidence has been presented by Atkin[60] and Atkin and Pickthall[61] showing that the heteromorphism of chromosome 1 predisposes towards ovarian and other forms of malignancy including lymphomas. An increased frequency of prominent secondary constrictions of chromosomes 1, 9 and 16 has been observed in individuals with congenital malformation[62,63], psychiatric disorders[64] and low IQ[65]. Nielsen et al.[51] observed an increased frequency of 9qh+ variant from 0.1% in normal populations to 2.8% in relatives of individuals with chromosomal aberrations. According to Palmer and Schroder[66], persons with 9qh+ variant might have an increased risk of progeny with chromosome abnormalities and/or congenital abnormalities. These variants are also found in increased frequency in couples with recurrent spontaneous abortions[67–69]. Lubs et al.[65] observed the presence of 9qh+ in 20% of black children with low IQ. An association between the size of the 9h, cranial circumference and mental retardation has been found by Funderburk et al.[30]. Increased heart defects have been detected in many carriers of 16qh+ variant by Kelly and Almy[62]. Nielsen et al.[51], however, failed to observe the pathogenic effect of 16qh+ variant. Soudek and Sorka[70] found 9qh+ and 16qh− variant to be more common in the mentally retarded children as compared to normal controls.

An increase in the frequency of inversion in chromosome 9 has been observed from 0.84 % in consecutive newborns to 5.9 % in retarded or autistic children by Lubs and Lubs[71]. Increased frequency of inversion in chromosome 9 has also been observed by Ballesta and Serra[72], and Say et al.[64] in mentally retarded populations and children with psychiatric disorders. A significantly higher frequency of partial inversion in the C-band segment of chromosome 9 was observed in the groups of patients with Down's syndrome and with idiopathic mental retardation by Wang and Hamerton[73].

## Heteromorphism of fluorescent bands

Very few studies are available on the clinical significance of Q-band heteromorphisms. Mikelsaar et al.[39] found increased frequency of intensely fluorescent satellites of the acrocentric chromosome and centromere of chromosome 3 in female oligophrenics compared with newborns. Funderburk et al.[30] (1976) found increased frequency of these variants in certain clinical conditions. Lubs et al.[65] found no correlation between IQ and Q-band heteromorphism in their population of 7-year-old children. Schwinger and Wehner[74] found no difference in the frequency of Q-band heteromorphism between 89 normal individuals and 247 with a variety of clinical abnormalities. Tharapel and Summitt[75] found no difference in Q-band heteromorphisms between 200 normal individuals and 200 with mental retardation. Matsuura et al.[76] found a significantly larger band 13 cen/p 1 in the patients with sociofamilial retardation. Mikelsaar et al.[77] and Soudek and Sorka[78] (1978) found no difference in the frequency of the inversion of the fluorescent region of chromosome 3 in mentally retarded and normal children.

## AETIOLOGICAL SIGNIFICANCE OF CHROMOSOME HETEROMORPHISM

Unlike numerical and structural chromosome aberrations, chromosome heteromorphism is associated with malformations only in some carriers. This makes it difficult to find out such associations and to elucidate the mechanism involved in the production of deleterious effects on the phenotype. However, the chromosomal heteromorphisms are observed in low frequency in phenotypically normal individuals but are elevated in patients with developmental anomalies. It may be postulated that increased heterochromatin and developmental disorders are the result of a common mechanism which are not related as cause and effect but as coexisting phenomena.

It has been suggested by Stonova[79] that since chromosomal heteromorphisms always involved the heterochromatic region of a chromosome, it is the heterochromatin that is different in individuals with developmental disorders. According to Davidenkova[80], the quantitative changes in the heterochromatin content in the limits of the normal karyotype may be correlated with the phenotype in the apparently normal individuals in the general population, and the non-specific and quantitative features of human developmental disorders were probably determined by the effect of heterochromatin of additional chromosomes.

The presence of chromosome heteromorphism may interfere with normal cell division. German et al.[81] are of the opinion that these heteromorphisms can contribute to abnormal disjunction and can interfere with normal development in a manner similar to 'structural genetic load'. It has been opined that heteromorphisms of human karyotype may lead to developmental effects by several mechanisms. These may result from a change in gene position, a gene mutation at the site of chromosome break or small duplication and deficiencies[82]. The presence of large areas of heterochromatin close to the centromere may provide a background for disturbed segregation of these chromosomes during meiosis[57]. The minus variants may also influence segregation of chromosomes during meiosis in a similar way.

There are indications that the heterochromatic part of chromosome 9 is a remarkable segment which may be active in the first meiosis[83]. Variants of human chromosome 9 involving subcentromeric heterochromatin are claimed to be associated with reduced sperm counts and subfertility. There are evidences suggesting that the variants play a role in the failure of gametogenesis in both males and females, perhaps by altering meiosis and/or mitosis[84]. The minus variants of chromosome 9 may promote failure of segregation due to shortness of the interstitial segment, i.e. q11–a13 (reported by Lindenbaum and Bobrow[85]). The duplication of the heterochromatic region in a chromosome No. 9 might affect pairing and segregation and thus give an increased risk of non-disjunction, deletion, translocation or duplication[51].

Hamerton et al.[86] and Ballantyne et al.[87] suggested that there might be an increased risk of non-disjunction in the progeny of individuals with increase in short arms satellite material, which might be related to the increase in heterochromatin. Various other workers have associated heterochromatic markers with reproductive failure in man and other organisms. They have suggested that the markers probably influence meiosis by an effect on chromosome pairing[47,69,88,89]. Ford and Lester[84] are of the opinion that the variants may influence meiosis by reducing the efficiency of chromosome pairing and can also influence cell division in other ways. Presence of extremely heteromorphic variants may show a decreased relative reproductive fitness[90] and there may be an increased risk of chromosome abnormalities in the progeny[51].

Observations from both plants and lower animals support the view that structural changes of heterochromatin may produce phenotypic effects[91,92]. A correlation between leg length and the amount of heavy satellite DNA has been observed in nine different species of kangaroo rat by Hatch and Mazrimas[93]. It has been observed that extra heterochromatic chromosomes affect birth weight, body weight, immunoglobulin level and cell growth rate. This last fact is reflected in affected intelligence as even small changes in division rate could lead to larger changes in the relative proportion of differentiated cells and so result in abnormal brain structure[94].

A better approach to the problem of the chromosome heteromorphism in relation to developmental anomalies can be achieved if we take into account the studies conducted in lower animals with respect to functional aspects of heterochromatin. In lower animals like Drosophila, alterations in the amount of heterochromatin in a single chromosome are not only capable of exerting an influence on the recombinational properties of that chromosome, but may

B

in addition alter the cross-over potential of other members of the genome[95]. These mechanisms could well provide a sufficient selective force to account for the presence of the chromosomal heteromorphism in question. It has been observed recently by Yamamoto and Miklos[96] that as the amount of centric heterochromatin is reduced, so too is the amount of recombination in the proximal euchromatin, confirming earlier observations that recombination decreases progressively when euchromatic sections are moved closer and closer to the centromere either by inversion or by translocation[97,98]. This may well explain the deleterious effect of the heteromorphisms observed in human karyotypes.

Thus we find that the heteromorphisms exhibited by chromosomes, which are mainly due to change in the amount, size and position of heterochromatin, influence the meiotic system, which may have important adaptive or evolutionary consequences. The process of normal development may be affected by a change in the amount of heterochromatin which in turn influences the recombinational properties of that chromosome. Meiotic studies in individuals who exhibit pronounced heteromorphism of karyotype may help us in understanding the role these heteromorphisms play in normal development.

## References

1 Paris Conference (1971) (1972). Standardization in human cytogenetics. *Birth Defects: Orig. Art. Ser.* **8**(7), 1. (New York: National Foundation)

2 Paris Conference (1971), Supplement (1975) (197̃5). Standardization in human cytogenetics. *Birth Defects Orig. Art. Ser.* **11**(9), 1. (New York: National Foundation)

3 Arrighi, F. E. and Hsu, T. C. (1971). Localization of heterochromatin in human chromosomes. *Cytogenetics*, **10**, 81

3a Sumner, A. T. (1972). A simple technique for demonstrating centromeric heterochromatin. *Exp. Cell Res.*, **75**, 304

4 Sofuni, T. K., Tanabe, K., Ohtaki, K., Shimba, H. and Awa, A. A. (1974). Two new types of C-band variants in human chromosome (6p + and 12ph +). *Jpn. J. Hum. Genet.* **19**, 251

5 Crossen, P. E. (1975). Variation in the centromeric banding of chromosome 19. *Clin. Genet.*, **8**, 218

6 Conen, P. E., Bailey, J. D., Allemang, W. H., Thompson, D. W. and Ezrin, C. (1961). A probable partial deletion of the Y chromosome in an intersex patient. *Lancet*, **2**, 194

7 Muldal, S. and Ockey, C. H. (1961). Muscular dystrophy and deletions of Y chromosomes. *Lancet*, **2**, 601

8 Vaharu, T., Patton, R. G., Voorhess, M. L. and Gardner, L. I. (1961). Gonadal dysgenesis and enlarged phallus in a girl with 45 chromosomes plus fragment. *Lancet*, **1**, 218

9 Van Wijck, J. A. M., Tijdink, G. A. G. and Stelte, L. A. M. (1962). Anomalies in the Y chromosome. *Lancet*, **1**, 218

10 Nakagome, Y., Sasaki, M., Matusi, I., Kawazura, M. and Fakuyama, Y. (1965). A mentally retarded boy with a minute Y chromosome. *J. Pediatr.*, **67**, 1163

11 Nuzzo, F., Bompiani, A., Moneta, E., Caviezel, F. and Mussinelli, M. (1967). Observations on some cases of structural variations in the Y chromosome in man. *Alti Assoc. Genet. Ital.*, **12**, 1919

12 Surana, R. B., Forbath, P. and Conen, P. E. (1971). Minute Y chromosome. *Ann. Genet.*, **14**, 145

13 Meisner, L. F. and Inhorn, S. L. (1972). Normal male development with Y chromosome long arm deletion (Yq −). *J. Med. Genet.*, **9**, 373

14 Bender, M. A. and Gooch, P. C. (1961). An unusually long human Y chromosome. *Lancet*, **2**, 463

15 Bishop, A., Blank, C. E. and Hunter, H. (1962). Heritable variation in the human Y chromosome. *Lancet*, **2**, 18

16 Kallen, B. and Levan, A. (1962). Abnormal lengths of chromosomes 21 and 22 in four patients with Marfan's syndrome. *Cytogenetics*, **1**, 5

17 Makino, S., Sasaki, M. S., Yamada, K. and Kaju, T. (1963). A long Y chromosome in man. *Chromosoma*, **14**, 154

18 Tonomura, A. and Ono, H. (1963). Variation in length of the Y chromosome in man. *Jpn. Nat. Inst. Genet. Rep.*, **14**, 131

19 Makino, S. and Muramoto, J. (1965). Some observations on the variability of Y chromosome. *Proc. Jpn. Acad.*, **40**, 457

20 Sugahara, T. and Sakurai, M. (1967). Chromosome studies in a family with an abnormally large Y chromosome. *Jpn. J. Hum. Genet.*, **12**, 190

21 Nielsen, J. (1969). Large Y and enlarged short arms of a small acrocentric chromosome. *Hereditas*, **61**, 416

22 Jeske, J. and Huebner, H. (1970). A familial variant of the Y chromosome. *Endokrynal. Pol.*, **21**, 411

23 Kahn, T., Carter, W. I., Dernley, N. and Slater, E. T. O. (1969). Chromosome studies in remand home and prison populations. In West, D. J. (ed.) *Criminological Implications of Chromosome Abnormalities*, pp. 44–48. (Cambridge: Cambridge UP)

24 Huebner, H. (1971). The Y chromosome in selected groups of men of the Polish population. *Bull. Acad. Pol. Sci. Ser. Sci. Biol.*, **19**, 467

25 Nielsen, J. and Friedrich, U. (1972). Length of the Y chromosome in criminal males. *Clin. Genet.*, **3**, 281

26 Soudek, D. and Larya, P. (1974). Longer Y chromosome in criminals. *Clin. Genet.*, **6**, 225

27 Nielsen, J. and Norland, E. (1975). Length of Y chromosome and activity in boys. *Clin. Genet.*, **8**, 291

28 Martin-Lucas, M. A. and Abrisqueta, J. A. (1976). Yq+ anomaly in a criminal population. *In Fifth International Congress of Human Genetics*, Mexico, 140

29 Tajmirova, O. and Ondrejcak, M. (1976). Cytogenetics in forensic genetics. In *Fifth International Congress of Genetics*, Mexico, 156

30 Funderburk, S. J., Sparkes, R. S., Guthric, D. and Westlake, J. R. (1976). The significance of minor chromosome variants. In *Fifth International Congress of Human Genetics*, Mexico, 125

31 Benezech, M., Robert, G., Noel, B. and Zinlitini, R. (1973). Aggressivité, délinquence psychopathique et longueur du chromosome Y. *Nouv. Presse. Méd.* **2**, 583

32 Schwinger, E. and Wild, P. (1974). Length of the Y chromosomes and antisocial behaviour. *Humangenetik*, **22**, 67

33 Urdal, T. and Brogger, A. (1974). Criminality and the length of Y chromosome. *Lancet*, **1**, 626

34 Benezech, M., Noel, B. and Manelphe, C. (1975). Chromosome Y long (Y q+) et comportement délictueux. *Ann. Méd.-Psychol.* (Paris), **2**, 313

35 Benezech, M., Noel, B., Travers, E. and Mottet, J. (1976). Conduite antisociale et longueur du chrome Y. *Hum. Genet.*, **32**, 77

36 Genest, P. and Dumas, L. (1976). The length of the Y chromosome and antisocial behaviour in human beings *Can. J. Genet. Cytol.*, **18**, 560

37 Akesson, H. O. and Wahlstrom, J. (1977). The length of the Y-chromosomes in men examined by forensic psychiatrists. *Hum. Genet.*, **39**, 1

38 Christensen, K. R. and Nielsen, J. (1974). Incidence of chromosome aberrations in a children's psychiatric hospital. *Clin. Genet.*, **5**, 205

39 Mikelsaar, A. V. N., Kaosaar, M. E., Tuur, S. J., Viikmaa, M. H., Tabik, T. A. and Laats, J. (1975). Human karyotype polymorphism III. Routine and fluorescence microscopic investigation of chromosome, in normal adults and mentally retarded population. *Humangenetik*, **26**, 1

40 Kadotani, T., Ohama, K. and Sato, H. (1969). A chromosome survey in 71 couples with repeated spontaneous abortions and still births. *Proc. Jpn. Acad.*, **45**, 180

41 Patil, S. R. and Lubs, H. A. (1977). A possible association of long Y chromosomes and fetal loss. *Hum. Genet.*, **35**, 233

42 Zizka, J. (1972). Down's syndrome: a cytogenetic and clinical study of 134 patients with a special reference to translocation cases. *Collection of Scientific Works, Med. Fac. Hradec Kralove*, **15**, 1

43 Giraud, F., Mattei, J. F. and Mattei, M. G. (1975). Étude chromosomique chez les parents d'enfants trisomique 21. *Lyon Méd.*, **233**, 241

44 Koulischer, L. (1976). Polymorphism of the Y chromosome in a population of 1000 infertile male patients. *Clin. Genet.*, **10**, 363

45 Kadotani, T., Ohama, K., Takahara, H. and Makino, S. (1971). Studies on the Y chromosome in human males from fertile and infertile couples. *Jpn. J. Hum. Genet.*, **16**, 35

46 Dekaban, A. S., Bender, M. A. and Economos, G. E. (1963). Chromosome studies in mongoloids and their families. *Cytogenetics*, **2**, 61

47 Subrt, I. (1970). A further example of familal Gp+ associated with trisomy G. *Humangenetik*, **9**, 86

48 Starkman, M. N. and Shaw, M. W. (1967). Atypical acrocentric chromosomes in Negro and caucasian mongols. *Am. J. Hum. Genet.*, **19**, 162

49 Hamerton, J. (1970). *Human Cytogenetics*. Vol. 11. (New York: Academic Press)

50 Bott, C. E., Sekhon, G. S. and Lubs, H. A. (1975). Unexpectedly high frequency of paternal origin of trisomy 21. *Am. J. Hum. Genet.*, **27**, 20A

51 Nielsen, J., Friedrich, U., Hreidarsson, A. B., and Zeuthen, E. (1974). Frequency of 9qh + and risk of chromosome aberrations in the progeny of individuals with 9qh +. *Hum. Genet.*, **21**, 211

52 Lubs, H. A. and Ruddle, F. H. (1970). Chromosomal abnormalities in the human population: Estimation of rate based on New Haven newborn study. *Science*, **169**, 495

53 Funderburk, S. J., Goldenberg, I., Klisak, I., Sparkes, R. S. and Westlake, J. (1979). Prominent acrocentric chromosome satellites in child patients with mental retardation on psychiatric disorders: No IQ–satellite size correlation. *Hum. Genet.*, **50**, 179

54 Kulazhenko, V. P., Lazink, G. I. and Livchenko, N. N. (1972). Chromosome aberrations found in parents in cases of recurrent abortions. *Genetika (Moskva)*, **8**, 154

55 Kaosaar, M. E. and Mikelsaar, A. V. N. (1973). Chromosome investigation in married couples with repeated spontaneous abortions. *Humangenetik*, **17**, 277

56 Papp, Z., Gardo, S. and Dalhay, B. (1974). Chromosome study of couples with repeated spontaneous abortions. *Fert. Steril.*, **25**, 713

57 Tsenghi, C., Metaxoton-Stavridaki, C., Strataki-Beneton, M., Kalpini-Mavrov, A. and Matsaniotis, N. (1976). Chromosome studies in couples with repeated spontaneous abortions. *Obstet. Gynecol.*, **47**, 463

58 Gardner, R. J. M., McCreanor, H. R., Parslow, M. I. and Veale, A. M. O. (1974). Are lq + chromosomes harmless? *Clin. Genet.*, **6**, 383

59 Halbrecht, I. and Shabtay, F. (1976). Human chromosome polymorphism and congenital malformations. *Clin. Genet.*, **10**, 113

60 Atkin, N. B. (1977). Chromosome 1 heteromorphism in patients with malignant disease; a constitutional marker for a high risk group? *Br. Med. J.*, **1**, 358

61 Atkin, N. B. and Pickthall, V. J. (1977). Chromosome 1 in 14 ovarian cancers, heterochromatin variants and structural change. *Hum. Genet.*, **38**, 25

62 Kelly, S. and Almy, R. (1971). Chromosome 16 heterology; non-specific variations in karyotype? *N. Y. State J. Med.*, **71**, 2297

63 Kunze, J. and Mour, G. (1975). $A_1$ and $C_9$ marker chromosomes in children with combined minor and major malformations. *Lancet*, **1**, 273

64 Say, B., Carpenter, N. J., Lanier, P. R. and Banez, C. (1977). Chromosome variants in children with psychiatric disorders. *Am. J. Psychiatry*, **134**, 424

65 Lubs, H. A., Kimberling, H. J., Hecht, F., Patil, S. R., Brown, J., Gerald, P. and Summitt, R. L. (1977). Racial differences in the frequency of Q and C chromosomal heteromorphisms. *Nature (London)*, **268**, 631

66 Palmer, C. G. and Schroder, J. (1971). A familial variant of chromosome 9. *J. Med. Genet.*, **8**, 202

67 Holbek, S., Friedrich, U., Lowritsen, T. G. and Therkelsen, A. J. (1974). Marker chromosomes in parents of spontaneous abortions. *Hum. Genet.*, **25**, 61

68 Boué, J., Taillemite, J. L., Hazet Massieux, P., Leonard, C. and Boué, A. (1975). Association of pericentric inversion of chromosome no. 9 and reproductive failure in ten unrelated families. *Humangenetik*, **30**, 217

69 Ford, J. H. (1977). Cytogenetics of infertility and habitual abortion. *Proc. Symp. Genet. Dis. Rec. Adelaide Child Hosp.*, **1**, 287

70 Soudek, D. and Sorka, H. (1979). Chromosomal variants in mentally retarded and normal men. *Clin. Genet.*, **16**, 109
71 Lubs, H. A. and Lubs, M. L. (1972). New cytogenetic techniques applied to a series of children with mental retardation. In Caspersson, T. and Zech, L. (eds.). *Chromosome Identification*, pp. 241–250. (New York: Academic Press)
72 Ballesta, F. M. and Serra, M. (1976). Chromosomal polymorphism; frequency and clinical manifestations. *Clin. Genet.*, **10**, 349
73 Wang, H. S. and Hamerton, J. L. (1979). C-band polymorphisms of chromosomes 1, 9 and 16 in four subgroups of mentally retarded patients and a normal control population. *Hum. Genet.*, **51**, 269
74 Schwinger, E. and Wehner, H. (1976). Frequency of chromosomal fluorescence polymorphism in normal persons and in clinical patients with diagnosed chromosome aberrations. *Hum. Genet.*, **33**, 115
75 Tharapel, A. T. and Summitt, R. L. (1978). Minor chromosome variations and selected heteromorphisms in 200 unclassifiable mentally retarded patients and 200 normal controls. *Hum. Genet.*, **41**, 121
76 Matsuura, J., Mayer, M. and Jacobs, P. (1978). A cytogenetic survey of an institution for the mentally retarded II. C-band chromosome heteromorphisms. *Hum. Genet.*, **45**, 33
77 Mikelsaar, A. V. N., Ilus, T. and Kivi, S. (1978). Variant chromosome 3 (inv 3) in normal newborns and their parents, and in children with mental retardation. *Hum. Genet.*, **41**, 109
78 Soudek, D. and Sorka, H. (1978). Inversion of 'fluorescent' segment in chromosome 3: a polymorphic trait. *Hum. Genet.*, **44**, 109
79 Stonova, N. S. (1968). Possible role of heterochromatin in the appearance of human inherent malformations. *Genetika*, **4**, 90
80 Davidenkova, E. F., Verlinskaya, O. K. and Tysyachnyuk, S. F. (1974). Clinical aspects of human chromosomal diseases and the role of heterochromatin of additional chromosomes. *Genetika*, **10**, 152
81 German, J., Ehlers, K. H. and Engle, M. A. (1966). Familial congenital heart disease II. Chromosomal studies. *Circulation*, **34**, 517
82 Jacobs, P. A. (1974). Correlation between euploid structural chromosome rearrangements and mental subnormality in humans. *Nature (London)*, **249**, 164
83 Stahl, A., Luciani, J. M., Dwictor, M., Capodano, A. M. and Gagne, R. (1975). Constitutive heterochromatin and micronuclei in the human oocyte at the diplotene stage. *Humangenetik*, **26**, 315
84 Ford, J. H. and Lester, P. (1978). Chromosomal variants and nondisjunction. *Cytogenet. Cell Genet.*, **21**, 300
85 Lindenbaum, C. and Bobrow, M. (1975). Reciprocal translocation in man, 3:1 disjunction resulting in 47- or 45- chromosome offspring. *J. Med. Genet.*, **12**, 29
86 Hamerton, J. L., Gianelli, F. and Polani, P. E. (1965). Cytogenetics of Down's syndrome (mongolism) I. Data on a consecutive series of patients referred for genetic counselling and diagnosis. *Cytogenetics*, **4**, 171
87 Ballantyne, G. H., Parslow, M. I., Veab, A. M. O. and Pullon, D. H. H. (1977). Down's syndrome and deletion of short arms of a G-chromosome. *J. Med. Genet.*, **14**, 147
88 Thomas, J. B. and Kaltsikes, P. J. (1974). Possible effect of heterochromatin on chromosome pairing. *Proc. Natl. Acad. Sci. USA*, **71**, 2787
89 Miklos, G. L. and Nankivell, R. N. (1976). Telomeric satellite DNA functions in regulating recombination. *Chromosoma*, **56**, 143
90 Jacobs, P. A., Frankiewicz, A., Law, P., Hilditch, C. J. and Morton, N. E. (1975). The effect of structural aberrations of the chromosomes on reproductive fitness in man II. Results. *Clin. Genet.*, **8**, 169
91 Quiros, C. F. (1976). Meiotic behaviour of extra heterochromatin in the tomato: Effects on several vital processes. *Can. J. Genet. Cytol.*, **18**, 325
92 Clarke, C. A., Sheppard, P. M. and Mittwoch, U. (1976). Heterochromatin polymorphism and colour pattern in the tiger swallowtail butterfly Papilio glauaes L. *Nature (London)*, **263**, 585
93 Hatch, F. T. and Mazrimas, J. A. (1977). Satellite DNA and cytogenetic evolution. In Sparkes, R. S., Comings, D. E. and Fox, C. F. (eds.) *Molecular human cytogenetics*, pp. 395–410. (New York: Academic Press)

94 Barlow, P. (1973). The influence of inactive chromosomes on human development. *Hum. Genet.*, **17**, 105
95 Miklos, G. L. G. and John, B. (1979). Heterochromatin and satellite DNA in man: Properties and prospects. *Am. J. Hum. Genet.*, **31**, 264
96 Yamamoto, M. and Miklos, G. L. G. (1978). Genetic studies on heterochromatin in Drosophila melanogaster and their implications for the functions of satellite DNA. *Chromosomo*, **66**, 71
97 Beadle, G. W. (1932). A possible influence of the spindle fibre on crossing over in Drosophila. *Proc. Natl. Acad. Sci. USA*, **18**, 160
98 Mather, K. (1939). Crossing over and heterochromatin in the X-chromosome of Drosophila melanogaster. *Genetics*, **24**, 413
99 Craig-Holmes, A. P. and Shaw, M. W. (1971). Polymorphism. of human constitutive heterochromatin. *Science*, **174**, 702
100 Craig-Holmes, A. P., Moore, F. B. and Shaw, M. W. (1973). Polymorphism of human C-band heterochromatin I. Frequency of variants. *Am. J. Hum. Genet.*, **25**, 181
101 Iinuma, K., Matsunaga, E. and Nakagoma, Y. (1973). Polymorphism of C- and Q-bands in human chromosomes. *Ann. Rep. Nat. Inst. Genet. (Mishima)*, **23**, 112
102 Park, J. and Antley, R. M. (1974). C-band chromosomal polymorphism in Orientals. *Am. J. Hum. Genet.*, **26**, 65A
103 Ferguson-Smith, M. A. (1974). Autosomal polymorphisms. *Birth Defects: Orig. Art. Ser.*, **11**(10), 19. (New York: National Foundation)
104 McKenzie, W. H. and Lubs, H. A. (1975). Human Q and C chromosomal variations: Distribution and incidence. *Cytogenet. Cell Genet.*, **14**, 97
105 Craig-Holmes, A. P., Moore, F. B. and Shaw, M. W. (1975). Polymorphism of human C-band heterochromatin II. Family studies with suggestive evidence for somatic crossing over. *Am. J. Hum. Genet.*, **27**, 178
106 Mueller, H. J., Klinger, H. P. and Glasser, M. (1975). Chromosome polymorphism in human newborn population II. Potentials of polymorphic chromosome variants for characterizing the idiogram of an individual. *Cytogenet. Cell Genet.*, **15**, 239
107 Nand, R. (1979). Chromosome polymorphism in North Indian populations. *Thesis submitted in partial fulfilment of MPhil*, University of Delhi (unpublished)
108 Geraedts, J. P. M. and Pearson, P. L. (1974). Fluorescent chromosome polymorphism: Frequencies and segregations in a Dutch population. *Clin. Genet.*, **6**, 247
109 Buckton, K. E., O'Riordan, M. L., Jacobs, P. A., Robinson, J. A., Hill, R. and Evans, H. J. (1976). C- and Q-band polymorphisms in the chromosomes of three human populations. *Ann. Hum. Genet.*, **40**, 99
110 Lin, C. C., Gedeon, M. M., Griffith, P., Smink, W. K., Newton, D. R., Wilkic, L. and Sewell, L. M. (1976). Chromosome analysis on 930 consecutive newborn children using quinacrine fluorescent banding technique. *Hum. Genet.*, **31**, 315
111 Verma, R. S., Dosik, H., Scharf, T. and Lubs, H. A. (1978). Length heteromorphism of fluorescent (bf) and non-fluorescent (ng) segments of human Y chromosome classification, frequencies and incidence in normal Caucasians. *J. Med. Genet.*, **15**, 277
112 Verma, R. S. and Dosik, H. (1979). Frequencies of centromeric heteromorphisms of human chromosomes 3 and 4 as detected by QFQ technique: Can they be identified by RFA technique. *Can. J. Genet. Cytol.*, **21**, 109

# 3
# Embryonic development in patients with a high risk of early pregnancy loss

BETTY J. POLAND

## INTRODUCTION

Spontaneous abortion or unavoidable pregnancy loss before 20 weeks of gestation occurs in about 15 % of all recognized conceptions. The association of abnormal development of the aborted conceptus in these cases has been well documented since the time of Mall in 1908[1] who also observed that the frequency of anomalies is directly proportional to the developmental age of the conceptus. In the aborted embryo the incidence is over 80 %, declining through the fetal stage of pregnancy to an incidence of 2–3 % in the term infant. In the past, aborted products of conception were frequently given little more than a cursory examination. The main reasons for this were twofold. Firstly, the products of conception were often incomplete when they arrived in the laboratory and the tissue poor in quality and difficult to evaluate. Secondly, the significance of the observations was not appreciated as a basis for subsequent counselling and management of the patient. It is as a result of the work done in the 1960s by the research groups led by Carr[2], Boué[3], Shepard[4], Nishimura[5], and their associates that the detailed diagnostic study of abortuses is gradually becoming an accepted part of the work of a service laboratory. The proper documentation of the results of the examination of the abortus is of particular importance to the patient who has repeated early pregnancy loss.

## THE PATIENT AT RISK

While spontaneous abortion is a common single event in the reproductive history, three or more consecutive abortions are uncommon and are likely to occur in not more than 4 % of the fertile couples[6]. In a follow-up study of 638 subsequent pregnancies from 472 women who had had at least one spontaneous abortion, it was shown that the risk for a second after a first spontaneous abortion was little raised on the empirical risk of 15 to 17%[7]. The same study showed that after two abortions, the empirical risk is doubled and continues to rise so long as the couple do not have a live child. In the event that a pregnancy is successful, the risk after two spontaneous abortions remains at 30%. A further study of the products of conception from 52 women who had at least two consecutive abortions showed that in over 75% of cases, the consecutive abortuses from one patient were of the same developmental stage, that is embryo or fetus[8]. Furthermore, in this study over 80% of the embryos were abnormal, all but one of the fetuses were normal. Recurrent abortion suggests a continuing problem, either in the quality of the conceptus or the competence of the uterine environment. Persistent abortion of abnormal embryos in the first trimester suggests a regimen of investigation different to that of a patient who aborts successive, normal fetuses in the second trimester. It is therefore of prime importance to examine the aborted products of conception in these cases and, from a careful evaluation, obtain a clue as to the aetiological factor.

Systemic anomalies in embryos or fetuses may be genetic, environmental, or a sporadic developmental event. The first may identify an increased risk for the child of a subsequent pregnancy. The diagnosis can only be made by an examination of the embryo or fetus, and correlation of the observations with a complete reproductive, medical and family history.

Cytogenetic anomalies occur prior to or at conception, and Edwards and Fowler give an excellent account of the current hypothesis on the origin of the chromosomal imbalance[9]. There is evidence from the animal experiments of Fugo and Butcher[10] that disturbance of the mechanism of ovulation is associated with an increased frequency of cytogenetic anomalies in the conceptus and it has been postulated by Mikamo[11] and others that inadequate gonadotrophin levels may have a similar effect in human reproduction.

Uterine anomalies, such as a septum, or pathology, such as fibroids or synechae or incompetent internal cervical os, may lead to disturbance of the uterine environment. As the fetus enlarges it can no longer be accommodated safely within the uterine cavity with the result that the abortion of a normal fetus and placenta in mid-trimester is a not uncommon result.

The results of examination of the abortus correlated with a detailed maternal history may identify the patient at risk and result in a solid basis for counselling and management of a subsequent pregnancy.

## EXAMINATION OF THE ABORTED PRODUCTS OF CONCEPTION

In order to carry out a meaningful examination of the products of conception it is essential that all the aborted material is made available whether passed spontaneously by the patient or curetted from the uterus. It should be sent to the laboratory without fixation. Spontaneous abortion does not necessarily occur immediately on embryonic or fetal death and the products of conception may be retained intact in the uterus for up to 6 weeks or more after embryonic development has ceased. This results in specific changes and distortion of the conceptus, and increases the difficulties of precise morpho-logical and cytogenetic evaluation. Although it may be impossible to be precise as to the specific abnormality, the time at which development ceased and the category of normal or abnormal embryo or fetus can be diagnosed in the majority of cases.

## DETERMINATION OF DEVELOPMENTAL AGE

This can be done most easily by reference to the standard information on normal growth and development. The developmental age is calculated from the duration of pregnancy counting from the first day of the last menstrual period minus 2 weeks to allow for the interval prior to ovulation. The difference between the expected developmental age calculated from the pregnancy dates and the actual developmental age, calculated from observ-ations made on the embryo or fetus, indicates uterine retention time. In this way it is possible to identify the approximate time during pregnancy when growth ceased and the conceptus died. If the estimated and actual develop-mental age coincide, this suggests that death is most likely related to a recent event.

The developmental age of the abortus should be determined by measure-ment of the crown–rump length. This is less than 30 mm in the embryo and between 30 and 180 mm in the early fetus. The developmental age at the end of the previable fetal stage is 20 weeks. In addition, an embryo is staged developmentally by external features based on the horizons described by Streeter in 1951[12,13], but using the simpler term 'stage' as recommended by O'Rahilly and Muecke in 1972[14]. The developmental age of the fetus in days can be obtained by reference to a standard graph correlating gestational age with the crown–rump length. Such a graph is used in the central laboratory of embryology in the University of Washington[15]. Identification of the sex of an embryo from external examination is impossible as the external genitalia remain in the undifferentiated stage until early fetal life, that is after 56 days. However, histological section of the gonad which lies in the urogenital ridge will provide an immediate diagnosis. The external genitalia of the fetus are fully differentiated and the sex is easily identified in normal cases. The embryo and fetus are examined internally by macro and micro dissection. There are several textbooks available on internal dissection and the staging of embryos[16,17]. Internal examination of the fetus is carried out as for a miniature autopsy, using the same guidelines. Graphs relating organ weights in the fetus to developmental age may be helpful, in the diagnosis of intrauterine growth retardation suggesting that maturity of organs and longitudinal growth are not synchronized[18].

A new and accurate source of information in recent years has resulted from serial examination by ultrasound in normal pregnancy. Graphs relating the crown–rump length of the embryo and fetus in the uterus and the size of the chorionic sac to the developmental age are now available from the 7th week of pregnancy, from the work of Piiroinen[19] and Robinson[20]. An accurate developmental age may be impossible to calculate if the specimen is severely abnormal.

## CATEGORY OF DEVELOPMENT

An abnormality in an embryo is sometimes impossible to recognize. For example, palatal fusion occurs at the end of the embryonic stage and a cleft palate is normal up to this time. However, the neural tube is closed by the 28th–30th day and non-closure after this time can be recognized as a defect. Extrusion of coils of intestine is normal between days 31 and 35, when the embryo grows from 18.5 mm to 45 mm, but after this time the intestine is retracted into the enlarged abdominal cavity. The cervical flexure in the embryo is acute so that the face is at right angles to the chest wall. Obliteration of the flexure in the embryonic stage is abnormal.

Evidence of morphological abnormality in embryos is present in at least 80% of cases. Because the stage of development is so sensitive and the differentiation interdependent, the result is frequently a disorganized specimen showing little evidence of normal organization. This lack of organization of development can result either in an empty chorionic sac or in a small embryo up to 10 mm in size in which it may not be possible to identify cephalic or caudal pole. These specimens were described as 'stunted embryos' by Mall in 1917[21]. In an attempt to bring some order to this group, we have categorized the degree of growth disorganization (GD) with respect to the amount of differentiation of tissue. An empty chorionic sac we have designated GD1. This diagnosis can only be made if the chorionic sac is intact. Frequently on opening the chorion, the amnion is found to be closely applied to the chorion and of very poor quality, impossible to separate as a complete layer. GD2 consists of a small 'nubbin' of embryonic tissue in which no differentiation with respect to external features, such as limb buds, face, tail etc., can be recognized. GD3 describes an embryo in which it is possible because of the appearance of retinal pigment or very imperfect limb buds to determine which is the cephalic pole. The difficulties of evaluating these embryonic specimens are exemplified by the intact chorionic sac containing amniotic fluid, amnion, yolk sac and body stalk, but no evidence of embryo. Undoubtedly, in these cases one must assume that the embryo has disintegrated as a result of postmortem autolysis. A full account of the growth disorganized embryo and the associated chromosomal anomalies was published in 1973[22].

Figure 3.1 is a composite photograph of GD1, GD2 and GD3 from the Human Embryology Study, University of British Columbia.

Focal anomalies of specific systems are less easy to recognize in an embryo because of the incomplete stage of organogenesis. However, both exencephaly and cyclopia were described by Mall in 1917. These anomalies are relatively common and easily identified.

Figure 3.2 shows an embryo in an intact amniotic sac with exencephaly.

Abnormality is present in approximately 25% of aborted fetuses. The abnormalities consist of either single system, focal anomalies or multiple anomalies which may represent a syndrome. Maceration may be a problem in aborted fetuses, but while this may make the specimen difficult to dissect and although it may result in severe distortion, maceration does not produce morphological anomalies. The presence of maceration identifies the fetus

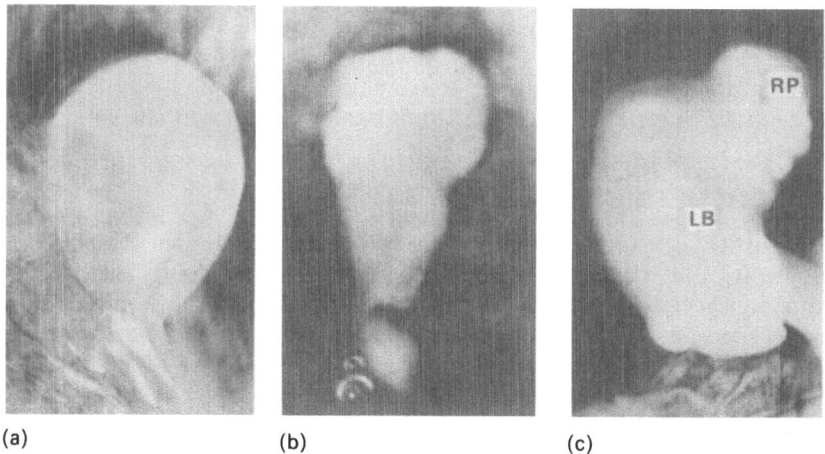

(a)                                    (b)                                    (c)

**Figure 3.1** Three categories of growth disorganization. a, Empty amniotic sac; b, GD2 – formless embryonic tissue; c, GD3 – some evidence of organization (RP = retinal pigment, LB = limb bud)

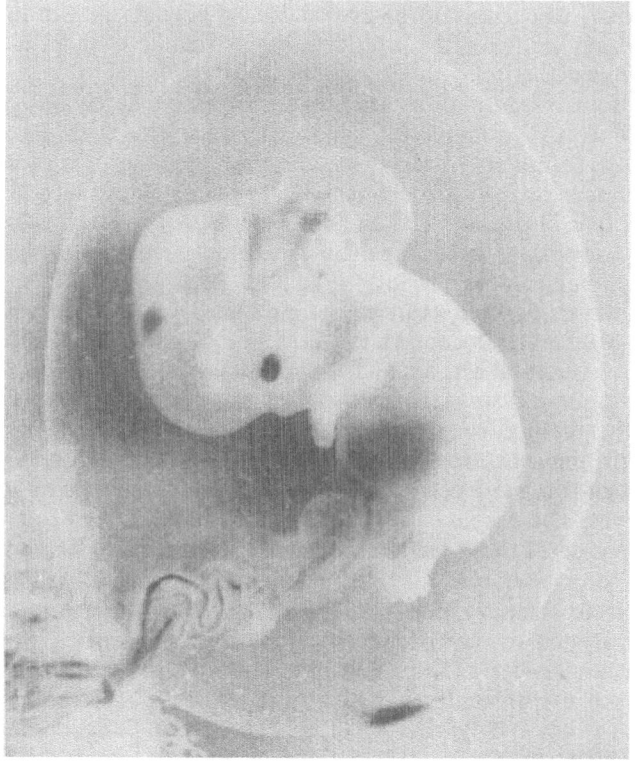

**Figure 3.2** Embryo with occipital encephalocoele, in intact amiotic sac. Stage 21. Developmental age 48 days. Crown–rump length 25 mm

dead *in utero* before abortion occurred. This can be readily distinguished from the fetus that died during the birthing process. Even this observation alone has some clinical significance. If a focal systemic anomaly is present, then it is mandatory that an effort be made to examine the family history so that if there is evidence of genetic disease the family can obtain appropriate counselling.

## CYTOGENETIC STUDIES

Exhaustive cytogenetic studies have been carried out in thousands of abortuses and there is no doubt that chromosomal anomalies are the commonest cause of death of the embryo. It was clearly shown by Boué and co-workers in 1973[23] that the frequency of chromosomal anomalies was directly proportional to the developmental age of the embryo with an overall incidence of 62 %. Successful tissue culture of abortuses is difficult because of the quality of the material. Amnion is the tissue usually used, as it leaves the embryo or fetus available for more detailed study. Because the culture of aborted tissue is unreliable, there has been considerable effort made to correlate the phenotype of the embryo with a specific chromosomal abnormality. No consistent phenotype has been associated with a trisomic anomaly, which is the most common anomaly in abortuses[23]. Harris[24], in a comprehensive discussion of the association of phenotype with the triploid karyotype, has emphasized the difficulties. No embryo or fetus has been described in association with tetraploidy.

Chromosomal abnormalities are present in approximately 20 % of fetuses prior to 20 weeks. Although the syndromes associated with chromosomal abnormalities identifiable in newborns are incompletely expressed in the early fetal stage, many features are identical. Descriptions are available of the early fetal phenotype in trisomy 18[25] and 45X[26]. It is of interest, however, that a trisomy 21 has none of the distinguishing features which make this anomaly so readily identifiable in the newborn. Undoubtedly in the fetus it is frequently missed; however, Bersu has carried out a detailed study of the early fetus in Down syndrome and identified the presence of extra facial muscles and variations in vertebral arteries and spinal nerves[27].

Tissue culture of consecutive abortuses reveals that the karyotypes are usually either both normal or both abnormal[28]. In the cases in which the karyotype is abnormal, the chromosomal anomalies are not necessarily the same, though it is more common for trisomy to follow trisomy than other combinations. There is no evidence that a triploid karyotype in an abortus, which is considered to be due to accident at the time of fertilization, has any increased recurrence risk. Similarly, the karyotype 45X has neither an increased recurrence risk nor is its frequency related to maternal age. The concern is related to the trisomic abortus, and the question whether this patient has an increased risk of non-disjunction in a subsequent pregnancy, and therefore an increased risk of a baby with Down syndrome. Counselling in this area varies and Warburton *et al.*[28] recommend amniocentesis in the subsequent pregnancy.

Structural chromosomal anomalies such as translocations, if identified in the karyotype of the abortus, while they may appear 'de novo' are an indication for chromosomal analysis in the parents. A balanced translocation in the parents may lead to an error at segregation and an unbalanced gamete. Structural chromosomal anomaly occurs more frequently in couples who have had two or more consecutive early spontaneous abortions[29]. Identification of a parent with a translocation is not only a possible explanation of repeated pregnancy loss, but an indication that the karyotype of a subsequent fetus should be monitored in an ongoing pregnancy, by amniocentesis.

The value of histological sections of the chorionic villi as a method of determining the developmental age of the conceptus was first described by Phillippe et al.[30]. By further study of the histology, he was able to obtain some indication of the karyotype of the abortus. This work was continued by Honore et al.[31] in 1978, and with very careful examination of multiple histological sections of the villi, he showed that it was possible to predict the presence of a chromosomal anomaly in over 50 % of cases.

The clinical significance of the findings in the examination of the abortus is suggested in the flow chart in Figure 3.3.

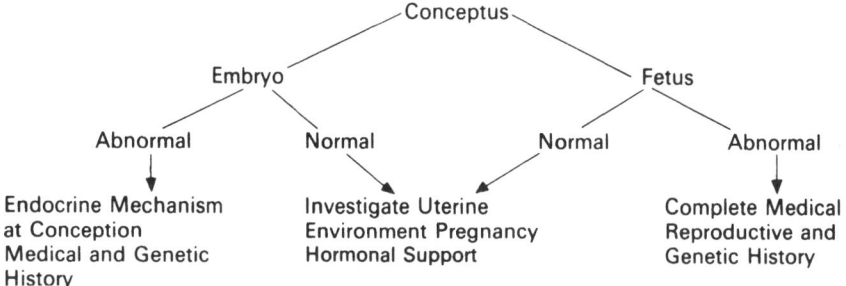

**Figure 3.3**  Morphology of conceptus related to maternal reproductive history

## References

1 Mall, F. P. (1908). A study on the causes underlying the origin of human monsters. *J. Morphol.*, **19**, 9

2 Carr, D. H. (1970). Chromosome abnormalities and spontaneous abortion. In Jacobs, P., Price, W. H. and Law, P. (eds.) *Human Population Cytogenetics*, p. 103. (Pfizer Medical Monograph 5, Edinburgh) (Baltimore: Williams & Wilkins)

3 Boué, J. and Boué, A. (1973). Anomalies chromosomiques dans les avortements spontanés. In Boué, A. and Thibault, C. (eds.) *Chromosomal Errors in Relation to Reproductive Failure*, p.29. (Paris: Inserm)

4 Shepard, T. H., Nelson, T., Oakley, G. P. Jr. and Lemire, J. (1971) Collection of human embryos and fetuses: seven years experience. In Hook, E. B., Janerich, D. T. and Porter, I. H. (eds.) *Monitoring Birth Defects and Environment. The Problem of Surveillance*, p. 29. (New York: Academic Press)

5 Nishimura, H., Takano, K., Tanimura, T. and Yasuda, M. (1968) Normal and abnormal development of human embryos. First report of the analysis of 1213 intact embryos. *Teratology*, **1**, 281

6 Hertig, A. T. and Livingstone, R. G. (1944). Spontaneous, threatened and habitual abortion: their pathogenesis and treatment. *N. Engl. J. Med.*, **230**, 797

7 Poland, B. J., Miller, J. R., Jones, D. C. and Trimble, B. K. (1977). Reproductive counselling in patients who have had a spontaneous abortion. *Am. J. Obstet. Gynecol.*, **127**, 685

8 Poland, B. J., Ho Yuen, B. (1978). Embryonic development in consecutive specimens from recurrent spontaneous abortion. *Am. J. Obstet. Gynecol.*, **130**, 512
9 Edwards, R. G. and Fowler, R. E. (1970). The genetics of human pre-implantation development. In Emery, A. E. H. (ed.) *Modern Trends in Human Genetics.* Vol 1. (London: Butterworth)
10 Fugo, N. W. and Butcher, R. L. (1966). Overripeness and the mammalian ova. 1. Overripeness and early embryonic development. *Fertil. Steril.*, **17**, 814
11 Mikamo, K. and Hamaguchi, H. (1975). Chromosomal disorder caused by pre-ovulatory overripeness of oocytes. In Blandau, R. J. (ed.) *Aging Gametes, International Symposium, Seattle,* 1973. (Basel: Karger)
12 Streeter, G. L. (1951). Developmental horizons in human embryos. Washington, D.C., Carnegie Institute Embryology. Description of age groups XV, XVI, XVIII. *Contrib. Embryol.*, **32**, 133
13 Streeter, G. L. (1948). Developmental horizons in human embryos. Description of age groups XIX, XX, XXI, XXII, XXIII, *Contrib. Embryol.*, **34**, 165
14 O'Rahilly, R. and Muecke, E. C. (1972). The timing and sequence of events in the development of the human urinary system during the embryonic period proper. *Z. Anat. Entwick.-lungsgesch.*, **138**, 99
15 Shepard, T. H., Nelson, T., Oakley, G., Lemire, R. J. (1971). A centralized laboratory for collection of human embryoss and fetuses. In Hook, E. B., Jamerich, D. T. and Porter, I. H. (eds.) *Monitoring Birth Defects and Environment. The Problem of Surveillance*, pp. 41–42. (New York: Academic Press)
16 Gasser, R. F. (1975). *Atlas of Human Embryos.* (Hagerstown, Maryland: Harper & Row)
17 Moore, K. L. (1973). *The Developing Human – Clinically Oriented Embryology.* (Toronto: Saunders)
18 Tanimura, T., Nelson, T. and Hollingsworth, R. R. (1970). Weight standards for organs from early fetuses. *Anat. Rec.*, **171**, 227
19 Piiroinen, O. (1975). Studies in diagnostic ultrasound. *Acta Obstet. Gynecol. Scand.*, **46** (Suppl.)
20 Robinson, H. (1978). Normal development in early pregnancy. In Villeger, M. (ed.) *Handbook of Clinical Ultrasound.* (New York: Wiley)
21 Mall, F. P. (1917). On the frequency of localized anomalies in human embryos and infants at birth. *Am. J. Anat.*, **22**, 49
22 Poland, B. J. and Miller, J. R. (1973). Les accidents chromosomiques de la reproduction. In Boué, A. and Thibault, C. (eds.) *Effects of Karyotype on Zygotic Development* (Paris: Inserm)
23 Boué, J. G., Boué, A. and Lazar, P. (1975). The epidemiology of human spontaneous abortions with chromosomal anomalies. In Blandau, R. J. (ed.) *Aging Gametes, International Symposium, Seattle,* 1973. (Basel: Karger)
24 Harris, M., Dill, F. J. and Poland, B. J. (1980). Triploidy in 40 human spontaneous abortions. A phenotype in the embryo. *Teratology*, (in press)
25 Baillie, J. E. and Poland, B. J. (1976). Developmental anomalies in a human fetus of 17 weeks gestational age. *Teratology*, **13**, 15
26 Poland, B. J., Dill, F. and Paradise, B. (1980). A Turner-like phenotype in the aborted fetus. *Teratology*, **21**, 46
27 Bersu, E. T. (1980). Anatomical analysis of the developmental effects of aneuploidy in man. The Down syndrome. *Am. J. Med. Genet.*, **5**, 399
28 Warburton, D., Susser, M., Stein, Z. and Kline, J. (1979). Genetic and epidemiologic investigation of spontaneous abortion. Relevance to clinical practice. In *Birth Defects: Orig. Art. Ser.*, **15**(5A), 127
29 Hevitage, D. W., English, S. C., Yound, R. B. and Chen, A. T. L. (1978). Cytogenetics of recurrent abortions. *Fertil. Steril.*, **29**
30 Phillippe, E. J., Ritter, J. M., Dehalleux, R., Renaud, R. and Gandor, R. (1968). De la pathologie des avortements spontanés. *Gyn. Omt. (Paris)*, **67**, 97
31 Honore, L. H., Dill, F. J., Poland, B. J. (1976). The placental morphology in spontaneous abortuses with normal and abnormal karyotypes. *Teratology*, **14**, 151

# 4
# Human chromosome polymorphism and congenital defects

## F. S. SHABTAI AND I. HALBRECHT

## INTRODUCTION

The polymorphism, or heteromorphism (Paris Conference, 1975) of human chromosomes is a fascinating field of cytogenetics. The most used techniques for the detection of human chromosome polymorphism are the QFQ (Q-bands by fluorescence using quinacrine), the CBG (C-bands by barium hydroxide staining with Giemsa), RFA (R-bands by fluorescence using acridine orange), and the silver staining[1] for the nucleolar organizer regions (NOR) of acrocentric chromosomes.

Due to recent advances in banding techniques an increasing number of heteromorphic sites are being recognized in the human genome and probably even more will be recognized in the near future.

Multiple consecutive banding techniques contribute to a better elucidation of the morphology of each chromosome, even if sometimes there is no direct relationship between a polymorphism identified by one technique and that identified by another. The common heteromorphic sites of human chromosomes are the short arms of the acrocentrics (13–15, 21 and 22), the centromeric area of chromosome 3 and 4, the long arm of the Y and the secondary constriction regions of chromosomes 1, 9, 16. Nevertheless each chromosome has a certain degree of polymorphism at its centromeric banding, either in size or in position[2]. Only extreme forms are probably noted and referred. Unusual variants have been reported for chromosome 6[3,4], for chromosome 11[5], for chromosomes 12 and 17[4,6], for chromosome 19[7], etc., apparently all of them without phenotypic effects.

The polymorphism of human chromosomes being so variable, the karyotype of an individual can be so well characterized using multiple techniques that it will be almost unique as an identity card. The polymorphisms are consistent within a person[8] (only few exceptions of somatic mosaicism have been reported) and are inherited in a Mendelian fashion (except for a few documented cases of '*de novo*' variations[9]). They may be used for several studies, e.g. on the mechanism producing triploidy in human fetuses, on the mechanism involved in the production of mosaics and for studying chimeras. Maternal cell contamination of amniotic fluid culture can be detected and the fate of transfused or transplanted cells may be followed.

Polymorphisms have been used to establish paternity and they have particular importance in forensic medicine. Zygosity testing in twins is another application of chromosomal polymorphisms. They may also be used extensively in gene mapping in somatic cell hybridization, as genetic markers in linkage studies. By using heteromorphic markers the origin of the extra chromosome in trisomies can be determined.

An important aspect of human chromosome polymorphisms is their clinical significance.

## THE CLINICAL SIGNIFICANCE OF POLYMORPHISMS

The polymorphic sites of the genome are usually sites of heterochromatin containing varying amounts of different classes of highly repetitive DNA sequences, some of them called satellites DNA[10-12]. These DNA sequences are believed to lack structural genes and not to be translated. For this reason so great a variability is tolerated in heterochromatic areas. On the other hand, everything in nature has its reason to be. The function of heterochromatin is only poorly understood; many hypotheses have been proposed and all reviewed by Hsu[13] (who sustained the 'bodyguard' hypothesis), by Miklos and John[14], and others. Recently, Varley *et al.*[15] communicated that satellites DNA on lambrush chromosomes is transcribed in RNA. It is unlikely that it is translated but the transcripts of the satellite DNA may regulate in some way the expression of other translatable sequences. Such control would be stage specific. Doolittle and Sapienza[16] and Orgel and Crick[17] discussed the idea of a 'selfish DNA' which makes no contribution to the phenotype and may have no specific function at all, being like a non-harmful parasite symbiotically living in the genome.

The subject is stimulating and the need to understand the clinical significance of the polymorphisms has encouraged a lot of investigations. Many authors have discussed the involvement of particular chromosomal variants in infertility[18-25], malformations, retardation and congenital defects[26-34], and cancer[40-43]. The conclusions are not unanimous or even controversial, the mechanisms not clear. Single case reports have no statistical significance. When different groups are compared the groups are usually too small for good statistical analysis. There are very few possibilities to compare one investigation with another, because different techniques or different criteria may be used in different laboratories for the evaluation of the variants. Each investigation has its own results. We have the impression that a long-lasting retrospective experience in cytogenetic work may contribute to a better understanding than that offered by limited programmed prospective investigations.

The problem of the clinical significance of human chromosome polymorphisms and their possible association with congenital defects has been considered in our laboratory since 1973. Three thousand index patients belonging to different clinical groups and 800 normal controls were comprehensively examined by multiple, consecutive banding techniques. The controls were of different ethnic groups (East European, North African and several groups of Asian Jews and Arabs), so that ethnical variations could also be considered.

It is not the purpose of this paper, however, to make numerical statistical analyses. We want to express some general concepts conceived by the observation of our material, although we are fully aware of the modesty of our experience.

## MULTIPLE VARIANTS. CHROMOSOMAL INSTABILITY. POSSIBLE GENE MUTATIONS, DELETIONS, DUPLICATIONS

Sometimes we are concerned with the significance of a variant of a particular chromosome and we are inclined to concentrate our attention on it, overlooking other variants of the karyotype in question for evaluation. This may be a source of error. The different classes of repetitive DNA are unequally distributed in several different chromosomes. It seems reasonable to think that each class has a particular function and we must consider the possibility of a threshold effect for every class. A better understanding in this field will probably be achieved only when more accurate determinations become feasible at a molecular level. For other reasons also, such as possible interactions between chromosomes[44-46], the karyotype in our opinion must be considered only as a whole. On the other hand, so many polymorphisms must be concomitantly evaluated, that comprehensive comparison is almost impossible until automated analyses become available.

We are fully aware of the limitation of our observations. Therefore we wish to express only some general considerations. We have the impression that the presence of unusual variants may by itself be the evidence of a certain inherited chromosomal instability.

Relatively increased breakage rate has been observed in individuals carrying unusual polymorphisms, particularly of the constriction regions of chromosome 1, 9 and 16[43]. Increased sensibility *in vitro* to chemicals or radiations has been also reported[47,48], for such individuals.

One explanation might be that a certain defect in DNA repair causes chromosomal instability which may be greater in hetero- than euchromatic sites of the genome. Mismatching of the repetitive DNA sequences with subsequent unequal crossing-over may easily happen and may be well tolerated because heterochromatin is not translated. A variety of polymorphisms may be found.

Interestingly, many families have been reported where multiple unusual variants[6,30,49] or unusual variants and structural rearrangements[50,51] were concomitantly segregating. Reports in the literature may be incomplete concerning the polymorphisms, but each laboratory has surely its own records. We have examined many individuals presenting multiple unusual variants, sometimes in association with structural rearrangements. For example, Figure 4.1 shows a metaphase of an individual presenting, a: A1 with increased heterochromatin (1h+); b: partial pericentric inversion in C9, inv

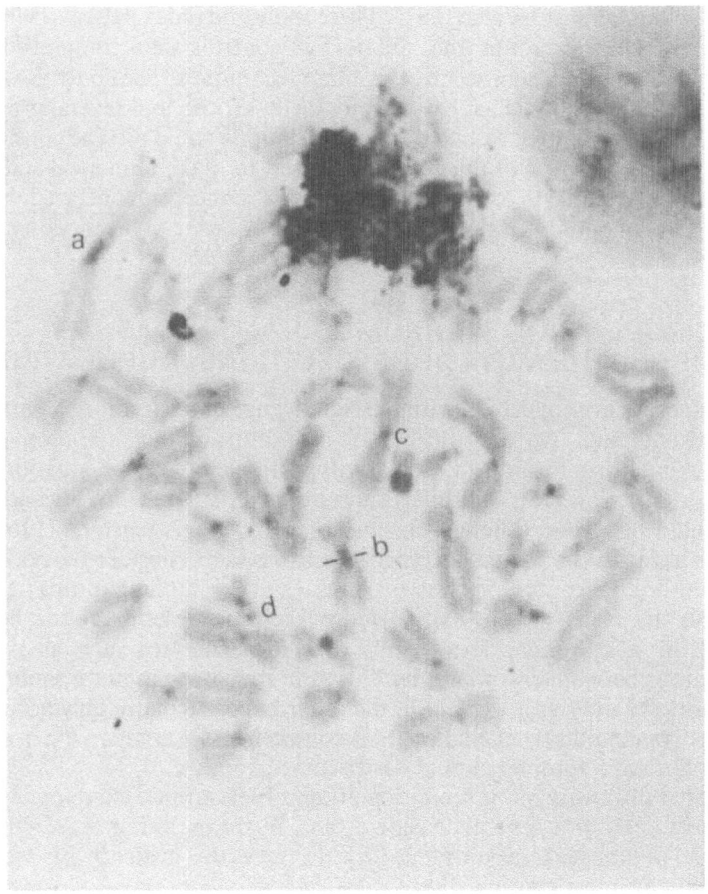

**Figure 4.1** Multiple heteromorphisms and structural rearrangements. a = A1h+; b = partial pericentric inversion of C9; c = pericentric inversion in the Y; d = prominent satellites in G22

9(p11 q11); c: pericentric inversion in the Y, and d: prominent satellites in G22.

The polymorphisms seem to be inherited without variations from one generation to the next, although it is difficult to rule out that the inborn instability might cause indetectable differences within the heterochromatin itself, which may change the clinical significance of the apparently same variant in different members of the same family. The chromosomal instability may eventually cause also minimal, submicroscopical undetectable deletions or duplications of genetic active material, of structural genes, accounting for the congenital defects and malformations frequently found in individuals carrying heteromorphic variants[29,43]. The polymorphism in this case is casual and not causal to the defect: both are due to the chromosomal instability. Being the heterochromatin–euchromatin junction sites most susceptible to this instability, it is reasonable to think that when euchromatic mutations, deletions or duplications happen, they frequently involve the sites next to the heterochromatic area. This may explain why particular genetic syndromes have been sometimes found in association with particular variants[26,30], although not necessarily[52]. Another explanation for such association may be the direct influence of the repeated sequences (when in particular quantity or composition) on the adjacent structural genes without any real change in the euchromatic material.

Mismatching of the repetitive DNA sequences with subsequent unequal crossing-over may happen during meiosis as well as in somatic cells. Somatic mosaicism of heterochromatic variants is very rare but it has been reported[41,53]. In our material we have at least three well documented examples of mosaicism for the heteromorphisms of chromosome 9. The first example was in a 60-year-old woman examined when developing acute myeloproliferative disease, 3 years after mastectomy for adenoma which was treated by chemotherapy and radiation. The mosaicism was thought to be a possible consequence of the treatment. The second and third examples were in young apparently healthy infertile males, suspected of being carriers of latent viruses. Virological studies indeed revealed herpes simplex infection in one (Figure 4.2), and a combined herpex simplex–adenovirus infection in the other (Figure 4.3).

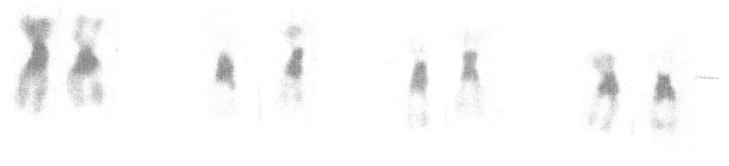

**Figure 4.2** Somatic mosaicism of a 9h variant

**Figure 4.3** Different heteromorphisms of the chromosomes 9 in the same individual

## THE BODYGUARD HYPOTHESIS. VIRUS SUSCEPTIBILITY. VIRUS INTEGRATION. VIRAL INFECTIONS AS CAUSE OF CONGENITAL DEFECTS

Hsu[13] proposed that the large areas of non-translated repeated sequences of DNA have the function of protecting the more important early replicating euchromatin by adsorbing the insults of the environment: the bodyguard hypothesis. This may be true for insults caused by chemicals, but an increased quantity of these repeated sequences may increase the risk when viral infections are considered. The repetitive sequences may be a good hiding site for viral DNA[54], thus permitting its integration in the genome. Indirect evidence is supplied by the study of Narovlyansky et al.[55], who found a relationship between the quantity of particular heterochromatic variants and resistance or susceptibility to Coxsackie B virus. The increased breakage rate frequently found in carriers of heterochromatic variants[43] instead of, or besides, being the consequence of an inherited chromosomal instability and general genetic susceptibility may also be the result of latent persistent infections of slow viruses. Furthermore, chronic infections suppress immunocompetence and susceptibility to other infections may thus be increased.

It is of interest that fragility at the constitutive heterochromatin of chromosomes 1, 9 and 16 has been found associated with immunodeficiency[56,57]. In our own material we found good confirmation that fragility at the heterochromatic blocks of chromosomes 1, 9 and 16 is constantly associated with selective or combined immunodeficiencies, IgM deficiency being the most common.

Virological studies frequently revealed a significant titre of antibody to viruses, especially of the herpes group. Another chromosome that may be involved in this question of viral integration is the Y chromosome. Personal epidemiological studies have shown that particular viruses preferentially damage the heterochromatic part of the Y long arm (Figure 4.4).

Many different classes of repetitive DNA are present in this area. On the other hand, only one repeated DNA sequence is involved in the polymorphism of the human Y chromosome. Differences in the Y's polymorphic size are mirrored by differences in the size of the restriction peak deriving from the Hae III fragment[14,58]. Increased reproductive risk and congenital defects have been found associated with large variants of the Y chromosome[21,23,24].

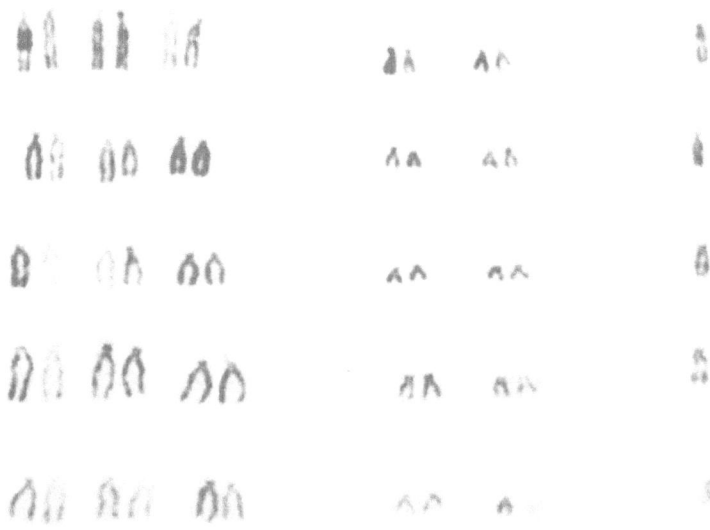

**Figure 4.4** Long damaged Y in the father of a child with Down syndrome

The increased quantity of the particular repetitive DNA involved in the polymorphism may be responsible for the reproductive risk, in a way that is still unclear. But the hypothesis that the risk is through the favoured eventual integration of viral DNA, and not merely by the quantity of the repeated sequences, may explain why other morphologically similar heteromorphic Ys are not associated with reproductive risk[24].

The Y chromosome loss or gain in leukaemias is a puzzling problem[59-61]. Integration of a viral DNA in the Y's long arm heterochromatin has not been proposed but may be a fact. It is worth mentioning that a Y chromosome structural rearrangement has been found in a case of chronic myelocytic leukaemia[62].

Polymorphisms involved in resistance or susceptibility to certain viruses seem to be also the NORs of the acrocentric chromosomes. The mechanism must be clarified. Unpublished results of our study on 102 families with central nervous system malformations show a very significant incidence of polymorphisms in the acrocentrics' short arms. Malformations of the central nervous system probably have a multifactorial origin: environmental factors have been considered and epidemiological studies seem to support also a viral involvement. Are the viruses in question RNA viruses? Is the genetic factor involved in the central nervous system malformations the individual genetic susceptibility to those viruses?

In conclusion, congenital defects are eventually caused by viral infections *in utero*. The teratogenic effect of the viral infection may be different according to the genetic susceptibility (or resistance) of the embryo and to its immunocompetence. All these factors seem to be related to the chromosomal polymorphisms in multiple ways.

## THE FRAGILE SITES ON HUMAN CHROMOSOMES

The fragile sites on human chromosomes have also been considered as a sort of polymorphism[63-66]. Sutherland[66] and others considering various hypotheses proposed among others that these sites are sites of viral modification or integration. We share and warmly support this hypothesis[67,68]. The congenital defects found in families where such polymorphisms are segregating are not specific. The hypothesis of a viral involvement, with all that implies (activation, replication, host's susceptibility and immunocompetence etc.) may fit for answering many of the puzzling questions on the fragile sites.

## POLYMORPHISMS AND ANEUPLOIDY

The tendency to aneuploidy in individuals carrying polymorphic variants has been evaluated and discussed[27,35-39]. Although the results of different studies are sometimes controversial, there is a feeling that at least some of the polymorphisms indeed increase the risk of aneuploidy[38,39]. We have this feeling too, especially for the 15p+ variants (Figures 4.5, 4.6).

Due to possible interactions between several different chromosomes[44-46] the tendency to aneuploidy may not be specific. On the other hand, several reports suggest that the chromosome homologous of the polymorphic one

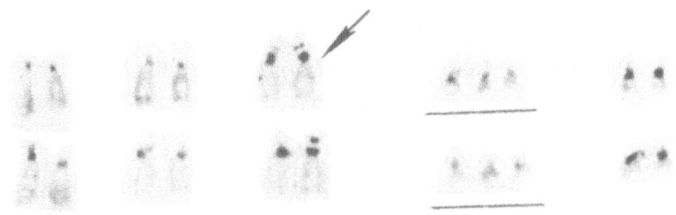

**Figure 4.5** Chromosome 15 with prominent heterochromatic satellites in a Down syndrome patient

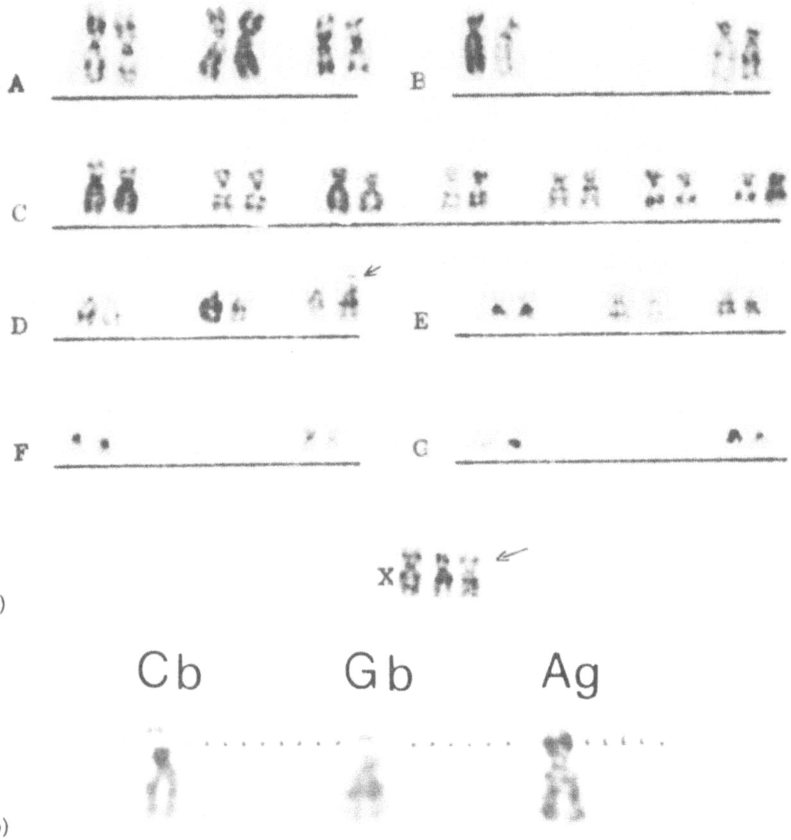

**Figure 4.6** (a) Long stalk and prominent satellites in a chromosome 15 of a XXX female
(b) Variant Dp$^+$ by C-banding (Cb), G-banding (Gb) and Ag staining (Ag)

may be preferentially involved. Partial and complete trisomy 9 has been frequently reported in patients with a polymorphic 9[69-73] (Figure 4.7). Trisomy 22 has also been found associated with variants of chromosome 22[74,75]. Significantly increased incidence of polymorphic chromosome 21

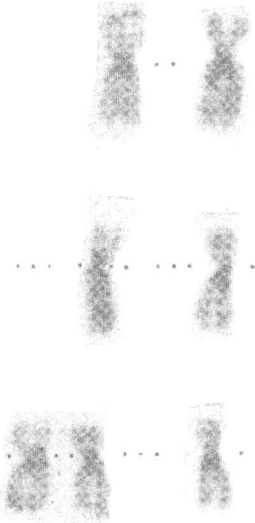

**Figure 4.7** Partial pericentric inversion of chromosome 9 in mother (upper row) and daughter with mosaic trisomy 9 (Ref. 73)

presenting a large NOR has been found in our group of Down's patients (Figure 4.8). When the satellites are small these kinds of variants may be overlooked without Ag staining. Interestingly a very high incidence of 21p+ has also been found in a group of patients with otherwise normal karyotype

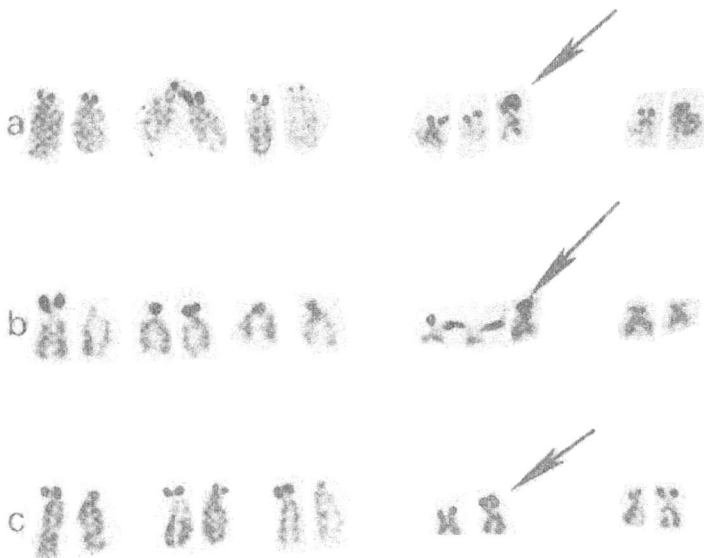

**Figure 4.8** Variant chromosome 21 with large NOR in two Down patients and one patient with suspected trisomy 21

but presenting some clinical features of Down's syndrome. It is difficult to exclude *a priori* a phenotypic influence of the polymorphism, but it seems not very likely.

We must have in mind the possibility of a low undetectable mosaicism for trisomy 21. Low mosaicisms are often very difficult to detect. Sometimes the mosaicism rapidly decreases and even disappears shortly after birth[76], and later in life it may be undetectable at all. A karyotypic analysis in a defective child must be as early as possible and on a large number of cells in multiple tissues[75].

Trabalza *et al.*[77] reported on a malformed child presenting a normal karyotype with a variant 22p+ which was inherited from the normal mother and normal maternal grandfather. Although a non-specific effect of the heterochromatic material was claimed by the authors as cause of the malformations, these are very suggestive of a trisomy 22 mosaicism. An undetected mosaicism seems to us a more logical explanation for the child's abnormal phenotype compared with the normal phenotype of his mother and grandfather.


## CONCLUSIONS

There is the feeling that polymorphisms of human chromosomes have a clinical significance and that they may be sometimes responsible for congenital defects. The mechanisms proposed are only conjectural and difficult to verify. We have tried to point out a few of them.

(1) A direct non-specific involvement of heterochromatic polymorphisms may be their eventual influence on metabolic processes[15,17], particularly important in the developing embryo.

(2) Involvement of the heterochromatin in the meiotic recombination mechanisms may influence the development and the incidence of aneuploidy[15,32].

(3) The polymorphisms may, by themselves, be the evidence of an inborn chromosomal instability and therefore susceptibility to environmental damage, which may be detrimental to the developing embryo.

(4) The polymorphic sites seem to be involved, probably in several different ways which must be still clarified, in virus susceptibility, resistance and perhaps integration. A possible relationship between viral infections and congenital defects or malformations is known.


### Acknowledgements

The authors warmly thank Mrs Neli Halperin, Sara Hadar, Shifra Golan for skilful technical assistance, Mr Joseph Sadovnik and Mrs Miriam Kleiner for the photographic work, and the Departments of Hasharon Hospital for providing the material of study.

# References

1 Tantravahi, R., Miller, D. A. and Miller, O. J. (1977). Ag staining of nucleolus organizer regions of chromosomes after Q, C, G or R banding procedures. *Cytogenet. Cell Genet.*, **18**, 364

2 McKenzie, W. H. and Lubs, H. A. (1975). Human Q and C chromosomal variations: distribution and incidence. *Cytogenet. Cell Genet.*, **14**, 97

3 Madan, K. and Bruinsma, A. H. (1979). C-band polymorphism in human chromosome No. 6. *Clin. Genet.*, **15**, 193

4 Sofuni, T., Tanabe, K., Ohtaki, K., Shimba, H. and Awa, A. A. (1974). Two new types of C-band variants in human chromosomes (6ph+ and 12ph+). *Jpn. J. Hum. Genet.*, **19**, 251

5 Simola, K., Karli, P. and De la Chapelle, A. (1977). Two pericentric inversions of human chromosome 11. *J. Med. Genet.*, **14**, 371

6 Mayer, M., Matsuura, J. and Jacobs, P. (1978). Inversions and other unusual heteromorphisms detected by C-banding. *Hum. Genet.*, **45**, 43

7 Crossen, P. E. (1975). Variation in the centromeric banding of chromosome 19. *Clin. Genet.*, **8**, 218

8 Hoehn, H., Au, K., Karp, L. E. and Martin, G. M. (1977). Somatic stability of variant C-band heterochromatin. *Hum. Genet.*, **35**, 163

9 Robinson, J. A., Buckton, K. E. and Evans, H. J. (1978). A possible mutation of a fluorescence polymorphism. *Ann. Hum. Genet.*, **41**, 323

10 Corneo, G., Ginelli, E. and Polli, E. (1970). Repeated sequences in human DNA. *J. Mol. Biol.*, **48**, 319

11 Jones, K. W. and Corneo, G. (1971). Location of satellite and homogenous DNA sequences in human chromosomes. *Nature (New Biol.)*, **233**, 286

12 Manuelidis, L. (1978). Chromosomal localization of complex and simple repeated human DNAs. *Chromosoma*, **66**, 23

13 Hsu, T. C. (1975). A possible function of constitutive heterochromatin: the bodyguard hypothesis. *Genetics*, **79** (Suppl.), 137

14 Miklos, G. L. G. and John, B. (1979). Heterochromatin and satellite DNA in man. Properties and prospects. *Am. J. Hum. Genet.*, **31**, 264

15 Varley, J. M., MacGregor, H. C. and Erba, H. P. (1980). Satellite DNA is transcribed on lambrush chromosome. *Nature (London)*, **283**, 686

16 Doolittle, W. F. and Sapienza, C. (1980). Selfish genes, the phenotype paradigm and genome evolution. *Nature (London)*, **284**, 601

17 Orgel, L. E. and Crick, F. H. C. (1980). Selfish DNA: the ultimate parasite. *Nature (London)*, **284**, 604

18 Nielsen, J., Friedrich, U., Hreisersson, A. and Zeuthen, E. (1974). Frequency of 9qh+ and risk of chromosome aberrations in the progeny of individuals with 9qh+. *Humangenetik*, **21**, 211

19 Boué, J., Taillemite, J. L., Hazael-Massieux, P., Leonard, C. and Boué, A. (1975). Association of pericentric inversion of chromosome 9 and reproductive failure in ten unrelated families. *Humangenetik*, **30**, 217

20 Howard-Peebles, P. N. and Stoddard, G. R. (1977). Inversions of chromosome 9 and associated risk for reproduction. *Am. J. Hum. Genet.*, **79**, 56A

21 Patil, S. R. and Lubs, H. A. (1977). A possible association of long Y chromosomes and fetal loss. *Hum. Genet.*, **35**, 233

22 Shabtai, F., Bichacho, S. and Halbrecht, I. (1977). Cytogenetics of male infertility. *Abstracts of the Second International Congress of Human Reproduction*, October 24–28, Tel Aviv, Israel, p. 90

23 Nielsen, J. (1978). Large Y chromosome (Yq+) and increased risk of abortion. *Clin. Genet.*, **13**, 415

24 Genest, P. (1979). Chromosome variants and abnormalities detected in 51 married couples with repeated spontaneous abortions. *Clin. Genet.*, **16**, 387

25 Serra, A., Bova, R., Neri, G., Brahe, C. and Tedeschi, B. (1980). Potential effects of pericentric inversion of the heterochromatic region of chromosome 9 on reproductive fitness. *Clin. Genet.*, **17**, 87 (Abstr.)

26 Gardner, R. J. M., McCreanor, H. R., Parslow, M. I. and Veale, A. M. D. (1974). Are 1q+ chromosomes harmless? *Clin. Genet.*, **6**, 383

27 Nielsen, J., Friedrich, U., Hreidersson, A. and Zeuthen, E. (1974). Frequency and segregation of 16qh+. *Clin. Genet.*, **5**, 316

28 Grosset, L., Jotterand, M. and Catti, A. (1975). A propos de quelques anomalies chromosomiques mineures. *J. Genet. Hum.*, **23** (Suppl.), 115

29 Kunze, J. and Mau, G. (1975). $A_1$ and $C_9$ marker chromosomes in children with combined minor and major malformations. *Lancet*, **1**, 273

30 Halbrecht, I. and Shabtai, F. (1976). Human chromosome polymorphism and congenital malformations. *Clin. Genet.*, **10**, 113

31 Lubs, H. A., Patil, S. R., Kimberling, W. J., Brown, J., Cohen, M., Gerald, P., Hecht, F., Myrianthopoulos, N. and Summitt, R. L. (1977). Q and C banding polymorphism in 7 and 8 year old children: Racial differences and clinical significance. In Hook, E. B. and Porter, I. H. (eds.) *Population Cytogenetics: Studies in Humans*, pp. 133–159. (New York: Academic Press)

32 Lubs, H. A. (1977). Occurrence and significance of chromosome variants. In Sparkes, R. S., Comings, D. E. and Fox, C. F. (eds.) *Molecular Human Cytogenetics*, pp. 443–455. (New York: Academic Press)

33 Funderburk, S. J., Sparkes, R. S., Klisak, I. and Goldenberg, I. T. (1978). Inherited acrocentric satellite variants: no correlation between satellite size and IQ scores. *Am. J. Hum. Genet.*, **30**, 82A

34 Tharapel, A. T. and Summitt, R. L. (1978). Minor chromosome variations and selected heteromorphisms in 200 unclassifiable mentally retarded patients and 200 normal controls. *Hum. Genet.*, **41**, 121

35 Therkelsen, A. J. (1964). Enlarged short arm of a small acrocentric chromosome in grandfather, mother and child, the latter with Down's syndrome. *Cytogenetics*, **3**, 441

36 Cote, G. B., Tsomi, K., Papadakou-Lagoyanni, S. and Petmezaki, S. (1978). Oligohydramnios syndrome and XYY karyotype. *Ann. Genet.*, **21**, 226

37 Faed, M. J. W., Lamont, M., Morton, H. G., Robertson, J. and Smail, P. (1978). An XYY boy with short stature and a case of Klinefelter's syndrome (XXY) in a family with inversion 9. *Clin. Genet.*, **14**, 241

38 Howard-Pebbles, P. and Stoddard, G. (1979). Pericentric inversions of chromosome number 9: benign or harmful? *Hum. Heredity*, **29**, 111

39 Lopetegni, P. H. (1980). 1, 9 and 16 C-band heteromorphisms in parents of Down's syndrome patients: distribution and etiological significance. *Jpn. J. Hum. Genet.*, **25**, 29

40 Atkin, N. B. and Baker, M. C. (1977). Pericentric inversion of chromosome 1: frequency and possible association with cancer. *Cytogenet. Cell Genet.*, **19**, 180

41 Atkin, N. B. (1977). Chromosome 1 heteromorphism in patients with malignant diseases: a constitutional marker for a high-risk group?. *Br. Med. J.*, **1**, 358

42 Berger, R., Bernheim, A., Le Coniat, M. and Vecchione, D. (1979). C-banding studies in various blood disorders. *Cancer Genet. Cytogenet.*, **1**, 95

43 Shabtai, F. and Halbrecht, I. (1979). Risk of malignancy and chromosome polymorphism: a possible mechanism of association. *Clin. Genet.*, **15**, 73

44 Gagne, R., Laberge, C. and Tanguay, R. (1973). Aspect cytologique et localisation intranucléaire de l'hétérochromatine constitutive des chromosomes $C_9$ chez l'homme. *Chromosoma*, **41**, 159

45 Steffensen, D. M., Duffy, P. and Prensky, W. (1974). Localization of 5S ribosomal RNA genes on human chromosome 1. *Nature (London)*, **252**, 741

46 Stahl, A., Luciani, J. M., DeVictor, M., Capodano, A. M. and Gagne, R. (1975). Constitutive heterochromatin and micronucleoli in the human oocytes at the diplotene stage. *Humangenetik*, **26**, 315

47 Meist, H. (1975). Der Clastogene Effekt von Trenimon auf die Chromosomen in vitro von phenotypisch gesunden Personen mit morphologisch Anomalea Karyotyp. *Acta Genet. Med. Gemell.*, **24**, 269

48 Seabright, M. (1976). Patterns of induced aberrations in humans with abnormal autosome complements. In Pearson, P. L. and Lewis, K. R. (eds.) *Chromosomes Today*. Vol. 5, pp. 293–298. (New York: Wiley)

49 Howard-Peebles, P. N. and Stoddard, G. R. (1976). A satellited Yq chromosome associated with trisomy 21 and an inversion of chromosome 9. *Hum. Genet.*, **34**, 223

50 Schmid, W. (1969). Satellites on the long Y chromosome arm: a familial autosome translocation in man. *Cytogenetics*, **8**, 415

51 Nora, A., Heideman, R., Peakman, D. and Morse, H. (1979). Cytogenetic studies in a acute leukemia patient following cerebellar astrocytoma. *Hum. Genet.*, **50**, 157

52 Bellesta, F., Ferrer, A. and Serra, M. (1974). 1q + chromosome in two families with Smith–Lemli–Opitz and Sedkal's syndrome. In Pearson, P. L. and Lewis, K. R. (eds.). *Chromosomes Today.* Vol. 5, p. 448 (New York: Wiley)

53 Craig-Holmes, A. P., Moore, F. B. and Shaw, M. W. (1975). Polymorphism of C-band heterochromatin. II. Family studies with suggestive evidence for somatic crossing over. *Am. J. Hum. Genet.*, **27**, 178

54 Lovinger, G. G. and Schochetman, G. (1980). 5′ terminal nucleotide sequences of type C retroviruses: features common to noncoding sequences of eucaryotic messenger RNAs. *Cell*, **20**, 441

55 Narovlyansky, A. N., Amchenkova, M. M., Stonova, N. S., Shatalin, K. Y., Gulevich, N. E. and Khesin, Y. E. (1980). Heterochromatin in cultured human chromosomes both susceptible and resistant to Coxsackie B viruses. *Tsitologiya*, **22**, 74

56 Hulten, M. (1978). Selective somatic pairing and fragility at 1 q12 in a boy with common variable immunodeficiency. *Clin. Genet.*, **14**, 294 (Abstr.)

57 Tiepolo, L., Maraschio, P., Gimelli, G., Cuoco, C., Gargani, G. F. and Romano, C. (1979). Multibranched chromosomes 1, 9, 16 in a patient with combined IgA and IgE deficiency. *Hum. Genet.*, **51**, 127

58 McKay, R. D. G., Bobrow, M. and Cooke, H. J. (1978). The identification of a repeated DNA sequence involved in the karyotype polymorphism of the human Y. *Cytogenet. Cell Genet.*, **21**, 19

59 Sakurai, M., Oshimura, M., Kakati, S. and Sandberg, A. A. (1974). 8–21 translocation and missing sex chromosomes in acute leukaemia. *Lancet*, **2**, 227

60 Padre-Mendoza, T., Farnes, P., Barker, B. E., Smith, P. S. and Forman, E. N. (1977). Y chromosome loss in childhood leukemias. *Am. J. Hum. Genet.*, **29**, 84A (Abstr.)

61 Berger, R. and Bernheim, A. (1979). Y chromosome loss in leukemias. *Cancer Genet. Cytogenet.*, **1**, 1

62 Verhest, A. and Lustman, F. (1980). Y chromosome structural rearrangement in Ph′ chromosome negative chronic myelogenous leukemia. *N. Engl. J. Med.*, **303**, 53

63 Schmid, E. and Bauchinger, M. (1969). Structural polymorphism in chromosome 17. *Nature (London)*, **221**, 387

64 Magenis, R. E., Hecht, F. and Lovrien, E. W. (1970). Heritable fragile site on chromosome 16: probable localization of haptoglobin locus in man. *Science*, **170**, 85

65 Giraud, F., Ayme, S., Mattei, J. F. and Mattei, M. G. (1976). Constitutional chromosomal breakage. *Hum. Genet.*, **34**, 125

66 Sutherland, G. R. (1979). Heritable fragile site on human chromosomes. II. Distribution, phenotypic effects and cytogenetics. *Am. J. Hum. Genet.*, **31**, 136

67 Shabtai, F., Bichacho, S. and Halbrecht, I. (1980). The fragile site on chromosome 16 (q21 q22). Data on 4 new families. *Hum. Genet.*, **55**, 19

68 Shabtai, F., Bichacho, S. and Halbrecht, I. (1981). Fragile sites on human chromosomes. (Proceedings of the Dubrovnik Symposium of the European Society of Human Genetics.) *Clin. Genet.* (to be published)

69 Bowne, P., Ying, K. and Chung, G. (1974). Trisomy 9 mosaicism in a newborn infant with multiple malformations. *J. Pediatr.*, **85**, 95

70 Schinzel, A., Hayashi, K. and Schmid, W. (1974). Mosaic-trisomy and pericentric inversion of chromosome 9 in a malformed boy. *Humangenetik*, **25**, 171

71 Francke, W., Benirsohke, K. and Jones, O. (1975). Prenatal diagnosis of trisomy 9. *Humangenetik*, **29**, 243

72 Seabright, M., Gregson, N. and Mould, S. (1976). Trisomy 9 associated with an enlarged 9 qh segment in a liveborn. *Hum. Genet.*, **34**, 323

73 Friedman, M., Shabtai, F., Halbrecht, I. and Elian, E. (1981). Normal psychomotor development in a child with mosaic trisomy and pericentric inversion of chromosome 9. *J. Med. Genet.* (to be published)

74 Schmid, L., Hepple, L. and Nitowsky, H. (1977). Maternal bisatellited chromosome 22 associated with the trisomy 22 syndrome. *Am. J. Hum. Genet.*, **29**, 95A (Abstr.)

75 Dulitzky, F., Shabtai, F., Zlotogora, J., Halbrecht, I. and Elian, E. (1981). Unilateral radial aplasia and trisomy 22 mosaicism. *J. Med. Genet.* (In press)

76 La Marche, P. H., Heisler, A. B. and Kronemer, N. S. (1967). Disappearing mosaicism. *RI Med. J.*, **50**, 184

77 Trabalza, N., Furbetta, M., Rosi, G., Donti, E., Venti, G. and Migliorini Bruschelli, G. (1978). Etude d'un sujet avec caryotype 46,XY,22p+. *J. Genet. Hum.*, **26**, 177

# 5
# H-Y antigen and the aetiology of genetic disorders in sex determination

A. SHALEV

## INTRODUCTION

### The process of sex determination

Mainly for historical reasons, the term 'sex determination' has occasionally been confused with the process of gonadal differentiation, the determination of the body sex (also referred to as 'secondary sex determination') or has been considered from one point of view, e.g. chromosomal. It is therefore essential to define sex determination as the complex of molecular (nucleic acids and their products), cellular and histological events which combine to establish testicular cords or follicular structures. Though the current state of knowledge does not permit drawing a sharp line between the nature of 'determining' and 'differentiation' events, it can be assumed with confidence that independent genes and processes control these stages. Once the essential cellular components of the gonad have been established, gonadal growth and function as well as secondary sex characteristics are perpetuated by hormonal feedback mechanisms controlled by an as yet undefined array of genes[1]. Thus, the present definition of sex determination includes both the genetic 'programme' for the establishment of the gonadal sex and the material execution of this programme at the morphogenic level. Assuming that a healthy set of genes for sex determination are transmitted by the parents, particular cistrons are soon activated, probably in an irreversible fashion, to produce sex-specific substance(s) which direct certain embryonic cells to organize the primary structures of the gonad. H-Y antigen has now been identified as such a sex-specific product in mammalian males. The existence of a female-specific substance has not been recognized as yet and it is possible that ovarian organogenesis simply proceeds in the absence of the male (H-Y) inducer[2].

The organogenesis of the human gonad has previously been described in detail by Jirasek[3,4], and is briefly summarized here. The undifferentiated

51

mammalian gonad, also referred to as gonadal blastema, comprises meso-blastic (primitive coelomic) cells, immigrating primordial germ cells and mesenchymal cells derived from the interstitium of the urogenital ridge. The onset of testicular organogenesis, at day 46–49 of the human embryo, is marked by polarization of some mesoblastic cells and the appearance of thin septa that divide the blastema into testicular cords. This is followed by the incorporation of germ cells into the cords and the formation of seminiferous cords at the periphery of the testicular cords which then separates the tubule from the surface epithelium. The accumulating connective tissue forms the tunica albuginea and the mediastinum testes. Fibroblast-like cells around the seminiferous cords acquire typical Leydig cell shape at about day 60. The Leydig cells are capable of synthesizing testosterone and also bind hCG and LH and thereby mark the onset of the differentiation cycle.

The transformation of the gonadal blastema into an ovary starts at day 45–55 of the embryo's life with the appearance of irregular epithelial cords and groups which remain attached to the epithelium. The intensive proliferation of primordial germ cells at the periphery produces more cords consisting of primordial germ cells and primitive granulosa cells. The primitive granulosa cells organize around oocytes to form the primordial follicles. These in turn will gradually mature to become vesicular follicles. A cell type equivalent to the male's Leydig cells that marks the onset of sexual differentiation in females has not been identified.

A central aspect of repoduction is the relatively high incidence of genetic disorders in sex determination. The chromosomal karyotype or the phenotype of an individual *per se* can, however, not serve as direct measures for the operation of the sex determining mechanism, since in no case can they point to the exact defective locus. This is illustrated by the seemingly 'normal' karyotype of, and external phenotype of, some XX males as well as individuals with XY chromosomes and gonadal dysgenesis. Such cases nonetheless, provide important clues to the existence of mutations of sex determining loci and the possible interactions between such loci. The existence of diverse genetic disorders of sex determination (see further below) and the difficulty in identifying the substance(s) that are involved did not permit the construction of a satisfactory model of sex determination in mammals. It was the discovery of H-Y antigen[5] and its role in sex determination[6] that allowed for the first time a testable experimental model of sex determination at the molecular level. This advance was largely due to the contributions of several distinguished investigators: E. Witchi[7], A. Jost[2], J. L. Hamerton[8], R. V. Short[9], U. Mittwoch[10] and S. Ohno[11]. General reviews on the nature of sex determination in animals are provided elsewhere[12,13].

## H-Y antigen: historical perspective

The histocompatibility-Y (H-Y) antigen was discovered by Eichwald and Silmser[5] as a minor male-specific antigen. Using the skin grafting technique it was shown that within syngeneic strains of mice females rejected the male skin transplant, but rejection did not occur when males were transplanted with female grafts or grafts were exchanged within the same sex. The analogous phenomenon was soon demonstrated in several other species including rats, rabbits, chickens and fish[14]. It was then clearly established that the expression of the antigen depends on the presence of the Y chromosome and was unrelated to the number of X chromosomes. The presence of H-Y antigen was subsequently demonstrated on virtually all male tissues examined, including skin, lung, blood cells, thyroid, pituitary and salivary glands, embryonic cells and others. These and more data on the achievements during the years 1955–1972 were summarized in a comprehensive review by Gasser and Silvers[14].

A central issue still which has been debated until recently was whether H-Y antigen is a constitutive component on the cell surface membrane or is induced by male androgens[14–16]. The significance of this issue in relation to cause and effect in sex determination is evident. In an elegant experiment Wachtel et al.[17] demonstrated that male lymphoid cells resident in female (chimaeric) mice continued to express H-Y antigen. This strongly argued against a hormonal or environmental influence on the expression of H-Y antigen. Since then the hormone-independent expression of H-Y antigen in mammals has unequivocally been established in several studies. Moreover, the detection of H-Y antigen on the cells of pre-implantation mouse embryos[18,19] confirmed the direct genetic control on H-Y antigen synthesis.

A major reason that initially led Wachtel et al.[6] to propose the role of H-Y antigen in sex determination was the observation that the previously mentioned sex-specific histocompatibility antigen in the heterogametic sex of several species could be identified as one cross-reactive (H-Y) antigen when tested by the in vitro sperm cytotoxicity test[20]. Thus, only cells derived from the heterogametic sex of several species including the chicken (Gallus g. domestica) and two amphibian species were capable of removing H-Y antibodies from the mouse anti-H-Y serum. These findings not only suggested the high evolutionary conservation of H-Y antigen, but were also in keeping with the expectations from cytogenetic and functional studies that pointed to the existence of a conserved and common Y-chromosome dependent sex-determining mechanism in mammals[6,11]. Furthermore, the generally accepted notion that the sex of the homogametic sex of non-mammalian species can readily be reversed to that of the heterogametic one, but not vice versa, was also compatible with the presence of a gonad inducer (H-Y) only in the heterogametic sex. Since those early observations, the evolutionary conservation of an H-Y cross-reactive antigen has been indicated by serological analysis in a wide range of species and classes including invertebrates[21–27]. Attempts to detect allelism of the H-Y structural gene among strains of mice have not been successful[28].

H-Y antigen also attracted the interest of scientists from various disciplines because of its interesting characteristics such as: association with other molecules on the plasma membrane[22,29], the genetic control on the cell-mediated cytotoxic response to H-Y antigen[30,31], its role in fetal–maternal immune relations[32,33] and its possible effect on sex ratio[32,34]. These subjects are, however, beyond the scope of this chapter.

## H-Y ANTIGEN AND GENETIC DISORDERS OF SEX DETERMINATION AND DIFFERENTIATION

### Human

The X-linked *Tfm* mutation in both humans and mice results in insensitivity to androgen-dependent differentiation of the urogenital tract during embryogenesis. The molecular mechanism of this mutation has been discussed elsewhere[1,29]. Since a product engaged in sex determination should be expressed independently of this mutation, the finding that in both species affected individuals with a female phenotype and XY chromosomes normally express H-Y antigen on the cell surface was concordant with the suggested role of H-Y antigen in sex determination[35,36].

The rarely detectable (sex reversed) males with 46,XX chromosomes have a rather normal male phenotype but small azoospermic testes[37]. These infertile males occur at an estimated incidence of 1/25 000 infants. Three theories on the origin of the XX male syndrome have been proposed[37] based on the premise that the Y chromosome normally carries the trigger gene(s) for testicular organogenesis. These theories are: (1) undetectable mosaicism for Y chromosome or the loss of Y chromosome following testicular organogenesis, (2) translocation of the male-determining gene(s) from the Y chromosome to other chromosomes and (3) activation of non-Y-linked sex-determining gene(s). There is experimental evidence to support each of these aetiologies and it is possible that more than a single mutation can give rise to XX males. Whatever the case may be, the mere presence of testicular tissue or the ability to detect testicular organogenesis at early embryonic states must imply the operation of the male-determining mechanism. However, the reverse does not follow, since testicular organogenesis may require the presence of more than one specific gene and product and all these products should cooperate to assure testicular organogenesis. If H-Y antigen is assigned the primary role in testicular organogenesis one can expect to detect it in all disorders where testicular tissue is detectable, including XX males. Indeed, both the familial and the sporadic cases of XX males are H-Y antigen positive[38,39]. Furthermore, it was possible to follow the inheritance of H-Y gene(s) from the mothers of such individuals, who also expressed H-Y antigen to a low extent and it was suggested that the familial sex reversal syndrome has an autosomal–recessive mode of inheritance[38].

Another disorder in sex determination related to the aetiologies mentioned above is the formation of mixed gonads or ovotestes in some individuals. These true hermaphrodites (to distinguish from pseudohermaphrodites with one type of gonads but ambiguous genitalia) display wide polymorphism in regard to gonadal structure[37]. Again, H-Y antigen typing established that these cases were H-Y antigen positive[39–42]. The question then arises: if H-Y antigen is organizing the testicular structure, will it be also present on the ovarian part of an ovotestis? In a direct study of tissues derived from the testicular and ovarian parts of an ovotestis, only testicular cells were found to carry H-Y antigen[42], suggesting that the ovotestis originated from a primordium which was H-Y$^+$/H-Y$^-$ mosaic. Can this explanation account for the occurrence of ovotestes in other cases including those with undetectable chromosomal mosaicism? And why do some cells lose the capacity to express H-Y antigen? These problems remain to be studied. Three cases of pseudohermaphroditism with testicular tissue have also been described, all of which were H-Y antigen positive[43-45].

A third category of mutations that bears great interest to the understanding of male sex determination occurs in individuals with a 46,XY karyotype but degenerated gonads. The syndrome is known as XY gonadal dysgenesis or Swyer syndrome. Individuals with 46,XY gonadal dysgenesis suffer from gonadal arrest at a very early stage in embryonic life and the gonadal tissue frequently develops dysgerminomas or gonadoblastomas. Occasionally such individuals may exhibit markers of Turner syndrome (XO females) but otherwise their phenotype is quite normal. Pedigree studies have suggested an X-linked recessive or male-limited autosomal dominant mode of inheritance[46,47]. Testing for H-Y antigen in cases of XY gonadal dysgenesis disclosed three different genotypes: (1) those with a normal density of H-Y antigen on the cell surface, (2) those with a subnormal level of H-Y antigen and (3) H-Y negative cases. Thus, the classification by the H-Y antigen criteria indicated at least two if not three different aetiologies in the origin of this syndrome. It is noteworthy that the classification of individuals with gonadal dysgenesis based on their karyotype or phenotype was not possible prior to the development of *in vitro* methods for H-Y antigen typing. The possible mutation sites which can produce these three categories and the current literature data on such cases are summarized in Table 5.1.

**Table 5.1** Classification of the 46,XY gonadal dysgenesis syndrome and its possible aetiologies according to H-Y antigen typing

| H-Y antigen finding | No. of cases reported (%) | Possible aetiology | Reference |
|---|---|---|---|
| Negative | 6 (24%) | Mutation in structural or regular H-Y locus | 40, 48 |
| Subnormal level* | 2 ( 8%) | Partial mutation in H-Y structural locus | 48 |
| Positive | 17 (68%) | (1) Mutation (absence) of the gonadal H-Y antigen specific receptor (2) Mutation in putative H-Y operator gene (3) Structural mutation which causes loss of binding capacity to the gonadal receptor | 32, 48, 49–51, 90 |

* The actual deficiency in H-Y antigen on the cell surface is yet to be demonstrated by more precise quantitative methods

## Other species (mouse, goat, cow, dog)

Considering the high evolutionary conservation of H-Y antigen as well as the sex determining mechanism in mammals, it is reasonable to draw some general conclusions on the function of H-Y antigen by examining mutations of sex determination in various mammalian species. The analogy of the *Tfm* mutation in humans and mice has been mentioned above. Another mutation (*Sxr*) that causes sex reversal in XX or XO mice has been described by Cattanach *et al.*[52]. As in other mutations affecting sex determination, the karyotype of the *Sxr* mice seems normal but the mutation is clearly inherited in an autosomal dominant mode. XX,*Sxr* males are sterile and azoospermic, but XO,*Sxr* males produce spermatozoa. The presence of the second X chromosome in the XX,*Sxr* mice impairs spermatogenesis as a result of the inability of XX germ cells to complete meiosis in the testicular milieu[9]. As could be expected, *Sxr* mice express H-Y antigen and the possibility of a hidden translocation of Y chromosome material to an autosomal has been proposed[53].

Contrasting with the situation in *Sxr* mice, in polled goats a mutation for sex reversal of XX animals is inherited in an autosomal recessive mode. Interestingly, the mutation is exclusively found in polled (hornless) animals and therefore must either be closely linked to the dominant gene for polledness (*P*) or otherwise the phenomenon is a sex-limited pleiotropic effect of *P* itself[9]. Heterozygous (*P*/±) animals are normal but homozygous XX animals are hermaphrodites or pseudohermaphrodites. Testicular tissue or its remnants are always found in such intersex goats[54]. In addition, intersex XX, *P*/*P* goats display large variability in gonadal differentiation and extent of masculinization. In one study[55], in which five animals (two intersexes, two heterozygous females and one heterozygous male) were tested for H-Y antigen, the male and the intersex animals were shown to express H-Y antigen. Assuming that the H-Y gene comprises a group of supernumerary H-Y loci, the authors postulated that a minor translocation of subcritical dose of H-Y loci to an autosome is responsible for the recessive mode of sex reversal in polled goats[55]. To account for the autosomal dominant mode of sex reversal, the same authors suggested that translocation of a larger (critical) number of H-Y loci is responsible for the induction of testicular organogenesis in the mouse. This theory underlines the quantitative significance of H-Y density on the cell surface. It would be interesting to see whether the subnormal dose of H-Y is also correlated to lower excretion of H-Y antigen by Sertoli cells. In a second study of polled goats, attempts were made to quantitate H-Y antigen on cultured fibroblasts derived from seven animals[56]. The results were in essence similar to those in the first study; three males (heterozygous or normal) and two intersex animals were positive for H-Y antigen but in addition one heterozygous female (mother of an intersex) was also found to be slightly positive for H-Y antigen. It was inferred that the *P*-linked sex reversal gene in polled goats is probably an H-Y antigen locus and that the abnormal inheritance of H-Y genes may be responsible for the intersexuality in XX,*P*/*P* goats. The tight association between the traits for polledness and sex reversal (H-Y antigen expression) remains to be investigated. An interesting prediction which in fact holds true for all conditions of sex reversal is that the density of H-Y antigen should quantitatively correspond to the number of H-Y loci. For instance, if one assumes complete linkage between *P* and the H-Y locus on the same autosome and that this particular H-Y locus codes for only 20 % of the H-Y surface molecules usually found in XY males, the following quantitative differences should be measurable in various genotypes of goats:

| Genotype | XX,+/+ | XX,P/+ | XX,P/P | XY,+/+ | XY,P/+ | XY,P/P |
|---|---|---|---|---|---|---|
| Relative density (%) of H-Y antigen on the cell membrane | 0% | 20% | 40% | 100% | 120% | 140% |

The confirmation of the correctness of this hypothetical scheme awaits the development of quantitative methods for H-Y antigen determination. A more detailed discussion on the quantitative inheritance of H-Y loci is also provided elsewhere[57].

The bovine freemartin is a unique example of non-genetic sex reversal in a mammalian species. Chorionic vascular anastomosis between heterosexual twins allows the masculinization of the female fetus, known as 'freemartin'. An analogous situation is not known for other mammalian species such as humans or marmoset monkeys[29]. Extensive cytogenetic studies on the possible presence of XY cells in the masculinized gonad of the freemartin were controversial[9,29] and when detected XY cells were in a minority of less than 10%. Thus, transformation of the XX female gonad into a small atrophic testis by the transposed XY cells is questionable[29]. Short[58] has pointed to the need to postulate the transfer of a male-specific testis inducer substance from the male co-twin since male hormones are totally ineffective in inducing gonadal transformation in mammals. The finding that the freemartin's gonad is indeed H-Y positive[59] was well anticipated based on the assigned role of H-Y antigen in mammalian testicular organogenesis. If H-Y antigen is the sole testis organizer then the condition of the freemartin could have been explained by transmission of a blood-borne H-Y antigen from the male to the female co-twin or by the active secretion of H-Y antigen by the XY cells settled in the female's gonad. In this case the XX gonadal cells do not produce H-Y antigen themselves and therefore they are also not expected to be H-Y positive once the putative supply of H-Y antigen from the male twin is terminated at birth. Also, if the minority of XY cells were the only active supplier of H-Y antigen the amount of H-Y antigen in the gonad should be reduced or negligible as compared to the amount detected in normal testicular cells. Since none of the these possibilities has been ascertained in the single study on H-Y antigen in the freemartin[59], an alternative hypothesis may be considered. It is tempting to postulate that expression of H-Y antigen in the freemartin's gonad is induced by a blood-borne H-Y antigen inducer substance that is transmitted from the male and activates an autosomal or X-linked structural H-Y gene(s).

The study of a family of cocker spaniel dogs shed some light on the aetiology of sex reversal in XX males and hermaphrodites[60]. A sire of a hermaphrodite XX female carried, in addition to a normal Y chromosome, a translocation of Y-chromosome material onto an autosome. This sire appeared to have an elevated level of H-Y antigen. His hermaphrodite daughter that carried the same translocation was also positive for H-Y antigen. Despite having bilateral ovotestes, this female gave birth to two pups of which one was a normal XY male and the other was an XX male and H-Y antigen positive. The fact that the same translocation resulted in a fertile hermaphrodite (mother) in one case and an XX male (pup) in the second case suggested to the authors that these two syndromes have a common aetiology.

The reason for the outcome of a different phenotype with seemingly the same genotype might be related to the association of H-Y antigen with cell surface molecules of the major histocompatibility complex[29].

## THE OPERATION OF H-Y ANTIGEN IN SEX DETERMINATION

### The genetics of H-Y antigen

Inasmuch as ample evidence supports the role of H-Y antigen in testicular organogenesis, the exact location of the structural H-Y gene and other genes that interact in the process of sex determination remains controversial. Clinical studies using cytogenetic and morphological analysis suggest that the human testis determining gene (TDG) is mapped close to the centromere of the Y chromosome and most probably on the short arm of the Y[12]. The extensive testing for H-Y antigen of cases with a structural aberration of the Y chromosome suggested the same location for the H-Y gene[61-65]. Some evidence also indicates the location of the TDG and the H-Y gene on the long arm of Y chromosome[12,61]. The alternative to this hypothesis is that the Y chromosome carries a regulator gene while structural H-Y loci are situated on the X chromosome or the autosomes, or are distributed on more than one chromosome. If the last possibility is correct, then to account for the increased level of H-Y antigen in human subjects with two Y chromosomes[62], a quantitative effect of presumably Y-linked H-Y inducer loci on the synthesis of H-Y antigen must be considered. Thompson[66] has further speculated on this possibility. Factors like the high evolutionary conservation of the X chromosome in mammals[29], the conservation of the structural H-Y gene (inferred from serological cross-reactivity of the antigen) and the predominance of the XX/XY sex chromosomes mechanism in mammals favour a common chromosomal location for the genes involved in mammalian sex determination. Nonetheless, this might not apply to *all* mammalian species and particularly not to those with irregular sex chromosomes. For instance, expression of H-Y antigen in the gonad of the freemartin can be taken as an argument against the location of the H-Y structural gene on the Y chromosome of the genus *Bos* (see above) and the same can be stated for XX human males and XX polled goats in which evidence for a translocation of a Y-chromosome derived material has not been presented. In the male vole (*Ellobius lutescens*) XO sex chromosomes are found in both males and females, but expression of H-Y antigen is confined to males only[67]. This implies that in this species the structural H-Y gene is located either on the X chromosome or on an autosome. In birds, the W chromosome is the equivalent of the mammalian Y chromosome and is present in the heterogametic females. If the H-Y structural gene in birds is located on the W chromosome one would not expect sex-reversed male testis (ZZ chromosomes) to express H-Y antigen. However, it was shown that hormonally sex-reversed male chicks do express H-Y antigen[68] and therefore the location of the structural H-Y gene on the W chromosome of this species is incompatible.

The wood lemming (*Myopus schistocolor*) has an unusual mechanism of sex determination that furnished important insight into the genetics of sex determination – not only that some females regularly carry XY chromosomes, but these females mother only female offspring[69]. Transfer of only X chromosomes to the oogonia of these XY females is made possible through early elimination of the Y chromosome by a double non-disjunction process. Later analyses of the X chromosomes of this species revealed that two types of X chromosomes can be identified, one of which is always present in XY females[70]. How do XY females escape the male determination axis when all somatic cells (including the gonadal) carry the Y chromosome? Upon testing for H-Y antigen, Wachtel *et al.*[71] detected the presence of H-Y antigen only in normal XY males but not XY females or XX females. The authors suggested that the most plausible explanation for this situation is that an X-linked gene (mutation?) regulates or suppresses the expression of H-Y antigen in XY females. This explanation is in accord with the finding of two types of X chromosomes and the conclusion that the condition is inherited in an X-linked mode[70]. These findings provided the first evidence of an X-linked H-Y regulator gene and established a basis for a general concept that at least the following three genes might be necessary for testis determination in mammals: the H-Y structural gene, an X-linked regulator gene and a gene that specifies a gonad receptor for H-Y antigen. The detection of a gonad-specific receptor for H-Y antigen in both testis and ovary[29,72] provides that it is not coded by a Y-linked gene. Figure 5.1 shows the possible organization, location and relation between the presumed sex-determining genes. This model assumes a typical XX/XY sex chromosomes mechanism and follows the operon model. Possible sites of mutations and the resulting phenotypes are indicated. Genes controlling the fate of the germ line cells are not included in this model.

## Testicular organogenesis

The strong circumstantial evidence supporting the role of H-Y antigen in testicular organogenesis has further been corroborated by direct experimentation *in vitro*[73]. When gonadal cells from a fetus or neonate were enzymatically dissociated and then allowed to reaggregate in the presence of H-Y antigen or after they were lysostripped of H-Y antigen they reassociated in the form of testicular cords or follicular structures, respectively. This system was originally used by Moscona[74] to demonstrate specific organogenic interactions between cells endowed with common cell surface antigens. Organization of ovary-like structures by testicular cells were demonstrated in the mouse[75] and rat[76] and organization of testicular cords by ovarian cells were shown in the rat[77] and cow[78,79]. How does H-Y antigen exercise its organogenic function on gonadal cells? Early experiments in which heterosexual fetal gonads were cotransplanted into ectopic sites of an animal's body or were cocultured together *in vitro* indicated the capacity of the mammalian testicular tissue to transform early mammalian ovary into a testis-like organ[81,82] and an avian testis into an ovary-like structure[80]. The transformed tissue was in all cases derived from the homogametic sex and the transformation capacity was also distance dependent, i.e. when gonads were transplanted beyond a certain distance from

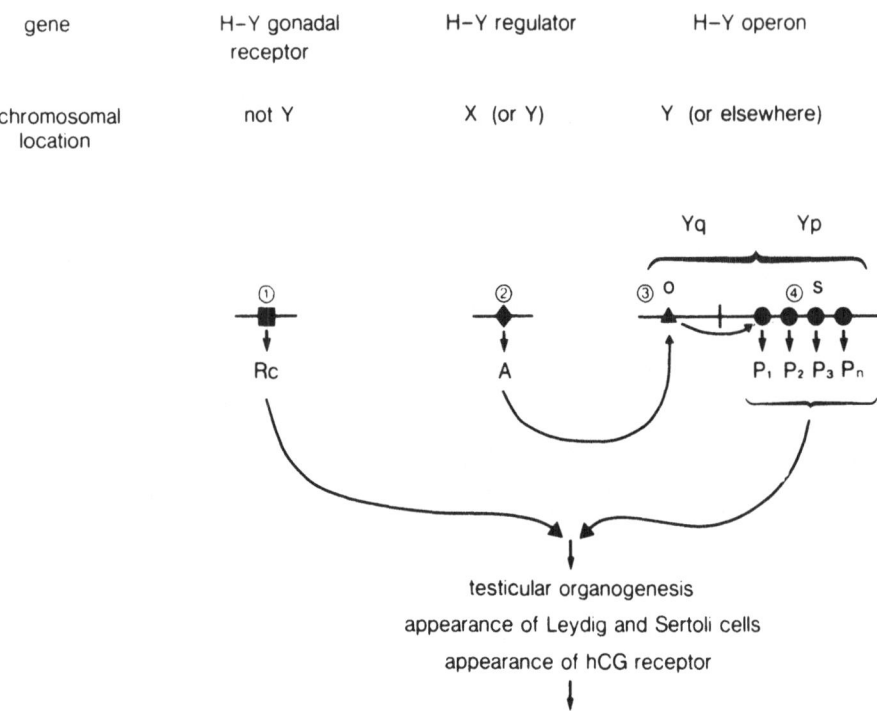

| gene | H–Y gonadal receptor | H–Y regulator | H–Y operon |
|---|---|---|---|
| chromosomal location | not Y | X (or Y) | Y (or elsewhere) |

**Figure 5.1** Model for operation of mammalian sex-determining genes and effect of possible mutations at these loci. Rc = gonadal specific receptor for H-Y antigen, A = activator molecule produced by the H-Y regulator gene, activates the H-Y operator gene (0), p,q = these letters correspond to the short arm (P) and long arm of the human Y chromosome, S = the structural H-Y gene which is comprised of supernumerary cistrons each coding for an H-Y molecule ($P_1$, $P_2$ . . . ) which can have exactly the same or similar structure, O = the operator gene that must be activated prior to translation of S; The H-Y operon might reside on the Y. *Mutations*: (1) = Mutation of the receptor locus will prevent testicular organogenesis and secondary male characteristics leading to a female phenotype with dysgenetic gonadal tissue. H-Y antigen expression is unaffected. This condition may be manifested in the H-Y positive 46,XY gonadal dysgenesis syndrome. (2) = Mutation of the regular gene may either reduce or completely eliminate the expression of H-Y antigen. This might be the case in XY females of the wood lemming (see above). (3) = Dysfunction of the operator gene entails shut-off in the translation of the H-Y structural gene(s) and no synthesis of H-Y antigen. (4) = A mutation in the structural H-Y cistrons can lead to partial or complete shut-off in H-Y antigen synthesis. This depends on the nature (local, chromosomal deletion etc.) of the mutation and the number and specificity of $P_1$ . . . $P_n$. Most of the quantitative differences in H-Y density may be interpreted by this type of mutation

each other, no transformation occurred. Since transformation occurred irrespective of the hormonal (host) environment, a specific testis inducer factor which is responsible for the sex reversal was implicated[82]. These experiments coincide very well with the recent studies on H-Y antigen showing it to be present in the heterogametic sex of vertebrates and capable of *in vitro* sex reversal of the homogametic gonad. Since the mammalian and avian H-Y antigens are serologically cross-reactive, it can be predicted that ovary organogenesis in birds may also be mediated by H-Y antigen. To administer its control on the orientation and differentiation of gonadal cells, H-Y antigen must attach to a specific receptor on gonadal cells. Indeed, both testicular and ovarian cells seem to be the only cell population endowed with H-Y antigen receptors[29,72,76]. It was also suggested that H-Y antigen is actively secreted by Sertoli cells and is present in the epididymis[83], and because of its effect over distance its mode of action was termed a 'hormone-like role'[29,59].

I believe that the study of H-Y antigen[84] will eventually lead to a full understanding of the sex determination mechanism in mammals; however, in the meantime much of this mechanism remains enigmatic. Two recent studies provide some clues to unknown and yet unexplained phenomena that may prove to be of central significance: the detection of an antigen common and specific to both Sertoli and follicular cells[79] and the appearance of hCG/LH receptor in newborn ovaries after *in vitro* treatment with H-Y antigen solution[85].

## UNSOLVED PROBLEMS IN SEX DETERMINATION

In the introduction to his book[29], Dr S. Ohno asks: 'Why sexual dimorphism?' An even more baffling question that occupied the minds of scholars for decades is: 'Why sex?'[86]. Cannot species reproduce equally successfully by cloning their genome vegetatively? Yet, the virtually ubiquitous distribution of sexual reproduction (recombination) among living organisms leaves no doubt that sex has definite evolutionary advantages. One can then attempt to look for the evolutionary origins of sex and ask 'how did it start'? The most comprehensive data on the mechanism of sexuality in a primitive organism come from the bacterium *Escherichia coli*. It is of interest to note that a specific piece of DNA, the episome, codes for sex specific pili (F pili or sex pili) and a sex specific antigen, the f[+] antigen[87]. The fertility of *E. coli* depends on the presence of these structures. Virtually nothing is known about sex determination in other primitive organisms such as fungi and algae where sexes are simply classified as + and − strains[88]. Is it only accidental that a unique piece of DNA (Y or W chromosome) is also characteristic to most metazoa and that sex specific antigens are common in various species[27]? If the evolution of sexuality from primitive organisms like *E. coli* is to be examined scientifically, much effort will have to be invested in understanding the mechanisms of sex determination in a wide array of classes. Alternatively, it can be postulated that sex determination originated independently and in parallel form at different stages of evolution. The evidence that H-Y antigen and the sex hormones are commonly shared by all vertebrate species lends support to the theory of gradual evolution of the sex determination mechanism. Sex determination in mammals has reached perfection in terms of genetic control while in lower organisms sex determination is vulnerable to the influence of sex hormones and environmental factors. Thus, it was claimed that in mammals a simple genetic programme of major genes controls sex determination[29]. In so far as some central questions remain to be clarified, this supposition might be premature. Some of the unresolved questions are as follows.

(1) How many cistrones does the H-Y gene represent and where are they located? What are the interacting genes and where are they located?

(2) Accepting the role of H-Y antigen in testicular organogenesis, what is the sequence of events induced by it? What is the role of the constitutive as compared to the presumed Sertoli cell-secreted H-Y antigen?

(3) Is there an equivalent to H-Y antigen in ovarian determination? Some interesting data in this respect were published[89], showing the presence of a substance in female fetal ovary that blocked the binding of H-Y antigen to ovarian cells.

(4) What is the genetic mechanism[90] that controls the fate of the germ line, and how do the germ line and the genes controlling the fate of gonadal soma interact?

(5) Does H-Y antigen play yet another role besides testicular organogenesis?

## Acknowledgements

This work was supported by grants from the National Cancer Institute and The Children's Hospital of Winnipeg Research Foundation. The author is very grateful to Drs A. H. Greenberg, J. L. Hamerton and S. Ohno for critically reading the manuscript and offering their comments.

## References

1 Goldstein, J. L. and Wilson, J. D. (1975). Genetic and hormonal control of male sexual differentiation. *J. Cell. Physiol.*, **85**, 365

2 Jost, A. (1970). Hormonal functions in the sex differentiation of the mammalian foetus. *Philos. Trans. R. Soc. Lond. B.*, **259**, 119

3 Jirasek, J. E. (1977). Morphogenesis of the genital system in the human. *Birth Defects: Orig. Art. Ser.*, **13**, 13

4 Jirásek, J. E. (1976). Principles of reproductive embryology. In Simpson, J. L. (ed.) *Disorders of Sexual Differentiation*, pp. 52–110. (New York, San Francisco, London: Academic Press)

5 Eichwald, E. J. and Silmser, C. R. (1955). Communications. *Transplant. Bull.*, **2**, 148

6 Wachtel, S. S., Ohno, S. and Koo, G. C. (1975). Possible role for H-Y antigen in the primary determination of sex. *Nature (London)*, **257**, 235

7 Witchi, E. (1957). The induced theory of sex differentiation. *J. Fac. Sci. Hokk. Zool.*, **13**, 428

8 Hamerton, J. L. (1968). The significance of sex chromosome derived heterochromatin in mammals. *Náture (London)*, **219**, 910

9 Short, R. V. (1972). Germ cell sex. In Beatty, R. A. and Gluecksohn-Wallsch, S. (eds.). *The Genetics of the Spermatozoon*, pp. 325–345. (Edinburgh)

10 Mittwoch, V. (1973). *Genetics of Sex Differentiation*. (New York, London: Academic Press)

11 Ohno, S. (1967). *Sex chromosomes and Sex-linked Genes*. (Berlin, New York: Springer)

12 Simpson, J. L. (1976). The nature of sex determination. In Simpson, J. L. (ed.) *Disorders of Sexual Differentiation*, pp. 141–155. (New York, San Francisco, London: Academic Press)

13 McCarrey, J. R. and Abbot, U. K. (1979). Mechanisms of genetic sex determination, gonadal sex differentiation, and germ-cell development in animals. *Adv. Genet.*, **20**, 217

14 Gasser, D. L. and Silvers, W. K. (1972). Genetics and immunology of sex-linked antigens. *Adv. Immunol.*, **15**, 215

15 Silvers, W. K., Billingham, R. E. and Sanford, B. H. (1968). The H-Y transplantation antigen: a Y-linked or sex influenced factor? *Nature (London)*, **220**, 401
16 Erickson, R. P.(1977). Androgen-modified expression compared with Y-linkage of male specific antigen. *Nature (London)*, **265**, 59
17 Wachtel, S. S., Goldberg, E. H., Zuckerman, E. and Boyse, E. A. (1973). Continued expression of H-Y antigen on male lymphoid cells resident in female (chimaeric) mice. *Nature (London)*, **244**, 102
18 Krco, C. J. and Goldberg, E. H. (1976). H-Y (male) antigen: detection on eight cell embryo. *Science*, **193**, 1134
19 Epstein, C. J., Smith, S. and Travis, B. (1980). Expression of H-Y antigen on pre-implantation mouse embryos. *Tissue Antigens*, **15**, 63
20 Wachtel, S. S., Koo, G. C. and Boyse, A. E. (1975). Evolutionary conservation of H-Y antigen. *Nature (London)*, **254**, 270
21 Shalev, A., Berczi, I. and Hamerton, J. L. (1978). Detection and cross-reaction of H-Y antigen by haemagglutination. *J. Immunogenet.*, **5**, 303
22 Fellous, M., Gunther, E., Kemler, R., Wiels, J., Berger, R., Guenet, J. L., Jacob, H. and Jacob, F. (1978). Association of H-Y male antigen with $\beta_2$-microglobulin on human lymphoid and differentiated mouse teratocarcinoma cell lines. *J. Exp. Med.*, **148**, 58
23 Zaborski, P. (1980). Invariability of H-Y antigen expression in the heterogametic sex of some amphibians and evidence for a sexual dimorphism of this antigen in *Pelodytes punctatus D* (Amphibia, Anura). *CR Acad. Sci. Ser. D.*, **289**, 1153
24 Pechan, P., Wachtel, S. S. and Reinboth, R. (1979). H-Y antigen in the teleost. *Differentiation*, **14**, 189
25 Müller, V. and Wolf, V. (1979). Cross-reactivity to mammalian anti-H-Y antiserum in teleostean fish. *Differentiation*, **14**, 185
26 Shalev, A. and Huebner, E. (1980). Expression of H-Y antigen in the guppy (*Lebistes reticulatus*). *Differentiation*, **16**, 81
27 Shalev, A., Goldenberg, P. Z. and Huebner, E. (1980). Evidence for an H-Y cross-reactive antigen in invertebrates. *Differentiation*, **16**, 77
28 Simpson, E., Brunner, C., Hetherington, C., Chandler, P., Brenan, M., Dagg, M. and Bailey, D. W. (1979). H-Y antigen: no evidence for alleles in wild strains of mice. *Immunogenetics*, **8**, 21
29 Ohno, S. (1979). *Major Sex-Determining Genes. Monographs of Endocrinology*, 11. (Berlin: Springer)
30 Hurme, M., Hetherington, C. M., Chandler, P. R. and Simpson, E. (1978). Cytotoxic T-cell responses to the H-Y: mapping of the Ir genes. *J. Exp. Med.* **147**, 758
31 Simpson, E. and Gordon, R. D. (1977). Responsiveness to H-Y antigen: Ir gene complementation and target cell specificity. *Immunol. Rev.*, **35**, 59
32 Shalev, A. (1980). *Immunogenetic Studies on H-Y Antigen. PhD Thesis*, University of Manitoba, Winnipeg
33 Shalev, A. (1980). Pregnancy-induced H-Y antibodies and their transmission to the foetus in rats. *Immunology*, **39**, 285
34 Lappé, M. and Schalk, J. (1971). Necessity of the spleen for balanced secondary sex ratios following maternal immunization with male antigen. *Transplantation*, **11**, 491
35 Bennet, D., Boyse, E. A., Lyon, M. F., Mathieson, B. J., Scheid, M. and Yanagisawa, K. (1975). Expression of H-Y (male) antigen in phenotypically female Tfm/Y mice. *Nature (London)*, **257**, 236
36 Koo, G. C., Wachtel, S. S., Saenger, P., New, M. I., Dosik, H., Amarose, A. P., Dorus, E. and Ventrulo, V. (1977). H-Y antigen: expression in human subjects with the testicular feminization syndrome. *Science*, **196**, 655
37 Simpson, J. L. (1976). Sex reversal in man: 46,XX males and true hermaphrodites. In Simpson, J. L. (ed.) *Disorders of Sexual Differentiation*, pp. 225–258. (New York, San Francisco, London: Academic Press)
38 de La Chapelle, A., Koo, G. C. and Wachtel, S. S. (1978). Recessive sex determination in human XX male syndrome. *Cell*, **15**, 837
39 Wachtel, S. S., Koo, G. C., Breg, R. W., Thaler, H. T., Morris, D. G., Rosenthal, I. M., Dosik, H., Gerald, P. S., Saenger, P., New, M., Lieber, E. and Miller, O. J. (1976). Serologic detection of a Y-linked gene in XX males and true hermaphrodites. *N. Engl. J. Med.*, **295**, 750

40 Ghosh, S. N., Shah, P. N., Gharpure, H. M. and Athreya, U. (1978). H-Y antigen in human intersexuality. *Clin. Genet.*, **14**, 31
41 Saenger, P., Levine, L. S., Wachtel, S. S., Korth-Schultz, S., Doberne, Y., Koo, G. C., Lavengood, R. W. Jr., German, J. L. III and New, M. I. (1976). Presence of H-Y antigen and testis in 46, XX hermaphroditism, evidence for Y-chromosomal function. *J. Clin. Endocrin. Metab.*, **43**, 1234
42 Winters, S. J., Wachtel, S. S., White, B. J. W., Koo, G. C., Javadpour, N., Loriaux, L. and Sherins, R. J. (1979). H-Y antigen mosaicism in the gonad of a 46,XY true hermaphrodite. *N. Engl. J. Med.*, **300**, 745
43 Mailhes, J. B., Pittaway, D. A., Rary, J., Chen, H. and Grafton, W. D. (1979). H-Y antigen-positive male pseudohermaphroditism with 45,X,46,XYq – mosaicism. *Hum. Genet.*, **53**, 57
44 Forobosco, A., Cheli, E., Noel, B. and Torus, J. (1978). H-Y antigen in a male with a 45,X karyotype. *Lancet*, **2**, 313
45 Vague, J., Guidon, J., Mathei, J. F., Luciani, J. M., Jouve, R. and Le-Gall, F. (1977). Apparent internal male pseudo-hermaphrodite. Gynecomastia. Negative gonad surgical research. Karyotype 46,XX. Presence of H-Y antigen. *Ann. Endocrinol.*, **38**, 395
46 Simpson, J. L. (1976). Gonadal dysgenesis. In Simpson, J. L. (ed.) *Disorders of Sexual Differentiation*, pp. 259–302. (New York, San Francisco, London: Academic Press)
47 German, J., Simpson, J. L. and Changati, R. S. K. (1978). Genetically determined sex-reversal in 46,XY humans. *Science*, **202**, 53
48 Wolf, U. (1979). XY gonadal dysgenesis and H-Y antigen. Report on 12 cases. *Hum. Genet.*, **47**, 269
49 Dorus, E., Amarose, A. P., Koo, G. C. and Wachtel, S. S. (1977). Clinical, pathologic and genetic findings in a case of 46,XY pure gonadal dysgenesis (Swyer's Syndrome). II. Presence of H-Y antigen. *Am. J. Obst. Gynecol.*, **127**, 829
50 Nazareth, H. R. S., Moreira-Filho, C. A., Cunha, A. J. B., Vierira-Filho, J. P. B., Lengyel, A. M. J. and Lima, M. C. (1979). H-Y antigen in 46,XY pure testicular dysgenesis. *Am. J. Med. Genet.*, **3**, 149
51 Moreira-Filho, C. A., Toledo, S. P., Bagonolli, V. R., Fronta-Pessoa, O., Bisi, H. and Wajntal, A. (1979). H-Y antigen in Swyer's syndrome and the genetics of XY gonadal dysgenesis. *Hum. Genet.*, **53**, 51
52 Cattanach, B. M., Pollard, C. E. and Hawkens, S. G. (1971). Sex-reversed mice: XX and XO males. *Cytogenetics*, **10**, 318
53 Bennett, D., Mathieson, B. J., Scheid., M., Yanagisawa, K., Boyse, E. A., Wachtel, S. and Cattanach, B. M. (1977). Serological evidence for H-Y antigen in Sxr,XX sex-reversed phenotypic males. *Nature (London)*, **265**, 255
54 Hamerton, J. L., Dickson, J. M., Pollard, C. E., Grieves, S. A. and Short, R. V. (1969). Genetic intersexuality in goats. *J. Reprod. Fertil.* (Suppl.), **7**, 25
55 Wachtel, S. S., Basrur, P. and Koo, G. C. (1978). Recessive male-determining genes. *Cell*, **15**, 279
56 Shalev, A., Short, R. V. and Hamerton, J. L. (1980). Immunogenetics of sex determination in the polled goat. *Cytogenet. Cell Genet.*, **28**, 195
57 Wachtel, S. S. and Ohno, S. (1979). The immunogenetics of sexual development. *Prog. Med. Genet.*, **3**, 109
58 Short, R. V. (1970). The bovine freemartin: a new look at an old problem. *Philos. Trans. R. Soc. Lond. B.*, **259**, 141
59 Ohno, S., Christian, L. C., Wachtel, S. S. and Koo, G. C. (1976). Hormone-like role of H-Y antigen in bovine freemartin gonad. *Nature (London)*, **261**, 597
60 Selden, J. R., Wachtel, S. S., Koo, G. C., Haskins, M. E. and Patterson, D. F. (1978). Genetic basis of XX male syndrome and XX true hermaphroditism: evidence in the dog. *Science*, **201**, 644
61 Koo, G. C., Wachtel, S. S., Krupen-Brown, K., Mittl, L. R., Breg, W. R., Genel, M., Rosenthal, I. M., Borgankar, D. S., Miller, D. A., Tantravahi, R., Schreck, R. R., Erlanger, B. F. and Miller, O. J. (1977). Mapping of the locus of H-Y gene on the human Y chromosome. *Science*, **198**, 940
62 Wachtel, S. S., Koo, G. C., Breg, R. W., Boyse, E. S. and Miller, O. J. (1975). Expression of H-Y antigen in human males with two Y chromosomes. *N. Engl. J. Med.*, **293**, 1070
63 Hasen, J., Ross, H. and Boyar, R. M. (1978). H-Y antigen: localization of H-Y gene. *Hormone Res.*, **9**, 102

64 Rary, J. M., Cummings, D. K., Jones, H. W. Jr. and Rock, J. A. (1979). Assignment of the H-Y antigen gene to the short arm of chromosome Y. *Heredity*, **70**, 78

65 Rosenfeld, R. G., Luzzatti, L., Hintz, R. L., Miller, O. J., Koo, G. C. and Wachtel, S. S. (1979). Sexual and somatic determination of the human Y chromosome: studies in a 46, XYp – phenotypic female. *Am. J. Hum. Genet.*, **31**, 458

66 Thompson, F. H. (1978). H-Y antigen gene loci. *Science*, **201**, 842

67 Nagai, Y. and Ohno, S. (1977). Testis-determining H-Y in XO males of the male vole (*Ellobius lutescens*). *Cell*, **10**, 729

68 Müller, U., Zenzes, M. T., Wolf, U., Engel, W. and Weniger, P. J. (1979). Appearance of H-W (H-Y) antigens in the gonads of oestradiol sex-reversed male chicken embryos. *Nature (London)*, **280**, 142

69 Fredga, K., Gropp, A., Winking, H. and Frank, F. (1976). Fertile XX- and XY-type females in the wood lemming *Myopus schistocolor*. *Nature (London)*, **261**, 225

70 Herbst, E. W., Fredga, K., Frank, F., Winking, H. and Gropp, A. (1978). Cytological identification of two X-chromosome types in the wood lemming (*Myopus schistocolor*). *Chromosoma*, **69**, 185

71 Wachtel, S. S., Koo, G. C., Ohno, S., Gropp, A., Dev, V. G., Tantravahi, R., Miller, D. A. and Miller, O. J. (1976). H-Y antigen and the origin of XY female wood lemmings (*Myopus schistocolor*). *Nature (London)*, **264**, 638

72 Müller, U., Aschmoneit, I., Zenzes, M. T. and Wolf, U. (1978). Binding studies of H-Y antigen in rat tissues. Indications for a gonadal-specific receptor. *Hum. Genet.*, **43**, 151

73 Wachtel, S. S. (1977). H-Y antigen and the genetics of sex determination. *Science*, **198**, 797

74 Moscona, A. (1961). Rotation-mediated histogenic aggregation of dissociated cells. *Exp. Cell Res.*, **22**, 455

75 Ohno, S., Nagai, Y. and Ciccarese, S. (1978). Testicular cells lyso-stripped of H-Y antigen organize ovarian follicle-like aggregates. *Cytogenet. Cell genet.*, **20**, 351

76 Zenzes, M. T., Wolf, U., Gunther, E. and Engel, W. (1978). Studies on the function of H-Y antigen: dissociation and reorganization experiments on rat gonadal tissue. *Cytogenet. Cell Genet.*, **20**, 365

77 Zenzes, M. T., Wolf, U. and Engel, W. (1978). Organization in vitro of ovarian cells into testicular structures. *Hum. Genet.*, **44**, 333

78 Nagai, Y., Ciccarese, S. and Ohno, S. (1979). The identification of human H-Y antigen and testicular transformation induced by its interaction with the receptor site of bovine fetal ovarian cells. *Differentiation*, **13**, 155

79 Ohno, S., Nagai, Y., Ciccarese, S. and Smith, R. (1979). In vitro studies of gonadal organogenesis in the presence and absence of H-Y antigen. *In Vitro*, **15**, 11

80 Akram, H. and Weniger, J. P. (1968). Feminization en culture *in vitro* du testicule embryonnaire de poulet par le testicule embryonnaire de veau. *Arch. Anat. Micro. Morphol. Exp.*, **57**, 369

81 McIntyre, M. N. (1956). Effect of the testis on ovarian differentiation in heterosexual embryonic rat gonad transplants. *Anat. Rec.*, **124**, 27

82 Mangoushi, M. A. (1977). Contiguous allografts of male and female gonadal primordia in the rat. *J. Anat.*, **123**, 407

83 Müller, U., Sibers, J. W., Zenzes, M. T. and Wolf, U. (1978). The testis as a secretory organ for H-Y antigen. *Hum. Genet.*, **45**, 209

84 Ohno, S., Nagai, Y., Ciccarese, S. and Smith, R. (1979). In vitro studies of gonadal organogenesis in the presence and absence of H-Y antigen. *In Vitro*, **15**, 11

85 Müller, U., Zenzes, M. T., Baukencht, T. and Wolf, U. (1978). Appearance of hCG-receptor after conversion of newborn ovarian cells into testicular structures of H-Y antigen in vitro. *Hum. Genet.*, **45**, 203

86 Maynard Smith, J. (1971). What use is sex? *J. Theor. Biol.*, **30**, 319

87 Ørskov, I. and Ørskov, F. (1960). An antigen termed f$^+$ occurring in F$^+$ E. Coli strains. *Acta Pathol. Microbiol. Scand.*, **48**, 37

88 Crandall, M. (1977). Mating-type interactions in micro organisms. In Cuatrecasas, P. and Greaves, M. F. (eds.) pp. 46–98. (London: Chapman & Hall)

89 Wachtel, S. S. and Hall, J. L. (1979). H-Y binding in the gonad: inhibition by a supernatant of the fetal ovary. *Cell*, **17**, 327

90 Wachtel, S. S., Koo, G. C., de la Chapelle, A., Kallio, H. and Heyman, J. M. (1980). *Hum. Genet.*, **54**, 25

# 6
# Effect of chromosome changes on body and mind development

MARGARETE S. MATTEVI AND F. M. SALZANO

## INTRODUCTION

The beginning of what can be called the 'human cytogenetic era' can be traced to 1956, with the classical paper by Tjio and Levan[1], in which a new methodology of study was described and the correct chromosome number of our species established. The new techniques that developed soon afterwards opened a whole new field in human biology, in relation to both normal and abnormal variation. Up to 1970 several new syndromes had been delineated, the aetiology of others, already known, clarified, and the importance of chromosome studies firmly recognized in human pathology. About 10 years ago a new significant technical development took place, with the so-called 'banding' methods which made possible the full identification of not only all the chromosome pairs, but also specific segments of them.

In this quarter of a century anomalies had been related with changes in practically all chromosomes of the karyotype. Besides the syndromes discovered in the pre-banding period, others were described, and their number now is of the order to two dozen. In addition, the importance of cytogenetic investigations in the determination of causal factors in disturbances of sexual development, mental retardation, repeated abortions and malignant growth, as well as in the monitoring of the effects of increasing levels of chemical substances or ionizing radiation, is not an object of dispute. An additional technical advance, involving the possibility of obtaining somatic cell hybrids, coupled with the development of sophisticated physicochemical determinations, made possible the detailed mapping of the human genome. This is of obvious importance for the determination of linkages between normal and abnormal characteristics. On the non-applied side, this new knowledge considerably improved our understanding of the fine structure and the spatial arrangement of the genetic material in man, and contributed to elaborated comparisons between the chromosomes of our species and those of the lower primates.

In what follows we will present a short appraisal of the new cytogenetic techniques, followed by a survey of the main results concerning the better delineated clinical syndromes. The effect of the lack or excess of autosome or sex chromosome material will be evaluated, as well as their influence on mental growth. To balance the fact that reviews of this kind are generally made by scientists living outside Latin America we shall try to give some special attention to papers which originated from this region. A review will also be made of the cytogenetic studies performed to date in our laboratory. The areas related to the cytogenetic aspects of human reproduction, especially the relationship between chromosome changes and abortions, as well as those of karyotypic abnormalities on human cancers, though of evident medical interest, will not be covered here.

## RECENT PROGRESS IN THE FIELD OF HUMAN CYTOGENETICS AND REPRODUCTION

### Banding techniques[2-5]

Until 1970 the mitotic chromosome was seen as a predominantly uniform structure, differentiated only by the centromere (primary constriction). In some cases, secondary constrictions and satellites could also be observed. The development of special staining techniques led to the discovery of numerous transverse bands, which are arranged in patterns characteristic for specific chromosomes.

The first bands described were those obtained using a solution of quinacrine mustard or some other fluorescent dyes. They have been termed *Q-bands*; about 300 of them can be identified by eye. If the chromosome preparations are incubated in saline solution, with or without previous treatment in alkali, and then stained with Giemsa, the so-called *G-bands* appear. An alternative method now probably more widely used than the original technique involves a . mild protease treatment (generally with trypsin) followed by Giemsa staining. These darkly staining bands are equivalent to the Q-bands.

Hot alkali pretreatment produces a reverse G pattern in which the regions that stained pale with Giemsa were found to be darkly staining; these reverse bands were christened *R-bands*. Prolonging the procedure telomeric (*T-bands*) are revealed.

*C-bands* appear when chromosomes are subjected to a strong alkali solution followed by treatment with warm saline and then stained with Giemsa. The prominent blocks formed are located at the centromeric regions of all chromosomes, as well as at the distal portion of the Y.

Other techniques produce types of bands that are specific for certain areas of the genome (the *N-bands* for the nucleolar organizing regions; the *G-11 bands* for the secondary constriction of chromosome 9, and the *BrdU-bands* for the early-replicating DNA segments).

What do the bands represent in terms of chromosome structure? They reveal the occurrence of repetitive DNA and differences in the base composition of this molecule and of protein components, as well as the presence of variable packing degress of these structures.

The Q- and C-bands especially show considerable variability in size, originating variants which present high population frequencies. Comparison of these frequencies in different populations has been hampered by the fact that the scoring procedures used by different investigators have varied widely and involved subjective decisions. Recently more objective methods, which involve quantitative measurements, have been developed. A review of those related to the C-bands can be found in Erdtmann et al.[6].

## Other methods of chromosome analysis[7-17]

The ultrastructure of human chromosomes is being actively investigated using ultrathin sectioning, spreading, critical point drying and X-ray diffraction studies. In addition, methods have been developed for comparing cells in the light and electron microscope. These investigations have furnished important clues about the organization of mitotic and meiotic chromosomes.

The cytochemical properties of the chromosomes can be investigated through the quantitation by eye or flow cytometry of their fluorescence patterns. Korenberg et al.[14] investigated the possible effect of the so-called 'hot-spots' (areas in which there is a high number of breaks and rearrangements) on trisomic abortions, while Hoehn and Callis[9] tested the use of flow-fluorometric methods in the diagnosis of aneuploidy.

Chromosomes can also be studied through a variety of agents that denature or fraction their DNA. They can act in three ways: (1) through the reassociation or hybridization of single DNA or RNA strands that have complementary base sequences, (2) by exploring differences in the buoyant densities of DNAs with variable base composition or (3) by using restriction endonucleases that cleave DNA at specific sites. By using these techniques it was possible to characterize a Y-chromosome-specific reiterated DNA.

The use of autoradiography and 5-bromodeoxyuridine (BrdU) indicated that the chromatid consists of but one DNA duplex that runs along its whole length and shows specific polarity. Moreover, BrdU-labelled cells allow the investigation of sister chromatid exchanges, that proved to be highly sensitive indices of the interaction of clastogens with chromosomes.

The frequency of aneuploidy in the sperm of normal males has been the subject of debate. Recently, a new method for its study has been developed, that uses hamster eggs. They can activate the human sperm to the point where their chromosomes can be studied directly[15]. This opens a whole new field of investigation, allowing the assessment of the effect on the male gamete of a variety of natural and experimentally induced phenomena.

## EFFECT OF CHROMOSOME ABERRATIONS IN THE DEVELOPMENTAL PROCESS IN HUMANS

### General

A chromosome abnormality does not generally imply a defect in the quality of the genes, but rather a quantitative abnormality in gene functions. The basic problem is one of genetic imbalance. The delicate control mechanisms which manage the timeliness and sequential nature of morphogenesis are disturbed. The result is the occurrence of multiple malformations. Single defects do not allow the identification of chromosome changes; they can only be related to clusters of malformations or syndromes.

## Autosomes

Table 6.1 presents the distinctive features in the most common autosomal full trisomies. Only traits that occur in at least half of the patients have been listed. As can be seen, there is practically no overlap between the main symptoms in

**Table 6.1  Distinctive features in the most common autosomal full trisomies***

| Ten most frequent clinical signs | Full trisomies | | |
|---|---|---|---|
| | 21 | 18 | 13 |
| Oblique palpebral fissures | + | | |
| Flat facial profile | + | | |
| Lack of Moro reflex | + | | |
| Round head shape | + | | |
| High cephalic index | + | | |
| 10 ulnar loops on fingers | + | | |
| Hyperabduction of hip | + | | |
| Small or absent lobes | + | | |
| Abundant skin on neck | + | ± | |
| Flat occiput | + | | |
| Developmental and/or mental retardation | + | + | + |
| Arches on 3 or more fingers | | + | |
| Low-set, malformed ears | | + | + |
| Failure to thrive | | + | |
| Micrognathia | | + | |
| Ventricular septal defect | | + | ± |
| Congenital heart disease | | + | |
| Flexion deformity of fingers | | + | ± |
| Elongated skull | | + | |
| Short sternum | | + | |
| Undescended testes | | ± | + |
| Presumptive deafness | | | + |
| Distal axial triradius | | | + |
| Single palmar crease | ± | | + |
| Microcephaly | | | + |
| Cleft palate | | | + |
| Long hyperconvex nails | | | + |
| Fibular arch in hallucal area | | | + |
| Polydactyly | | | + |

* Traits are listed in order of decreasing frequency. A + sign indicates the occurrence in 75% or more of the patients; a ± sign, a frequency between 50% and 74%

*Sources*: Hamerton[18], De Grouchy and Turleau[19], Hodes *et al.*[20], Hecht[21], Magenis and Hecht[22] and Miller[23]

these three syndromes, with some exceptions in the 18/13 comparison. Of the 29 signs listed, 11 are expressed on the head; 5 dermatoglyphic features can help in the diagnosis.

The 21 trisomy (Down syndrome) is by far the most common of the autosome aberrations. In Porto Alegre, despite the fact that karyotypes are made only from patients born from young mothers (below 30 years of age), 135 cases have been detected out of a total of 235 persons with abnormal chromosomes (57 %). Of these, 128 are simple trisomics, 5 have arisen through translocations (D/G or G/G) and 2 are mosaics (see Table 6.9). Our experience with trisomies 18 (Edward syndrome) and 13 (Patau syndrome) is much less extensive; in relation to the former 6 cases (1 mosaic) and to the latter 3 cases (1 mosaic) have been studied (4 % of the abnormal karyotypes). Similar frequencies of mosaics and translocation patients have been found in another Brazilian series[24]. These authors have also investigated the possible occurrence of mosaicism among 190 parents of regular 21 trisomy children with negative results.

As an example of an interesting case involving a mosaic Down syndrome patient we can mention the report by Vianna-Morgante and Nunesmaia[25] who studied an individual with a 45,XY,t(15;21) (15qter → 15p13: : 21p11 → 21qter)/46,XY,i(21) (qter → cen → qter) karyotype. The two chromosome abnormalities can be explained by independent origins or through single events (either a chromatid translocation or dissociation of a dicentric translocation chromosome). Toledo and Wajntal[26] described another mosaic, involving in this case the 13 trisomy, and compared the clinical picture of regular 13 trisomics with those of normal/trisomy mosaics.

The association between the 21 trisomy and mother's age led to the assumption that almost all cases were due to non-disjunction in the female. The banding techniques allowed a more precise investigation of this question. For instance, Mattei et al.[27], using the Q polymorphisms, verified that out of 42 patients non-disjunction was of paternal origin in 8 (19 %). In these the event occurred with equal frequency in the first or second meiotic division. This was not true in the cases of maternal origin, in which non-disjunction occurred far more often during the first meiotic division.

The main clinical signs present in the full 8 and five partial trisomies are listed in Table 6.2. Thirty-nine of them are from the craniofacial region, 16 are thoracic, 9 involve the limbs and 3 the psychomotor system. The number of cases of these relatively new syndromes studied so far is not large, ranging from 12 to 59. Their most distinctive features are as follows:

(1) *Trisomy 8*: everted lower lip, short and crowded neck with cutaneous folds, skeletal malformations, narrow shoulders and '*plis capitonnés*' palms and soles,

(2) *4p*: 'boxer' nose aspect and wide-spaced nipples,

(3) *Partial 4q*: pursed mouth, large and low-set ears with protruding antihelix, folded helix, tragus and lobe little developed,

(4) *9p*: asymmetrical grin (anxious look), hands with palms too long in relation to the fingers showing a single crease, and brachymesophalangy,

Table 6.2  Clinical signs in the full 8 and other partial trisomies

| Clinical signs | Trisomies | | | | | |
| --- | --- | --- | --- | --- | --- | --- |
| | 8 (N=35) | 4p (N=40) | Partial 4q (N=13) | 9p (N=59) | Partial 11q (N=13) | Proximal 15q (N=12) |
| *Craniofacial* | | | | | | |
| High forehead | + | | | | | |
| Forehead low and down | | + | | | ± | |
| Microcephaly | | | + | M | ± | ±M |
| Brachycephaly | | | + | + | ± | |
| High arched palate | | + | | | | ± |
| Microretrognathia | | | | | +* | |
| Everted lower lip | +* | | | + | | |
| Asymmetrical grin | | | | +* | | |
| Pursed mouth | | | +* | | | |
| Protuberant and fissured upper lip | | + | | | | |
| Glossoptosis | | | | | + | |
| Palpebral fissures horizontal or slanted downward and outward | | + | M | + | + | |
| Hypertelorism | M | + | | M | ± | |
| Blepharophimosis | | ± | + | | | |
| Microphthalmia | | + | + | + | | ± |
| Strabismus | M | + | | + | | ± |
| Narrow palpebral fissures | | | + | | | |
| Epicanthus | | | + | | ± | |
| Ptosis | | | + | | | |
| Full or flesh nose | | | | + | + | + |
| Nares open downward | | | | + | | |
| Pronounced nasofrontal angle | | | | | + | |
| Nasal bridge shallow or absent | | | + | | | |
| 'Boxer' nose aspect | | +* | | | | |
| Upturned nose | + | | | | | |
| Low-set ears | + | + | +* | | + | + |
| Large ears | + | + | +* | + | + | |
| Protruding antihelix | | + | +* | | + | |
| Folded helix | | + | +* | | + | |
| Round ears | | | +* | | | |
| Tragus and lobe underdeveloped | | | +* | | | |
| Flat antihelix | + | | | | | |
| Detached ear | | | | + | | |
| Posteriorly rotated ear | | | | | + | |
| Blind fistula or pediculated tag | | | | | + | |
| Round chin | | + | | + | | |
| Small and slightly receding chin | | | + | | | |
| Protruding chin | | + | | ± | | |
| Cerebral malformation | ± | ± | ± | | + | |
| *Thorax and abdomen* | | | | | | |
| Short neck | +* | + | + | + | + | |
| Wide neck | +* | | | | | |
| Wide-spaced nipples | | +* | + | + | + | |
| Narrow chest | | | | | + | |
| Wide chest | | | | + | | |
| Funnel chest | | | | + | | |
| Cardiopathy | ± | ± | ± | | + | |

(*continued*)

72

**Table 6.2** (*continued*)

| Clinical signs | Trisomies | | | | | |
|---|---|---|---|---|---|---|
| | 8 (N=35) | 4p (N=40) | Partial 4q (N=13) | 9p (N=59) | Partial 11q (N=13) | Proximal 15q (N=12) |
| Skeletal malformations | +* | + | | | + | |
| Narrow shoulders | +* | | | | | |
| Diastasis or hernia of the recti | | | ± | + | + | |
| Umbilical hernia | | | | + | | |
| Abdominal striae atrophicae cutis | | | | + | | |
| Visceral malformations | ± | ± | ± | | + | |
| Hypogonadism | ± | ± | | + | + | |
| Delayed menarche | | | | + | | |
| Cryptorchidism | ± | ± | + | | | |
| *Limbs* | | | | | | |
| Bird's wing posture of fingers | | | | | +* | |
| Large palms | | | | +* | | |
| Single palmar crease | | | | +* | | |
| Brachymesophalangy | | | | +* | | |
| Clinodactily | + | | | ± | | |
| Thumb anomalies | | ± | + | | | |
| 'Plis capitonnés' palms and soles | +* | | | | | |
| Clubfoot | | | | + | | ± |
| Hallux valgus | | | | + | | |
| *Psychomotor* | | | | | | |
| Mental retardation | + | + | + | + | + | +* |
| IQ | 50–80 | ≤50 | ≤50 | 55 | | <50 |
| Language difficulties | + | ± | | + | | |

+*: characteristic feature; +: occurrence in the majority of the cases; ±: irregular occurrence; M: moderate
   *Sources*: Caspersson et al.[28], Crandall et al.[29], Tuncbilek et al.[30], Baccichetti et al.[31], Cassidy et al.[32,33], Centerwall et al.[34], Giraud et al.[35,36], De Grouchy and Turleau[19], Francke[37]
   N = Approximate number of cases studied[38]

(5) *Partial 11*: microretrognathia, pleading or bird's wing posture of fingers,

(6) *Proximal 15q*: extremely profound mental retardation.

Gagliardi et al.[39] have reported a Brazilian 21-year-old male patient with the complete trisomy 8 (Figure 6.1) and compared him with other cases known at the time. Penchaszadeh and Coco[40] described an Argentinian male infant with an extra derivative chromosome 9 resulting from a 3:1 meiotic segregation of a maternal balanced translocation involving the long arms of chromosomes 7 and 9. In addition to the features of the 9p trisomy syndrome he showed marked congenital myopia and extreme penis hypoplasia.

Table 6.3 lists the main clinical signs in the most studied chromosome deficiencies. Besides one general characteristic, the traits listed by region are as follows:

(1) craniofacial: 35,

(2) thorax and abdomen: 14 and

73

**Table 6.3  Clinical signs in the most studied deficiencies**

| Clinical signs | Partial deficiencies | | | | |
| --- | --- | --- | --- | --- | --- |
| | 4p−<br>(N=45) | 5p−<br>(N=150) | 13q−<br>(N=72) | 18p−<br>(N=85) | 18q−<br>(N=50) |
| *General* | | | | | |
| Short stature | | + | | +* | + |
| *Craniofacial* | | | | | |
| Greek warrior helmet appearance | +* | | | | |
| Moonlike face | | +* | | ± | |
| Greek profile and asymmetrical face | | | +* | | |
| Midface dysplasia | | | | | +* |
| Microcephaly | +* | +* | + | | + |
| Dolichocephaly | + | | | | |
| Midline scalp defect | + | | | | |
| Prominent forehead | +* | | | | |
| Trigonocephaly | + | | ± | | |
| Fish-shaped mouth | + | | | | +* |
| Cleft palate | +* | | | | ± |
| Upper lip short and protruding | | | | + | |
| Lower lip everted | | | | + | |
| Decayed teeth or severe caries | | + | | ± | |
| 'Rabbit' teeth | | | +* | | |
| Micrognathia | + | +* | | | |
| Retrognathia | + | | | | |
| Palpebral fissures slanted downward | +* | + | | | |
| Hypertelorism | +* | +* | + | +* | + |
| Epicanthal folds | + | + | + | +* | ± |
| Strabism | | + | | +* | ± |
| Ptosis | | | + | +* | |
| Ocular anomalies | | | +* | | + |
| Retinoblastoma | | | +* | | |
| Nystagmus | | | | | + |
| Nasal bridge flat and broad | +* | + | + | ± | |
| Aplastic anthelix | | | | + | |
| Floppy ears | + | | | + | |
| Low-set ears | + | + | | + | |
| Preauricular tag or pit | + | ± | | | |
| Narrow or atresic auditory canals | | ± | | | + |
| Large and prominent ears | | | + | | +* |
| Helix with deep sulcus | | | + | | |
| Small chin | | | + | + | |
| Receding chin | | ± | | + | |
| *Thorax and abdomen* | | | | | |
| Abnormal larynx and cry | | +* | | | |
| Long neck | + | | | | |
| Short and webbed neck | | | + | ± | |
| Wide-spaced nipples | | | | +* | +* |
| Wide and depressed chest | | | | +* | |
| Skeletal anomalies | ± | + | + | + | |
| Congenital heart disease | | ± | ± | | ± |
| Elongated chest | +* | | | | |
| Inguinal or umbilical hernia | | ± | | + | ± |
| Diastasis recti | | ± | | + | |
| Cryptorchidism | + | | ± | ± | ± |
| Hypospadia | +* | | ± | ± | ± |

*(continued)*

**Table 6.3** (*continued*)

| | Partial deficiencies | | | | |
| --- | --- | --- | --- | --- | --- |
| *Clinical signs* | *4p−* (N = 45) | *5p−* (N = 150) | *13q−* (N = 72) | *18p−* (N = 85) | *18q−* (N = 50) |
| Hypoplastic or bifid scrotum | | | ± | | ± |
| Genital hypoplasia | | | | ± | ± |
| *Limbs* | | | | | |
| 'Hanging water-drops' hands | | | | | +* |
| Clinodactyly | | | | + | ± |
| Broad and short hands | | | | + | |
| Malformed hands or feet | ± | | + | | ± |
| Transverse palmar crease | | + | | | |
| Distal axial triradius | | + | | | |
| Excess of whorls | | | | | + |
| Excess of arches | + | | | | |
| Clubfoot and abnormal position, 2nd toe | | | | | + |
| Dimples on the epitrochlea, hands and knees | | | | | +* |
| Slender limbs | +* | | | | |
| *Psychomotor* | | | | | |
| IQ | <20 | <20 | <50 | 25−75 | 30−70 |
| Seizures | + | | | ± | ± |
| Hypothonia | + | + | | | + |
| Frog-like position | | | | | +* |

+*: characteristic feature;   +: occurrence in the majority of the cases;   ±: irregular occurrence
*Sources*: Guthrie *et al.*[41], Hamerton[18], Schinzel *et al.*[42], De Grouchy and Turleau[19], Carlin and Norman[43], Breg[44], Miller[45,46], Warburton[47]
N = Approximate number of cases studied[38]

(3) limbs: 11.

Four of them affect the psychomotor system. The approximate number of cases described for these five syndromes range from 45 to 150. Their most characteristic symptoms (besides mental retardation, present in all) are the following:

(1) *4 p− (Wolf-Hirschhorn syndrome)*: Greek warrior helmet appearance, microcephaly, prominent forehead, cleft palate, palpebral fissures slanted downward, hypertelorism, flat and broad nasal bridge, elongated chest, hypospadia and slender limbs,

(2) *5p− (cri-du-chat syndrome)*: moonlike face, microcephaly, micrognathia, hypertelorism and abnormal larynx and cry,

(3) *13 q−*: Greek profile and asymmetric face, 'rabbit' teeth and ocular anomalies including retinoblastoma,

(4) *18 p−*: short stature, hypertelorism, epicanthal folds, strabism, ptosis, wide-spaced nipples, wide and depressed chest and skeletal anomalies,

(5) *18 q−*: midface dysplasia, fish-shaped mouth, large and prominent ears, wide-spaced nipples, 'hanging water-drops' hands, dimples on the epitrochlea, hands and knees, and frog-like position.

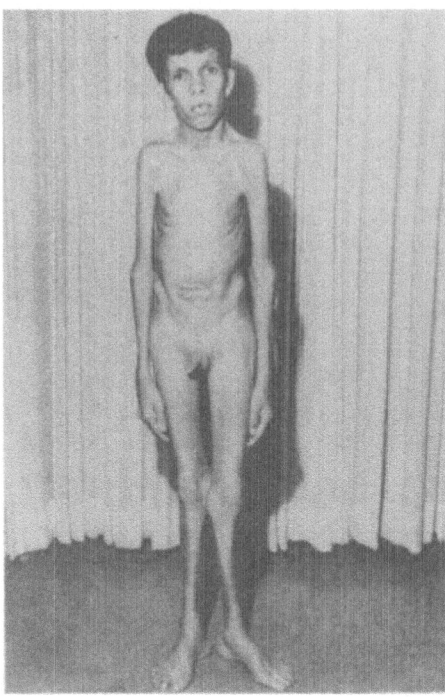

**Figure 6.1** Patient with the complete trisomy 8 studied by Gagliardi *et al.*[39]. (Reproduced with permission from the authors and the *Journal of Medical Genetics*)

In our laboratory we have found two cases of 4p−, three of 5p− and a r(13) which involved a small deletion of the long arm of this chromosome. Therefore, out of the 235 abnormal karyotypes observed, 2.5 % consisted of patients with these syndromes (see Table 6.9).

Marçallo *et al.*[48] described a 10-year-old Brazilian girl with a mental age of 7–8 years, several minor anomalies and body asymmetry. Her lymphocytes showed a balanced but very unequal translocation of most of 13q transferred to 7p, both translocation chromosomes being present, but all examined fibroblasts lacked the small translocation chromosome, hence being mono-somic for 13p, proximal part of 13q and a terminal portion of 7p. They suggested that the anomalies seen, as well as the borderline intelligence, are a result of deleterious effects during development of the unbalanced cell line.

As for chromosome 18, two reports in which abnormalities have been detected were presented by Brazilian investigators. Vianna-Morgante *et al.*[49] described a pericentric inversion in this chromosome in the mother of a patient with clinical diagnosis of the 18q− syndrome. The *propositus*'s complement included the recombinant 18 with deficiency of the distal third of the long arm and duplication of the terminal segment of the short arm. His sister carried a duplication of the distal third of the long arm and a deficiency of the terminal segment of the short arm. Almeida *et al.*[50] reported a patient with clinical signs of 18q– which was a carrier of a ring chromosome with breaks in positions p11 and q21.

## Sex chromosomes

Multiple malformations are much less frequent in persons with sex chromosome abnormalities. In general and as expected their effect is usually restricted to problems of sex differentiation. Therefore, these conditions are rarely diagnosed at birth; in addition, mental retardation is not a constant feature.

Table 6.4 lists the sexual abnormalities due to excess of sex chromosome material. X polysomy generally leads to triple or tetra-X, as well as Klinefelter syndromes. True hermaphroditism may also eventually occur. Y polysomy

**Table 6.4  Sexual abnormalities due to excess of X or Y chromosome material**

| | Phenotypes | |
|---|---|---|
| Karyotypes | Frequently | Possibly |
| *X polysomy* | | |
| 47,XXX | Triple-X syndrome | — |
| 46,XX/47,XXX | Triple-X syndrome | True hermaphroditism |
| 48,XXXX | Tetra-X syndrome | — |
| 47,XXX/48,XXXX | Tetra-X syndrome | — |
| 46,XX/47,XXX/48,XXXX | Tetra-X syndrome | — |
| 49,XXXXX | Penta-X syndrome | — |
| 46,XX/46,XY | True hermaphroditism | Pure gonadal dysgenesis |
| 47,XXY | Klinefelter syndrome | Male pseudoher-maphroditism |
| 46,XX/47,XXY | Klinefelter syndrome | True hermaphroditism |
| 46,XY/47,XXY | Klinefelter syndrome | — |
| 46,XX/46,XY/47,XXY | Klinefelter syndrome | True hermaphroditism |
| 48,XXXY | Klinefelter syndrome | — |
| 46,XY/48,XXXY | Klinefelter syndrome | — |
| 49,XXXXY | XXXXY-males | — |
| 48,XXXY/49,XXXXY | XXXXY-males | — |
| 48,XXXY/49,XXXXY/50,XXXXXY | XXXXY-males | — |
| *Y polysomy* | | |
| 47,XYY | YY-males | — |
| 46,XY/47,XYY | YY-males | — |
| 47,XYY/48,XYYY | YY-males | — |
| *X and Y polysomy* | | |
| 46,XY/47,XYY/47,XXY | Male pseudoher-maphroditism | — |
| 48,XXYY | XXYY-males | — |
| 46,XX/48,XXYY | True hermaphroditism | — |
| 46,XY/47,XXY/48,XXYY | Klinefelter syndrome | — |
| 47,XXY/48,XXYY | XXYY-males | — |
| 49,XXXYY | XXYY-males | — |
| 46,XX/47,XXY/49,XXYYY | True hermaphroditism | — |
| 48,XXXY/49,XXXXY/50,XXXXXY | XXYY-males | — |

Sources: Hamerton[18], Simpson[51], Federman[52], De Grouchy and Turleau[19], King and Cooke[53], Riccardi et al.[54]

does not lead to defined symptoms, while X *and* Y polysomy determine a large phenotypic array. Usually several karyotypes may determine what seems to be the same clinical entity. An extreme example is the Klinefelter syndrome; as is indicated in this table, and Table 6.5, eight distinct karyotypes may lead to it.

The cases of X polysomy studied in Porto Alegre included six 47,XXY, two 47,XXX, two 46,XY/46,XX with phenotypes of true hermaphrodites and one 48,XXY + G (see Table 6.9; the latter patient was described in detail by Erdtmann *et al.*[55]). As for the Y polysomies, just one individual has been detected with a 47,XYY karyotype that surprisingly showed multiple malformations (Figures 6.2, 6.3). The question of the identification of individuals with this chromosome constitution (such individuals generally show a normal phenotype) was considered by Vianna *et al.*[56]. They observed one XYY individual out of 14 with abnormal electrocardiograms (prolonged P–R intervals). This is of course much higher than the frequency of these individuals in the general population, indicating an association between these conditions. On the other hand, Varella-Garcia *et al.*[57] observed a patient with signs of pure gonadal dysgenesis (a condition in which the karyotypes are

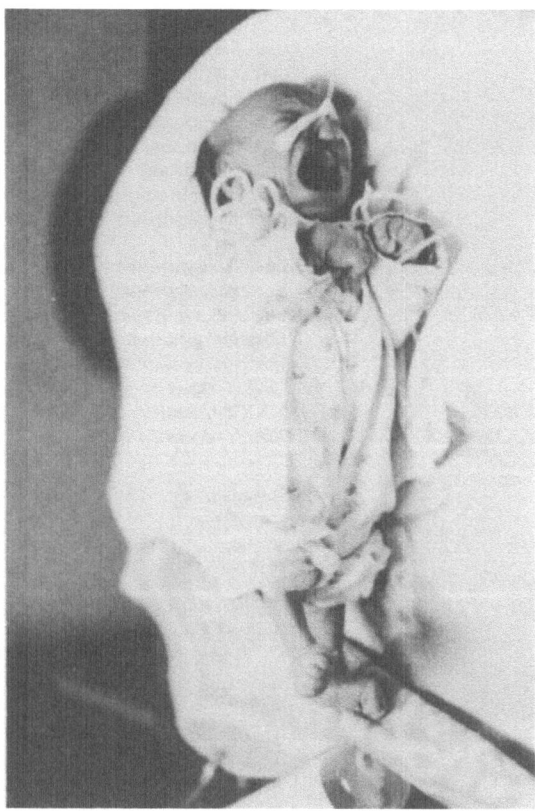

**Figure 6.2** 47,XYY patient studied in Porto Alegre. Overall view

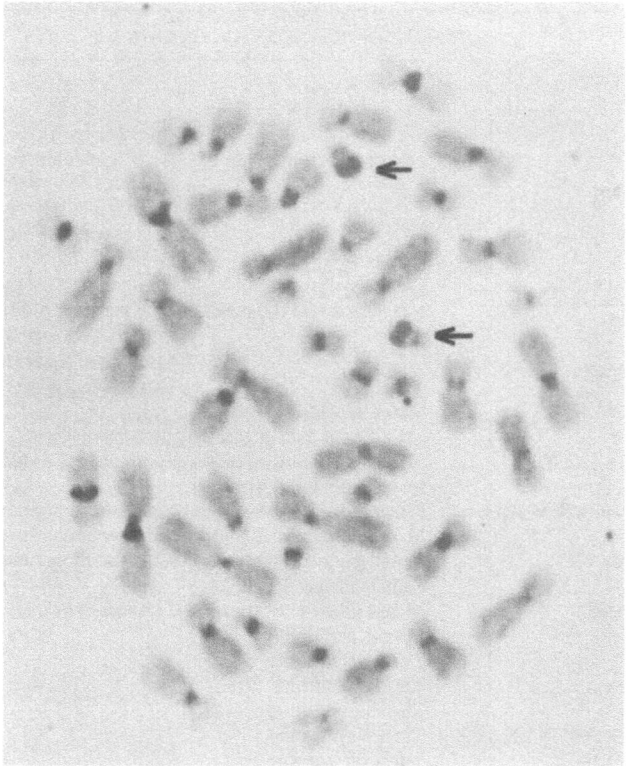

**Figure 6.3** 47,XXY patient shown in Figure 6.2. C-banded metaphase plate in which the Ys are indicated by arrows

generally normal) with a duplication of a large region of the long arm of the X.

The sexual abnormalities that can arise due to lack of X or Y chromosome material are listed in Table 6.5. As can be seen, the number of karyotypes observed is very large. However, a few generalizations are possible. X monosomy, as well as X or Y (short arm) deletions usually give rise to a Turner phenotype. X monosomy associated with the full Y or its short arm generally leads to mixed gonadal dysgenesis. Conditions that may also eventually occur are male pseudohermaphroditism, pure gonadal dysgenesis and true hermaphroditism. The most unpredictable karyotype is the 45,X/46,XY. No fewer than five clinical conditions can be associated with this type of mosaicism.

A total of 44 patients with these types of chromosome abnormalities was studied in Porto Alegre (see Table 6.9). Half of them are of the classical 45,X type and 12 are 45,X/46,XX. The majority of them have been described by Suñé et al.[58] and Mattevi et al.[59]. As for other Latin American contributions to these investigations, we should mention the series of 125 patients with Turner syndrome studied by Coco and Bergada[60]; the study of ovarian

**Table 6.5 Sexual abnormalities due to lack of X or Y chromosome material**

| Karyotypes | Phenotypes | |
|---|---|---|
| | Frequently | Possibly |
| *X monosomy* | | |
| 45,X | Turner syndrome | Mixed gonadal dysgenesis; male pseudohermaphroditism |
| 45,X/46,XX | Turner syndrome | Pure gonadal dysgenesis; male pseudohermaphroditism |
| 45,X/47,XXX | Turner syndrome | Triple-X syndrome |
| 45,X/46,XX/47,XXX | Turner syndrome | Pure gonadal dysgenesis |
| 45,X/46,XY | Mixed gonadal dysgenesis | True hermaphroditism; male pseudohermaphroditism; pure gonadal dysgenesis; Turner syndrome |
| 45,X/46,XY/46,XX | Mixed gonadal dysgenesis | — |
| 45,X/47,XYY | Mixed gonadal dysgenesis | Turner syndrome |
| 45,X/46,XY/47,XYY | Mixed gonadal dysgenesis | Pure gonadal dysgenesis |
| 45,X/48,XXXY | Mixed gonadal dysgenesis | — |
| 45,X/46,XY/46,XX/47,XXY | Mixed gonadal dysgenesis | Male pseudohermaphroditism |
| 45,X/46,X,i($Y_p$)/47,XY,i($Y_p$) | Mixed gonadal dysgenesis | Male pseudohermaphroditism |
| 45,X/46,X,del($Y_q$) | Mixed gonadal dysgenesis | Male pseudohermaphroditism |
| 45,X/46,XY/47,XXY | Klinefelter syndrome | — |
| 45,X/46,X dic(Y) | Mixed gonadal dysgenesis | Turner syndrome |
| 45,X/46,XX p− | Turner syndrome | — |
| 45,X/46,XX q− | Turner syndrome | — |
| 45,X/46,X,i($X_q$) | Turner syndrome | — |
| 45,X/46,X,i($X_p$) | Turner syndrome | — |
| 45,X/46,X,i($X_q$)/47,X,i($X_q$),i($X_q$) | Turner syndrome | — |
| 45,X/46,X,r(X) | Turner syndrome | — |
| 45,X/46,X,r(X)/47,X,r(X),r(X) | Turner syndrome | — |
| 45,X/46,X,fra+ | Turner syndrome | — |
| *X deletions* | | |
| 46,X,del($X_p$) | Turner syndrome | — |
| 46,X,i($X_q$) | Turner syndrome | — |
| 46,X,del($X_q$) | Turner syndrome | Pure gonadal dysgenesis |
| 46,X,i($X_p$) | Turner syndrome | — |
| *Y deletions* | | |
| 46,X,i($Y_q$) | Turner syndrome | — |
| 46,X,dic(Y) | Turner syndrome | — |

*Sources*: Hamerton[18], Simpson[51], Federman[52], De Grouchy and Turleau[19], King and Cooke[53], Riccardi et al.[54]

differentiation in 17 of these patients by Rivelis et al.[61]; the observation of Magnelli et al.[62] of a Turner syndrome patient with an isochromosome of the long arm of the Y, providing an elegant demonstration that the male determining factors are situated on the short arm of the Y chromosome; and the calculations made by Otto et al.[63] that the risk of abnormal offspring in mosaics 46,XX/45,X is very low.

Disorders of sexual development can arise in the presence of apparently normal karyotypes. A list of hereditary conditions that lead to such

Table 6.6   Hereditary disorders of sexual development with normal karyotypes

---

46,XX

*Usually*

Amastia, arcuate uterus, ataxia–hypogonadism syndrome; Cowden disease, hand–foot–uterus syndrome, isolated gonadotropin deficiency, Laurence–Moon–Biedl syndrome, Leprechaunism, Meckel syndrome, Noonan syndrome, panhypopituitary dwarfism, popliteal pterygium syndrome, polycystic ovarian disease, polymastia, Prader–Willi syndrome, Robinow syndrome, Rokitansky–Kuster–Hauser syndrome, sex-reversal syndrome (XX-males), Silver syndrome, steroid 11 β-hydroxylase deficiency, steroid 17 α-hydroxylase deficiency, steroid 21-hydroxylase deficiency, steroid 3 β-hydroxysteroid dehydrogenase deficiency, transverse vaginal septum, true hermaphroditism, ulnar–mammary syndrome, vaginal atresia

*Possibly*

Male pseudohermaphroditism, pure gonadal dysgenesis, Smith–Lemli–Opitz syndrome, Turner syndrome

46,XY

*Usually*

Ataxia–hypogonadism syndrome, congenital anorchia, cystic fibrosis, defective androgen synthesis, defective androgen action, defective Müllerian duct regression, G syndrome, hypertelorism–hypospadias syndrome, hypospadias, incomplete testicular feminization, isolated gonadotropin deficiency, Laurence–Moon–Biedl syndrome, Leprechaunism, Meckel syndrome, Noonan syndrome, panhypopituitary dwarfism, popliteal pterygium syndrome, polymastia, Prader–Willi syndrome, pseudovaginal perineoscrotal hypospadia, pure gonadal dysgenesis, Reifenstein syndrome, Robinow syndrome, Silver syndrome, Smith–Lemli–Opitz syndrome, steroid 17 α-hydroxylase deficiency, steroid 17,20-desmolase deficiency, steroid 3 β-hydroxysteroid dehydrogenase testicular feminization syndrome, tubular male pseudohermaphroditism syndrome, ulnar–mammary syndrome

*Possibly*

Amastia, mixed gonadal dysgenesis, true hermaphroditism

---

*Sources*: Bergsma[64,65]

abnormalities is given in Table 6.6. We were able to locate 31 of them that are usually or possibly associated with a 46,XX and 34 associated with a 46,XY constitution.

## CHROMOSOME BALANCE AND MENTAL RETARDATION

As has already been mentioned, mental retardation is almost invariably a consequence of an autosomal aberration. Therefore, many studies have been conducted in unselected series of such patients, to establish the role of chromosome abnormality in the aetiology of this rather heterogeneous condition. Most of the surveys conducted in the pre-banding era have been reviewed in Erdtmann *et al.*[66] . More recent evaluations have been made by Tharapel and Summitt[67], Jacobs *et al.*[68] and Ally and Grace[69]. The data are summarized in Table 6.7. For each mental retardate with a non-21-trisomy chromosome abnormality there are at least four Down patients. When only persons with unclassifiable syndromes and who showed besides the mental retardation multiple malformations are included, a frequency of 9.1 % of chromosome disorders is encountered. Full or partial trisomies are twice as frequent as deletions and rings.

Table 6.7 Chromosome surveys in mentally retarded individuals

| Type of chromosome aberration | | Nature of sample | |
|---|---|---|---|
| | | Mental retardates (institutionalized or not) (N = 6675)* | Mental retardation plus multiple malformations (N = 1424)† |
| Sex chromosome | N | 58 | 18 |
| abnormalities | % | 0.9 | 1.3 |
| 21 trisomy | N | 637 | — |
| | % | 9.5 | — |
| Full or partial | N | 44 | 66 |
| trisomies | % | 0.7 | 4.5 |
| Deletions and | N | 23 | 27 |
| rings | % | 0.3 | 1.9 |
| Apparently balanced | N | 20 | 18 |
| rearrangements | % | 0.3 | 1.3 |
| Total | N | 782 | 129 |
| | % | 11.7 | 9.1 |

* Series assembled or studied by Jacobs et al.[68] and Ally and Grace[69]
† Series assembled or studied by Tharapel and Summit[67] and Jacobs et al.[68]. The absence of 21 trisomy is due to an ascertainment effect (only persons with unclassifiable syndromes were included in the studies)

## THE PORTO ALEGRE STUDIES

Tables 6.8 and 6.9 present a summary of the human cytogenetic studies performed to date in Porto Alegre. Reference to some of these cases has already been made in previous sections; only an overall evaluation of the findings will therefore be presented here. As is indicated in Table 6.8, the subjects studied were ascertained in four ways. The largest category involves 741 normal individuals, investigated for the relationship between aneuploidy and age (Mattevi and Salzano[70,71]), as well as for C-band polymorphisms (Viégas and Salzano[72], Erdtmann et al.[6,73]). A total of 507 persons had been referred to us by physicians for cytogenetic investigation, the two other sources of ascertainment being a survey in an institution for mental retardates[66] and family follow-ups. The total frequency of abnormal karyotypes, considering only the 'high-risk' populations (ascertainment categories 2, 3 and 4) is $235/639 = 37\%$. It should be stressed that stringent selective criteria are used by us for the inclusion of a patient in the cytogenetic series. Obvious cases of 21 trisomy arising from old mothers are not included, and persons with problems of sexual development are first screened by the analysis of their X and Y chromatin. If this analysis agrees with the diagnostic hypothesis, no karyotype studies are performed. In the unclassified patients, only those with multiple malformations are chosen for further investigation. It comes as no surprise, therefore, that the proportion of autosome versus sex chromosome abnormal karyotypes is of the order of 3:1.

Details about the chromosomally abnormal individuals are given in Table 6.9. Changes were observed in chromosomes 2, 4, 5, 12, 13, 17, 18, 21, 22, X and Y. In addition, one case could be ascribed to the chromosome group only (F), and a centromeric fragment could not be classified.

**Table 6.8** **Frequency of subjects with abnormal karyotypes studied in Porto Alegre classified by type of ascertainment**

| Karyotype | Type of ascertainment | | | | |
|---|---|---|---|---|---|
| | 1 | 2 | 3 | 4 | Total |
| *Normal* | | | | | |
| Males | 355 | 30 | 35 | 132 | 552 |
| Females | 386 | 16 | 34 | 157 | 593 |
| Total | 741 | 46 | 69 | 289 | 1145 |
| *Abnormal* | | | | | |
| Autosomes | — | 4 | 12 | 153 | 169 |
| Sex chromosomes | — | 1 | — | 65 | 66 |
| Total | — | 5 | 12 | 218 | 235 |
| *Grand total* | 741 | 51 | 81 | 507 | 1380 |

*Type of ascertainment*: 1 = survey, general populations; 2 = survey, institutions or special populations; 3 = research studies, family follow-ups; 4 = clinical referral

**Table 6.9  Types of chromosome abnormalities observed in Porto Alegre**

| Chromosome identification | Type of abnormality* | Number of subjects | Frequency |
|---|---|---|---|
| 02p130 | 46,XY(4) or 46,XX(4), inv(2) (p13→qter)† | 8 | 3.4 |
| 04p100 | 46,XX,4p− | 2 | 0.9 |
| 05p130 | 46,XY(1) or 46,XX(2), 5p− | 3 | 1.3 |
| 05q000 | 46,XX,5q− | 1 | 0.4 |
| 05q350 | 46,XX,t(4;13) (q35;q21)† | 4 | 1.7 |
| 120000 | 46,XX,inv(12) (pq) | 2 | 0.9 |
| 130000 | 47,XY,+13; 46,XY/47,XY,+13; 47,XX,+13 | 3 | 1.3 |
| 130000 | 46,XX or 46, XY,−D, +t(Dq Gq)‡ | 2 | 0.9 |
| 130000 | 46,XX,r(13) (p11→q32) | 1 | 0.4 |
| 170000 | 46,XX/46,XX,r(17) | 1 | 0.4 |
| 180000 | 47,XX,+18(5); 46,XX/47,XX,+18(1) | 6 | 2.6 |
| 210000 | 47,XY,+21(70); 47,XX,+21(58) | 128 | 54.5 |
| 210000 | 46,XY/47,XY,+21 | 2 | 0.9 |
| 210000 | 46,XY(1) or 46,XX(2), −G,+t(Gq Gq)‡ | 3 | 1.3 |
| 220000 | 46,XX/46,XX,−G,+Ph$_1$ | 1 | 0.4 |
| 0X0000 | 45,X(22); 45,X/46,X,i(Xq) (1); 45,X/46,X,Xp−q−(1) | 24 | 10.2 |
| 0X0000 | 45,X/46,XY(4); 45,X/46,XX(12) | 16 | 6.8 |
| 0X0000 | 47,XXX | 2 | 0.9 |
| 0X0000 | 46,X,i(X) (qter→cen→qter) | 1 | 0.4 |
| 0X0000 | 46,XX/46,X,Xp−q− | 2 | 0.8 |
| 0X0000 | 46,XX, hermaphrodite | 2 | 0.8 |
| 0Y0000 | 47,XYY;45,X/47,XYY | 2 | 0.8 |
| XY0000 | 47,XXY(6); 48,XXY,+G(1) | 7 | 3.0 |
| XY0000 | 46,XY/46,XX | 2 | 0.8 |
| XY0000 | 46,XY, female (†or‡) | 8 | 3.4 |
| Group F | 46,XY/46,XY,+F | 1 | 0.4 |
| Unidentified | 47,XX,+mar | 1 | 0.4 |

\* When necessary the number of individuals in each category is indicated in parenthesis
† Familial
‡ Sporadic

## PERSPECTIVES

It can be confidently expected that great progress will occur in the field of cytogenetics and its relationship to pathology. This is because new techniques are being developed at a fast rate both at the cytological and physicochemical sides. On the other hand, the mastering of new ways of cultivation of somatic and germinal tissues is opening the possibility of meaningful comparisons between *in vitro* and *in vivo* conditions. The culture of early human embryos and the analysis of their growth requirements will certainly lead to a better understanding of the complex mechanisms of sexual differentiation.

These advances will provide us with the proper knowledge about chromosome structure and function that is vital for the solution of some basic problems in biology, such as the processes that change a normal into a malignant cell and those responsible for the ageing phenomena at the cellular and organismal levels. Clarification of the nature of the chromosome polymorphisms and the establishment of their frequencies in different populations will be necessary for the study of their linkage with pathological traits. It is estimated, for instance, that about 1000 bands can be identified in the uncondensed human chromosomes. This is still very far from the estimated number of genes that we possess, but a significant beginning has been made in the task of locating specific genes in specific chromosomes. Calculations indicate that about 2 % of the human genes have been identified but less than 1 % assigned to specific chromosomes (Evans et al.[74]). An important new technique in this field involves the use of restriction fragments and plasmid vectors containing inserted DNA sequences.

These developments will provide the basis for the delineation of new syndromes, the location of the responsible genes in given chromosomal sites and the establishment of recombination frequencies with normal, frequent traits. In this manner proper genetic counselling will be possible, and the way opened for a greater degree of intervention in the growth determinants that may lead to congenital malformations. It is already possible, through amniocentesis, to interrupt pregnancies when the embryo is abnormal. *In utero* treatment, however, has just begun. Such manipulations will undoubtedly become more common in the future.

We close by stressing the need for a better exchange of knowledge and experience between scientists in the clinical and non-clinical areas. Only by this interdisciplinary approach will we be able to understand the complex interactions that can lead to normal or abnormal patterns of development.

### Acknowledgements

Our researches are supported by the Conselho Nacional de Desenvolvimento Científico e Tecnológico (Programa Integrado de Genética), Fundação de Amparo à Pesquisa do Estado do Rio Grande do Sul and Pró-Reitoria de Pesquisa e Pós-graduação da Universidade Federal do Rio Grande do Sul.

### References

1 Tjio, J. H. and Levan, A. (1956). The chromosome number of man. *Hereditas*, **42**, 1
2 Schwarzacher, H. G. (1976). *Chromosomes in Mitosis and Interphase*, p. 182. (Berlin: Springer)

3  Evans, H. J. (1977). Some facts and fancies relating to chromosome structure in man. *Adv. Hum. Genet.*, **8**, 347

4  Yunis, J. J. (1977). *Molecular Structure of Human Chromosomes*, p. 336. (New York: Academic Press)

5  Bostock, C. J. and Sumner, A. T. (1978). *The Eukaryotic Chromosome*, p. 525. (Amsterdam: North-Holland)

6  Erdtmann, B., Salzano, F. M. and Mattevi, M. S. (1980). Quantitative analysis of C-band size in human chromosomes (Submitted for publication)

7  Ruzicka, F. (1974). Organization of mitotic chromosomes. *Humangenetik*, **23**, 1

8  Bahr, G. F. (1977). Chromosomes and chromatin structure. In Yunis, J. J. (ed.) *Molecular Structure of Human Chromosomes*, pp. 143–203. (New York: Academic Press)

9  Hoehn, H. and Callis, J. (1977). Flow-fluorometric DNA content differences as a function of chromosome constitution in resting human lymphocytes. In Sparkes, R. S., Comings, D. E. and Fox, C. F. (eds.) *Molecular Human Cytogenetics*, pp. 243–256. (New York: Academic Press)

10  Holm, P. B. and Rasmussen, S. W. (1977). Three-dimensional reconstructions of meiotic chromosomes in human spermatocytes. *Chromosomes Today*, **6**, 83

11  Rasmussen, S. W. and Holm, P. B. (1978). Human meiosis II. Chromosome pairing and recombination nodules in human spermatocytes. *Carlsberg Res. Commun.*, **43**, 275

12  Kunkel, L. M., Smith, K. D. and Boyer, S. H. (1977). Y-chromosome DNA. In Sparkes, R. S., Comings, D. E. and Fox, C. F. (eds.). *Molecular Human Cytogenetics*, pp. 305–313. (New York: Academic Press)

13  Latt, S. A., Allen, J. W., Shuler, C., Loveday, K. S. and Munroe, S. H. (1977). The detection and induction of sister chromatid exchanges. In Sparkes, R. S., Comings, D. E. and Fox, C. F. (eds.) *Molecular Human Cytogenetics*, pp. 315–334. (New York: Academic Press)

14  Korenberg, J. R., Therman, E. and Denniston, C. (1978). Hot spots and functional organization of human chromosomes. *Hum. Genet.*, **43**, 13

15  Rudak, E., Jacobs, P. A. and Yanagimachi, R. (1978). Direct analysis of the chromosome constitution of human spermatozoa. *Nature (London)*, **274**, 911

16  Adolph, K. W. (1980). Isolation and structural organization of human mitotic chromosomes. *Chromosoma*, **76**, 23

17  Langlois, R. G., Carrano, A. V., Gray, J. W. and Van Dilla, M. A. (1980). Cytochemical studies of metaphase chromosomes by flow cytometry. *Chromosoma*, **77**, 229

18  Hamerton, J. L. (1971). *Human Cytogenetics*. Vol. II, p. 545. (New York: Academic Press)

19  De Grouchy, J. and Turleau, C. (1977). *Clinical Atlas of Human Chromosomes*, p. 319. (New York: Wiley)

20  Hodes, M. E., Cole, J., Palmer, C. G. and Reed, T. (1978). Trisomy 18 (29 cases) and trisomy 13 (19 cases): a summary. In Summitt, R. L., Bergsma, D., Simpson, J. L. and Paul, N. W. (eds.) *Sex Differentiation and Chromosomal Abnormalities*, pp. 377–382. (New York: A. R. Liss)

21  Hecht, F. (1979). Chromosome eighteen trisomy syndrome. In Bergsma, D. (ed.) *Birth Defects Compendium*, pp. 201–202. (New York: A. R. Liss)

22  Magenis, E. and Hecht, F. (1979). Chromosome thirteen trisomy syndrome. In Bergsma, D. (ed.) *Birth Defects Compendium*, pp. 212–213. (New York: A. R. Liss)

23  Miller, O. J. (1979). Chromosome twenty-one trisomy syndrome. In Bergsma, D. (ed.) *Birth Defects Compendium*, pp. 215–216. (New York: A. R. Liss)

24  Kasahara, S., Viegas-Pequignot, E. M. and Frota-Pessoa, O. (1977). A search on karyotypic mosaicism in Mongoloid patients and their parents. *Rev. Bras. Pesq. Med. Biol.*, **10**, 225

25  Vianna-Morgante, A. M. and Nunesmaia, H. G. (1978). Dissociation as probable origin of mosaic 45,XY,t(15;21)/46,XY,i(21q). *J. Med. Genet.*, **15**, 305

26  Toledo, S. P. and Wajntal, A. (1977). 47,XX, + 13/46,XX mosaicism: a case report. *Acta Genet. Med. Gemellol.*, **26**, 71

27  Mattei, J. F., Mattei, M. G., Ayme, S. and Giraud, F. (1979). Origin of the extra chromosome in trisomy 21. *Hum. Genet.*, **46**, 107

28  Caspersson, T., Lindsten, J., Zech, L., Buckton, K. E. and Price, W. H. (1972). Four patients with trisomy 8 identified by the fluorescence and Giemsa banding techniques. *J. Med. Genet.*, **9**, 1

29  Crandall, B. F., Bass, H. N., Marey, S. M., Glovsky, M. and Fish, C. H. (1974). The trisomy 8 syndrome: two additional mosaic cases. *J. Med. Genet.*, **11**, 393

30  Tuncbilek, E., Halicioglu, C. and Say, B. (1974). Trisomy-8 syndrome. *Humangenetik*, **23**, 23
31  Baccichetti, C., Tenconi, R., Anglani, F. and Zacchello, F. (1975). Trisomy 4q32 → 4qter due to a maternal 4/21 translocation. *J. Med. Genet.*, **12**, 425
32  Cassidy, S. B., McGee, B. J., Van Eys, J., Nance, W. E. and Engel, E. (1975). Trisomy-8 syndrome. *Pediatrics*, **56**, 826
33  Cassidy, S. B., McGee, B. J. and Engel, E. (1979). Chromosome eight trisomy syndrome. In Bergsma, D. (ed.) *Birth Defects Compendium*, pp. 198–199. (New York: A. R. Liss)
34  Centerwall, W. R., Mayeski, C. A. and Cha, C. C. (1975). Trisomy 9q – . A variant of the 9p trisomy syndrome. *Humangenetik*, **29**, 91
35  Giraud, F., Mattei, J. F., Mattei, M. G., Ayme, S. and Bernard, R. (1975). La trisomie 4p. A propos de 3 observations. *Humangenetik*, **30**, 99
36  Giraud, F., Mattei, J. F., Mattei, M. G. and Bernard, R. (1975). Trisomie partielle 11q et translocation familiale 11–22. *Humangenetik*, **28**, 343
37  Francke, U. (1979). Chromosome eleven q partial trisomy syndrome. In Bergsma, D. (ed.) *Birth Defects Compendium*, pp. 203–204. (New York: A. R. Liss)
38  Yunis, J. J. (1977). *New Chromosomal Syndromes*, p. 404. (New York: Academic Press)
39  Gagliardi, A. R. T., Tajara, E. H., Varella-Garcia, M. and Moreira, L. M. A. (1978). Trisomy-8 syndrome. *J. Med. Genet.*, **15**, 70
40  Penchaszadeh, V. B. and Coco, R. (1975). Partial 9 trisomy by 3:1 segregation of balanced maternal translocation (7q + ;9q – ). *J. Med. Genet.*, **12**, 301
41  Guthrie, R. D., Aase, J. M., Asper, A. C. and Smith, D. W. (1971). The 4p – syndrome. *Am. J. Dis. Child.*, **122**, 421
42  Schinzel, A., Hayashi, K. and Schmid, W. (1975). Structural aberrations of chromosome 18. II. The 18q – syndrome. Report of three cases. *Humangenetik*, **26**, 123
43  Carlin, M. E. and Norman, C. (1978). Case report: partial trisomy 12p associated with 4p deletion due to paternal t(12p – ;4p + ) translocation. In Summitt, R. L. and Bergsma, D. (eds.) *Sex Differentiation and Chromosomal Abnormalities*, pp. 399–406. (New York: A. R. Liss)
44  Breg, W. R. (1979). Chromosome five p – syndrome. In Bergsma, D. (ed.) *Birth Defects Compendium*, pp. 205–206. (New York: A. R. Liss)
45  Miller, O. J. (1979). Chromosome eighteen p – syndrome. In Bergsma, D. (ed.) *Birth Defects Compendium*, pp. 199–200. (New York: A. R. Liss)
46  Miller, O. J. (1979). Chromosome eighteen q – syndrome. In Bergsma, D. (ed.) *Birth Defects Compendium*, pp. 200–201. (New York: A. R. Liss)
47  Warburton, D. (1979). Chromosome four p – syndrome. In Bergsma, D. (ed.) *Birth Defects Compendium*, pp. 207–208. (New York: A. R. Liss)
48  Marçallo, F. A., Werneck, L. C., Pilotto, R. F. and Opitz, J. M. (1977). Hemihypotrophy in a girl with a translocation t(13q;7p) *Eur. J. Pediat.*, **124**, 167
49  Vianna-Morgante, A. M., Nozaki, M. J., Ortega, C. C., Coates, V. and Yamamura, Y. (1976). Partial monosomy and partial trisomy 18 in two offspring of carrier of pericentric inversion of chromosome 18. *J. Med. Genet.*, **13**, 366
50  Almeida, J. C. C., Campos, J. M. S., Martins, R. R. S., Kayath, H. C. and Barbosa Neto, J. G. (1979). Anel de 18.46,XX,r(18). Apresentação de um caso. *J. Pediatr. (Rio de Janeiro)*, **46**, 377
51  Simpson, J. L. (1975). Gonadal dysgenesis and abnormalities of the human sex chromosomes: current status of phenotypic–karyotypic correlations. In Bergsma, D. (ed.) *Genetic Forms of Hypogonadism*, pp. 23–60. (New York: Stratton Intercontinental)
52  Federman, D. D. (1975). Mixed gonadal dysgenesis and other intersex states. In Bergsma, D. (ed.) *Genetic Forms of Hypogonadism*, pp. 61–62. (New York: Stratton Intercontinental)
53  King, C. R. and Cook, D. M. (1978). Bilateral gonadoblastoma in a phenotypic female with 45,X/46,X, dicentric iso Y[45,X/46,X,idic(Yq)] mosaicism. In Summitt, R. L. and Bergsma, D. (eds.) *Sex Differentiation and Chromosomal Abnormalities*, pp. 109–122. (New York: A. R. Liss)
54  Riccardi, V. M., Duck, S. and Katayama, O. (1978). The Turner syndrome and the Y chromosome: mechanisms of diminished Y-determined maleness. In Summitt, R. L. and Bergsma, D. (eds.) *Sex Differentiation and Chromosomal Abnormalities*, pp. 123–132. (New York: A. R. Liss)
55  Erdtmann, B., Gomes de Freitas, A. A., De Souza, R. P. and Salzano, F. M. (1971). Klinefelter's syndrome and G trisomy. *J. Med. Genet.*, **8**, 364

56 Vianna, A. M., Frota-Pessoa, O., Lion, M. F. and Decourt, L. (1972). Searching for XYY males through electrocardiograms. *J. Med. Genet.*, **9**, 165

57 Varella-Garcia, M., Tajara, E. H. and Gagliardi, A. R. T. (1980). Structural aberration of the X chromosome in a patient with gonadal dysgenesis: an approach to the karyotype-phenotype correlation. (Submitted for publication)

58 Suñé, M. V., Centeno, J. V. and Salzano, F. M. (1970). Gonadoblastoma in a phenotypic female with 45,X/47,XYY mosaicism. *J. Med. Genet.*, **7**, 410

59 Mattevi, M. S., Wolff, H., Salzano, F. M. and Mallmann, M. C. (1971). Cytogenetic, clinical and genealogical analyses in a series of gonadal dysgenesis patients and their families. *Humangenetik*, **13**, 126

60 Coco, R. and Bergada, C. (1977). Cytogenetic findings in 125 patients with Turner's syndrome and abnormal karyotypes. *J. Génét. Hum.*, **25**, 95

61 Rivelis, F., Coco, R. and Bergada, C. (1978). Ovarian differentiation in Turner's syndrome. *J. Génét. Hum.*, **26**, 69

62 Magnelli, N. C., Vianna-Morgante, A. M., Frota-Pessoa, O. and Taboada-Lopez, M. G. (1974). Turner's syndrome and 46,X,i(Yq) karyotype. *J. Med. Genet.*, **11**, 403

63 Otto, P. A., Kasahara, S., Nunesmaia, H. G. and Frota-Pessoa, O. (1977). Risk of 45,X karyotype in offspring of Turner's syndrome patients. *Lancet*, **2**, 257

64 Bergsma, D. (1975). *Genetic Forms of Hypogonadism*, p. 161. (New York: Stratton Intercontinental)

65 Bergsma, D. (1979). *Birth Defects Compendium*. 2nd Edn., p. 1183. (New York: A. R. Liss)

66 Erdtmann, B., Salzano, F. M. and Mattevi, M. S. (1975). Chromosome studies in patients with congenital malformations and mental retardation. *Humangenetik*, **26**, 297

67 Tharapel, A. T. and Summitt, R. L. (1977). A cytogenetic survey of 200 unclassifiable mentally retarded children with congenital anomalies and 200 normal controls. *Hum. Genet.*, **37**, 329

68 Jacobs, P. A., Matsuura, J. S., Mayer, M. and Newlands, I. M. (1978). A cytogenetic survey of an institution for the mentally retarded: I. Chromosome abnormalities. *Clin. Genet.*, **13**, 37

69 Ally, F. E. and Grace, H. J. (1979). Chromosome abnormalities in South African mental retardates. *South Afr. Med. J.*, **55**, 710

70 Mattevi, M. S. and Salzano, F. M. (1975). Senescence and human chromosome changes. *Humangenetik*, **27**, 1

71 Mattevi, M. S. and Salzano, F. M. (1975). Effect of sex, age and cultivation time on number of satellites and acrocentric associations in man. *Humangenetik*, **29**, 265

72 Viégas, J. and F. M. Salzano (1978). C-bands in chromosomes 1, 9 and 16 of twins. *Hum. Genet.*, **45**, 127

73 Erdtmann, B., Salzano, F. M. and Mattevi, M. S. (1980). Quantitative analysis of C bands in chromosomes 1, 9 and 16 of Brazilian Indians and Caucasoids. *Hum. Genet.* (In press)

74 Evans, H. J., Hamerton, J. L., Klinger, H. P. and McKusick, V. A. (1979). Human genetic mapping 5. *Fifth International Workshop on Human Gene Mapping. Cytogenet. Cell Genet.*, **25**, 1

# 7
# Early psychomotor development of children with sex chromosome aneuploidies

KATERINA HAKA-IKSE

## INTRODUCTION AND GENERAL CONSIDERATIONS

Present knowledge about early psychomotor development in aneuploidies results mainly from prospective studies of children detected through large newborn cytogenetic surveys over the past two decades. Such surveys were conducted either through examination of the chromosomes of the peripheral blood leukocytes, or through examination of X- and/or Y-chromatin from the amnion or from the cells of buccal mucosa, followed by karyotyping those children found to present chromatin anomalies. The first method was used in Århus, Denmark[1], Northeastern, USA[2,3], Scotland[4], Tokyo, Japan[5] and the collaborative study in the USA[6]. The Denver[7], Edinburgh[8] and Toronto[9] studies carried instead sex chromatin surveys.

The children who were found to present sex chromosome aneuploidies were subsequently followed up for various parameters of physical growth and psychomotor–behavioural development[5,6,10-16]. Several of these follow-up studies are still in progress[5,12,14,16] and at least one[14] will continue until the subjects reach maturity.

There has been considerable interstudy variation in research design, methodology, scope, use of controls, length of follow-up and even ethical considerations such as communication with the parents of affected children and obtaining informed consent. Despite such significant differences most of the conclusions arrived at by various research groups do seem to coincide.

One of the important contributions of these prospective studies in children has been the understanding that a certain chromosome constitution does not necessarily entail the presence of a particular developmental or behavioural phenotype. The latter has been suggested by earlier research[17-20], the subjects of which were adults drawn from skewed populations such as mental, penal or mental/penal institution inmates.

However, even with the knowledge that a sex chromosome aberration is not associated with any definite syndrome in all individuals presenting the aberration, the prospective studies have demonstrated that certain developmental and behavioural traits are prevalent amongst the various aneuploidy types[21]. It has long been known that the 45,XO karyotype is related with spatial cognitive deficits[22-25] and current results suggest that the presence of an additional X chromosome accounts for increased incidence of verbal deficits[26-29]. The presence of an extra Y chromosome is less clearly associated with particular deficits, although there is a tendency for the group as a whole to be slightly skewed to the left of the general intelligence distribution[21]. Boys with additional X chromosomes are generally passive and present low activity level[16,21,30], while no definitive data exist today about the early behaviour of extra Y chromosome children. Girls with additional X chromosomes may present increased risk for behavioural–psychiatric problems[14,31,32], although the pattern of such problems is not uniform. Children with mosaic karyotypes do not seem to differ significantly in any parameter from the general population[13].

There has been wide speculation about the factors that may be responsible for the tendency to particular cognitive (and behavioural in boys with additional X chromosomes) profiles in sex chromosome aneuploidies. Such differences were attributed[33] to sex hormone action, namely to the prenatal androgen/oestrogen balance which could influence the relative development of verbal and non-verbal abilities. This theory, however, could not explain the essentially similar verbal deficit pattern in 47,XXY and 47,XXX children who presumably have sufficient hormonal differences to develop into male and female respectively. Also, the hormonal hypothesis would not explain why the phenotypically female 45,XO and 47,XXX differ in their cognitive pattern. Obviously, the theory that intellectual differences between normal males (who are more gifted in spatial tasks) and normal females (who are more advanced in verbal abilities) are due to hormonal changes occurring in the period around puberty[34,35] is not relevant in aneuploids as the differences under discussion are noted much before the onset of puberty.

The spatial deficit of 45,XO subjects has been also attributed to parietal lobe dysfunction[36]; there is however no direct neuropathological evidence to support this conclusion or others which have attributed this deficit to frontal lobe[37] or generalized right hemisphere damage disorders[38,39].

Barlow's theory that additional X chromosomes impede the mitotic cell-division rates[40] opened the way for a more plausible explanation. This theory was used in conjunction with the hypothesis of hemispheric specialization[41] to explain why the presence of an additional X chromosome could lead to verbal deficits[29,42]. Children with additional X chromosomes present a delay in bone age[14], sulphation rates[43] and dentition[44]; they also have low dermal ridge counts[45], while the ridge counts of 45,XO girls are higher than normal[45]. Since the dermal ridge counts are considered as an index of growth during fetal life, one can hypothesize that the growth rate of certain tissues may be delayed in children with extra X chromosomes. Cerebral development growth rate could be delayed as well, particularly in sites processing language functions[29]. On the contrary, the 45,XO individuals would have faster growth rate of language processing cerebral sites and thus normal verbal abilities[46]. This was demonstrated in 45,XO subjects through the dichotic listening test which indicated that the right hemisphere of 45,XO is more involved in verbal processing than in normals[42]. Since, however, language and spatial abilities do compete for neural tissue sites in the course of the development of hemispheric specialization[41], the spatial abilities of the 45,XO subjects would be at a disadvantage as verbal abilities would have taken over the sites of the 'spatial' right hemisphere[46]. The conclusion is that the specific cognitive deficits observed in aneuploids are the result of neural processes which depend on chromosomal functions[29].

There is no evidence found that correlation exists between social class and the frequency of sex chromosome anomalies[7,47,48]. Also it has been demonstrated[14] that aneuploids differed in verbal and other parameters from their sibling controls although the two groups were obviously subjected to the same family and socioeconomic influences. Thus, neither cognitive nor behavioural characteristics in aneuploid states could be attributed to particular socioeconomic class distribution.

In prospective studies such as these, questions arise about the effect on the children of parental knowledge that their child presents an abnormality which may entail deviant mental and/or behavioural patterns. Such knowledge could possibly lead to parental underexpectations, overreactions to common behavioural manifestations, increased anxiety and overprotectiveness or rejection of the child. Also the child himself could conceivably be influenced directly by the study, being the subject of repeated examinations and scrutiny by professionals and also the cause of anxiety to parents of which he feels the impact without understanding the reason. None of the research groups, however, has reported any such measurable adverse effects on the parents or children; as a matter of fact some objective proof of low overall anxiety in parents has been reported[13]. Moreover, most study findings indicate that certain deviant cognitive or behavioural patterns are related to specific karyotypes rather than to other variables.

Certainly, children with aneuploidies would react to adverse environmental factors as any other child would[32]. It has been suggested[10,13] that aneuploid children may have particular vulnerability to environmental stresses and thus be more prone to develop reactive behavioural and/or cognitive problems.

## DEVELOPMENTAL CHARACTERISTICS OF THE MAIN ANEUPLOIDY GROUPS

### 47,XXX girls

*Physical characteristics*
Girls with 47,XXX karyotype have no physical characteristics that would distinguish them on casual observation. They do not present as a rule major congenital anomalies and have few minor dysmorphic signs, the commonest of which are clinodactyly and epicanthus[21].

After the age of 6 years these children tend to become taller than the mean or their predicted height from parental averages[14]. One study[12] has not noted such a tendency. Data from Denver[13] and Toronto[14] suggest that increased height is due to an U < L segment ratio.

Bone age development is slower than expected[12,14]. Head circumference distribution tends to be skewed to the left[21].

## Intelligence

This group presents mild cognitive deficits with the mean IQs being in the mid to upper 80s, regardless of the type of intelligence test used. Differences of 17[12] to 20 IQ points[13] between 47,XXX girls and controls are reported. These differences, measured by the Stanford–Binet and Wechsler scales respectively, are statistically significant.

Although there are reports of 47,XXX probands with average intelligence[31], others[32] found very narrow range of individual achievements with all the probands clustering in dull to low normal level. Five of 34 probands[13] were found to have IQs below 70.

Short term memory function weaknesses have been suggested as contributing to the low performance of 47,XXX[14], as well as low scores in Arithmetic and Object Assembly subtests of the Wechsler scales[11]. Tendency to do poorly in coding type tasks was also noted[13].

## Verbal development

Verbal abilities in 47,XXX probands are significantly lower than the non-verbal. Delayed onset of language both in acquisition of single words and word combinations is reported[10,12,21,32]. One study[11] reports that the mean age of saying the first understandable word is 22 months for a group of 47,XXX and 47,XXY children. Speech impediments are reported too[13,32]; these are severe enough to require speech and language remediation[12].

Both expressive and receptive language problems have been found. The Vocabulary subtest scores on the Weschsler scales were found to be significantly below norms[11]. The Verbal IQ was found to be below 80 in 42 % of probands[14]. It has been shown[29] that there is a verbal comprehension deficit in 47,XXX probands; by using a sentence verification task it was demonstrated that the probands have difficulty in analysing sentence components. Thus it was concluded that encoding for meaning is a major determinant in 47,XXX verbal deficit.

## Motor development

Motor incoordination[11,13], visuo-motor integration problems[10], slow gross and fine motor development[32], hypotonia and hyporeflexia[12] have been reported in 47,XXX probands. These deficits appear to be mild and perhaps transient in character.

## School performance

As expected from the overall cognitive and verbal deficits of this group there is increased incidence of school learning problems which appear to be sensitive to family environment[13]. Severity of problems ranges from attendance at school for educationally subnormal and need for remedial teaching[12] to grade repetition[32]. School progress is considered to be reasonably good considering the group's limitations in ability[14] and it should be kept in mind that there are reports of 47,XXX girls doing well at school[31,32]. Problems reported by teachers are inattentiveness, difficulties with language or speech, erratic performance[31], social immaturity[32], reading and spelling[14].

### Behavioural and emotional development
A higher incidence of behavioural–emotional problems in 47,XXX probands is reported. In 4 out of 12 subjects, problems such as extreme timidity, anxiety, antisocial behaviour and stubbornness were noted[14]. Tendency to withdraw from group activities has been noted and attributed in part to verbal deficits; also a tendency to passivity, general emotional immaturity and expressionless face[11]. Overanxiety, breath holding and temper tantrums, hyperactivity, elective mutism and extreme shyness were also noted[10].

This list of symptoms indicates that there is no uniformity in behavioural symptoms although the incidence of such problems is increased.

### Psychiatric problems
Notation of higher incidence of such problems has been made[11,13] and in one instance the incidence was found to be 25 % in 12 47,XXX probands[14] when the criterion of psychiatric referral was utilized.

### Summary
The 47,XXX group presents no distinguishing somatic phenotype although there is increased incidence of clinodactyly and epicanthus, the head size tends to be smaller and after 6 years of age these girls tend to be taller than average as they have longer legs.

The overall intelligence of the group is in the dull or low normal range. Speech and language deficits are prevalent and, to a lesser extent, mild cognitive (memory) deficits. Motor development is mildly impaired too. There is an increased incidence of school learning problems congruent with the group's below-average intelligence. Increased incidence of behavioural–emotional and psychiatric problems is known but no particular pattern of such problems has been established.

## 47,XXY children

### Physical characteristics
An increased incidence (18 %) of various major congenital malformations is reported together with higher incidence (26 %) of minor dysmorphic signs, the most consistent of the latter being clinodactyly[21].

After the age of 4 years there is a tendency for head circumference to be below the mean, although within normal limits. From the age of 3 years and over there is increased height growth due to an U < L segment ratio[11,13,14]. Bone age of the whole group is substantially below the mean for age[14]. Although significant as group characteristics these signs would not distinguish the 47,XXY individuals from their normal karyotype coevals.

## Intelligence

Mild cognitive deficits have been reported in children with 47,XXY karyotype. Infant scores of cognitive development were average at 5–6 months, and within normal but ranging from 78 to 133 DQ between 9 and 18 months of age.[30] Mean intelligence scores of 47,XXY boys on the Wechsler scales are lower than those of sibling controls; difficulties in coding type tasks, perhaps because of limited concentration, sequencing problems and poor fine motor skills are reported[13]. Although Bayley Mental Scale Scores did not distinguish between 47,XXY and control groups and although the 47,XXY children were consistently above their chronological age level on visual perception tasks, they showed on testing some lack of persistence, inattentiveness and distractibility[16]. Short term memory deficits are also reported.[14]

## Verbal development

A consensus seems to exist that the deficits in 47,XXY children are mainly verbal. Thus the lower Wechsler scores in 47,XXY children are attributed to low verbal scores; performance abilities in the subjects are highly similar to the average scores obtained by familial controls[29]. Significant discrepancies between verbal and performance scores were noted on Wechsler scales at 5–7 years of age; verbal scores were within normal limits but subaverage and tended to depress the full IQ scores[16]. Delay in early language development has been reported in children with 47,XXY chromosome constitution[10,12,30,32]. High frequency of auditory discrimination failures and poor response to Comprehension and Sentences subtests of the Wechsler Preschool and Primary Scale of Intelligence (WPPSI) are reported[11]. Others report immature articulation and syntax, difficulties in expressing ideas clearly, in verbal comprehension, abstraction and sequencing[10]. Language is slow to appear, verbalizations are decreased, voice is soft, enunciation poor, connected language close to normal for age[13]. Expressive language and articulation are the most frequent problems; some cases present both expressive–receptive language problems. Articulation problems are transient[13]. The Stanford–Binet scores were by 17 points lower than those of controls at 4 years; however the Wechsler scores were average for both Verbal and Performance skills in the same group at 7 years[12]. This difference may be explained by the nature of the Stanford–Binet test which is heavily weighted with verbal items. Syntax, word finding, articulation and in some cases voice production problems are also reported[16]. Some part of the verbal impairment must affect the comprehension of language, particularly the analysis of sentence components, although the existence of associated non-verbal deficits cannot be precluded[29].

## Motor development

Motor incoordination[11], gross and fine motor deficits[14,32] and gross and fine motor slowness and clumsiness[13] are frequent findings. Deficiency in motor organization and delay in inhibition of primary reflexes was found in infancy; later on, mild impairment in motor coordination and speed of movement was noted[30].

## School performance

There are many indications of frequent school learning problems for this group. Reading and spelling difficulties related to dysnomia and temporal sequencing, most of them severe, are reported[14,16]. Arithmetic attainments seem to be essentially identical with those of controls[14]. Other school difficulties are indicated by repeating grades, being assigned to special education class or showing disruptive classroom behaviour[13].

Amongst 12 probands in kindergarten grade 2, 4 were performing well, 3 were repeating a grade and 5 were placed in special class because of learning and/or behaviour problems[32].

## Behavioural–emotional development

This group is characterized by lower activity level, tendency to withdraw from novel stimuli, pliancy in social interaction and difficulty in organizing daily routines and responding to parental expectations[16]. Distractibility, short attention span and a lack of persistence were noted after 25–30 months of age[16]. Passivity, distractibility and expressionless face are noted as well as the child's being quiet and easy to take care of in infancy[11]. Low activity, bland facies and low emotional tone are also reported[30].

Parent rating of their children indicates that the 47,XXY boys are unassertive and tend to be anxious[14]. The anxiety and passivity may be due to direct chromosomal effect or to specific impairments in memory and motor function leading to educational and athletic failures that affect their self-esteem and competitiveness[14].

Emotional and/or behavioural problems are reported in as high as 27% of subjects studied; there was, however, a variety of symptoms and environmental factors were prominent[14]. The latter are reported elsewhere also[32].

## Psychiatric problems

An incidence of 13.7% of psychiatric problems defined by the number of 47,XXY children who have had psychiatric contacts is reported; there was no identifiable syndrome, however, with the symptoms varying widely[14]. Indications of mild psychiatric problems[11], severe disturbance of personality development, shyness, fearfulness and 'immaturity' in three of the subjects[10] are reported. One study[13] has found an incidence of 25% psychiatric problems in 47,XXY subjects compared to 13% in siblings.

## Summary

The 47,XXY children show an increased incidence of minor (mainly clinodactyly) and to lesser extent major (various types) congenital anomalies. They tend to have below mean head size and above average heights after the age of 3 years, the latter due to longer legs.

Mild cognitive deficits do seem to exist; verbal deficits, however, seem to be more important and affect both expressive and receptive language development.

Slow acquisition of gross and fine motor skills and some tendency to motor incoordination are noted.

The early behaviour of 47,XXY children indicates a tendency to passivity, unassertiveness and low activity level. Increased incidence of emotional–behavioural and psychiatric problems is suggested although no particular syndrome has been found. Finally, there are frequent school learning problems mainly in reading and writing.

As with the 47,XXX girls, the 47,XXY boys are not easily distinguishable amongst the normal population and there are individuals amongst this aneuploidy group who function normally in the cognitive and behavioural areas.

## 47,XYY children

### Physical characteristics
These children do not present an increased incidence of major congenital anomalies; there is a suggestion that they may have increased frequency of one or more minor dysmorphic signs such as clinodactyly, inguinal hernia or abnormal ears[6]. There has been no evidence that these boys are taller than controls, at least those who have been followed up to the age of 12 years[21]. One study[12] has suggested a height growth spurt in some boys at 2–6 years of age and a tendency for heights to be skewed to the right of the mean after 4 years of age.

The head size of 47,XYY children is normal[21].

### Intelligence
This group also shows a tendency for mild cognitive deficits. Their IQ scores present a mild skew to the left with 14 of 37 children (38 %) being in the 70–89 IQ group; IQ range was 78–145[6]. At 4 years of age the Stanford–Binet mean IQ was 98.4 in comparison to 115.7 for 39 male controls; Wechsler mean score was 100.9, essentially unchanged at 7 years for the 47,XYY group[12]. In three sets of dizygotic twins a difference of 13–23 IQ points was noted with the unaffected twins performing better[12]. Mean DQ of 3 47,XYY boys was 96.3 below 3 years, 91.6 at 3–6 years[32]. Amongst 3 other probands, 1 was functioning in the mildly retarded and 2 in the borderline range[10].

### Verbal development
Delay in early language development has been noted[12,30,32]. Speech impediments and language problems in older children with 47,XYY have also been reported[11,12,13,15]. No significant difference between verbal and non-verbal skills was reported[12].

### Motor development
There is no consensus about gross motor coordination problems in this group and one study describes the probands as agile[13]. Fine motor incoordination[12] and tremors were reported[49] by some authors but refuted by others[13,15].

### Behavioural–emotional development
Four of 11 probands exhibited difficulties in this area compared to two of 18 controls; temper tantrums and difficulty to relate to peers were described[12]. In another study only one of six boys had presented transient behavioural

problems, considered to be reactive[15]. All three boys in yet another group had behaviour problems ranging between passivity and unassertiveness to temper tantrums[10]. One of three probands was negativistic, resistive, provocative and a loner, while a boy with 48,XXYY was aggressive, destructive, irritable, restless and negativistic[32]. Distractibility in test situations was described[11].

*Summary*
No conclusions can be drawn from the present data other than that no specific 47,XYY phenotype seems to emerge, at least before puberty. There is some evidence that IQ scores tend to be skewed to the left; also that there is increased incidence of speech and language problems, possibly of fine motor incoordination and perhaps of behavioural problems of various descriptions.

## 45,XO girls

Only a few children with this chromosome constitution were detected in prospective studies, the reason probably being the high percentage of spontaneous abortions in the case of conceptuses with 45,XO karyotype[50].

This is the only sex aneuploidy group associated with a specific physical phenotype (including short stature, webbed neck, oedema of the dorsum of hand and foot, shortening of the 4th metacarpal or metatarsal, cubitus valgus, cardiac and renal anomalies and characteristic facies with epicanthus, small mandible and prominent ears as well as with specific cognitive deficits. The latter are well known[22-25,51] and consist of a space–form perception difficulty[22], problems in numerical manipulations[23] and word fluency deficiency[51].

The verbal abilities of 45,XO are known to be normal and this finding was confirmed in the few individuals followed up in the prospective studies[11,13]. There was one report that these girls present poor auditory discrimination and emotional immaturity[11]. One child was found to be borderline in intelligence and bland in personality; she was developmentally slow and has repeated one grade[10]. No significant difference was found in another study[13] between 45,XO females and their sibling controls; however, one proband in this study had an IQ below 70.

Comparison of 31 45,X0 subjects identified at later age with 31 matched controls on spatial[52] and 23 45,XO with 23 matched controls on verbal[53] information processing tasks indicated specific spatial processing deficits. Particularly affected are the processes dealing with transformation of spatial information while the encoding of such information is unimpaired[52]. Verbal processing is normal, albeit a little slower than in controls[53]. The 45,XO girls have a faster maturation rate than normals as shown by their dentition[44,54], dermal ridge counts[34,55] and sulphation rates[43]; since adolescents who enter puberty early have higher verbal than spatial abilities[56] it is possible that the 45,XO individuals develop verbal abilities to the detriment of their spatial skills[52]. The dichotic listening test indicates that indeed the 45,XO have less hemispheric specialization[42]. Thus the spatial deficit of 45,XO may be the result of their accelerated growth due to the absence of the X chromosome[53]. The opposite pattern is noted in boys and girls with additional X chromosomes who show more hemispheric specialization with normal spatial but impaired language functions.

Despite their spatial deficit, 45,XO girls are not at increased risk to present learning or behavioural–psychiatric problems.

## Mosaic karyotypes

There have been no indications from the prospective studies that children with mosaicism are at increased risk for developmental or behavioural anomalies[11,13].

## CONCLUSIONS

Prospective studies in children with aneuploidy have indicated that some of them do present specific behaviour profile and cognitive–verbal deficits of mild degree, also increased incidence of school learning problems and some tendency for emotional–behavioural difficulties. Not all children present such characteristics, however, and those affected function as a rule within the normal, albeit subaverage, range. Continuing studies of these children to puberty and maturity will provide the information whether these mild deficits may become more evident and impede the functioning of the individuals as they enter adulthood.

Much controversy still exists about the ethical aspects of such studies. It is all too clear that future research in this area should be undertaken only after obtaining informed consent from the parents, keeping them informed about the findings on an ongoing basis and giving parents the option to continue or not continue their participation in the study. Legal counsel should be obtained around this aspect prior to designing the study[57]. It remains one of the medical dilemmas whether such study should be undertaken at all. The rationale for doing so is that aneuploid children are at high risk for developmental problems; like any such group (prematures as an example) they should be followed for the early detection of cognitive, verbal, speech and behavioural difficulties and timely provision of the best available remedial help to them. In boys with extra X chromosomes hormonal therapy should be instituted, as there are reports of beneficial effects on their behaviour[58], penile size[12] and hypogonadism. Information, anticipatory counselling and continuing support should be also provided to parents, with the initial emphasis being on the generally optimistic outlook that the present studies suggest about the possibility of the child to develop in the normal range in early life.

A benefit that has derived from the prospective studies is that knowledge about the early development of aneuploids now allows the physician to provide information and counselling to parents if such diagnosis is made through amniocentesis or karyotyping the child later in life.

The experience gained from the present prospective studies will be most valuable in helping future research groups to coordinate efforts and arrive at uniform and appropriately designed research to maximize the scope and the validity of data thus obtained.

# References

1 Nielsen, J. and Sillesen, I. (1975). Incidence of chromosome aberrations among 11,148 newborn children. *Humangenetik*, **30**, 1

2 Lubs, H. A. and Ruddle, F. H. (1970). Applications of quantitative karyotype to chromosome variation in 4,400 consecutive newborns. In Jacobs, P., Price, W. H. and Law, P. (eds.) *Human Population Cytogenetics*, pp. 120–142. (Edinburgh: University of Edinburgh Press)

3 Walzer, S. and Gerald, P. S. (1977). A chromosome survey of 13,751 male newborns. In Hook, E. B. and Porter, I. H. (eds.) *Population Cytogenetics*, pp. 45–61. (New York: Academic Press)

4 Jacobs, P. A., Melville, M., Ratcliffe, S. G., Keay, A. J. and Syme, J. (1974). A cytogenetic survey of 11,680 newborn infants. *Ann. Hum. Genet.*, **37**, 359

5 Higurashi, M., Iijima, K. and Ikeda, U. (1979). Chromosome survey of newborn infants in Tokyo: Follow-up study for XYY. In Robinson, A., Lubs, H. A. and Bregsma, D. (eds.) *Sex Chromosome Aneuploidy: Prospective Studies on Children*, pp. 161–174. (New York: A. R. Liss)

6 Lubs, H. A., Patil, S. R., Kimberling, W. J., Brown, J., Cohen, M. M., Gerald, P. S., Hecht, F., Moorhead, P., Myrianthopoulos, N. C. and Summit, R. L. (1979). Chromosome abnormalities ascertained in the collaborative perinatal survey of 7- and 8-year old children. In Robinson, A., Lubs, H. A. and Bregsma, D. (eds.) *Sex Chromosome Aneuploidy: Prospective Studies on Children*, pp. 191–202. (New York: A. R. Liss)

7 Robinson, A., Goad, W. B., Puck, T. T. and Harris, J. S. (1969). Studies on chromosomal nondisjunction in man, III. *Am. J. Hum. Genet.*, **21**, 466

8 Court Brown, W. M. (1969). Sex chromosome aneuploidy in man and its frequency, with special reference to mental subnormality and criminal behavior. *Int. Rev. Exp. Pathol.*, **7**, 31

9 Bell, A. G. and Corey, P. N. (1974). A sex chromatin and Y-body survey of Toronto newborns. *Can. J. Genet. Cytol.*, **16**, 239

10 Leonard, M. F., Showalter, J. E., Landy, G., Ruddle, F. H. and Lubs, H. A. (1979). Chromosomal abnormalities in the New Haven newborn study: A prospective study of development of children with sex chromosome anomalies. In Robinson, A., Lubs, H. A. and Bregsma, D. (eds.) *Sex Chromosome Aneuploidy: Prospective Studies on Children*, pp. 115–160. (New York: A. R. Liss)

11 Nielsen, J., Sillessen, I., Sørensen, A. M. and Sørensen, K. (1979). Follow-up until age 4 to 8 of 25 unselected children with sex chromosome abnormalities, compared with sibs and controls. In Robinson, A., Lubs, H. A. and Bergsma, D. (eds.) *Sex Chromosome Aneuploidy: Prospective Studies on Children*, pp. 15–74. (New York: A. R. Liss)

12 Ratcliffe, S. G., Axworthy, D. and Ginsborg, A. (1979). The Edinburgh study of growth and development in children with sex chromosome abnormalities. In Robinson, A., Lubs, H. A. and Bregsma, D. (eds.) *Sex Chromosome Aneuploidy: Prospective Studies on Children*, pp. 243–260. (New York: A. R. Liss)

13 Robinson, A., Puck, M., Pennington, B., Borelli, J. and Hudson, M. (1979). Abnormalities of the sex chromosomes: A prospective study on randomly identified newborns. In Robinson, A., Lubs, H. A. and Bregsma, D. (eds.) *Sex Chromosome Aneuploidy: Prospective Studies on Children*, pp. 203–242. (New York: A. R. Liss)

14 Stewart, D. A., Netley, C. T., Bailey, J. D., Haka-Ikse, K., Platt, J., Holland, W. and Cripps, M. (1979). Growth and development of children with X and Y chromosome aneuploidy: A prospective study. In Robinson, A., Lubs, H. A. and Bregsma, D. (eds.) *Sex Chromosome Aneuploidy: Prospective Studies on Children*, pp. 75–114. (New York: A. R. Liss)

15 Valentine, G. H. (1979). The growth and development of six XYY children. In Robinson, A., Lubs, H. A. and Bregsma, D. (eds.) *Sex Chromosome Aneuploidy: Prospective Studies on Children*, pp. 175–190. (New York: A. R. Liss)

16 Walzer, S., Wolff, P. H., Bowen, D., Silbert, A. R., Bashir, A. S., Gerald, P. S. and Richmond, J. B. (1978). A method for the longitudinal study of behavioral development in infants and children: The early development of XXY children. *J. Child Psychol. Psychiatry*, **19**, 213

17 Jacobs, P., Brunton, M., Melville, M. M., Brittain, R. P. and McClemont, W. F. (1965). Aggressive behaviour, mental subnormality and the XYY male. *Nature (London)*, **208**, 1351

18 Casey, M. D., Segall, L. J., Street, D. R. K. and Blank, C. E. (1966). Sex chromosome abnormalities in two state hospitals for patients requiring special security. *Nature (London)*, **209**, 641

19  Jacobs, P.A., Price, W. H., Court Brown, W. M., Brittain, R. P. and Whatmore, P. B. (1968). Chromosome studies on men in a maximum security hospital. *Ann. Hum. Genet.*, **31**, 339
20  McLean, N., Court Brown, W. M., Jacobs, P. A., Mantle, D. J. and Strong, J. A. (1968). A survey of sex chromatin abnormalities in mental hospitals. *J. Med. Genet.*, **5**, 165
21  Robinson, A., Lubs, H. A., Nielsen, J. and Sørensen, K. (1979). Summary of clinical findings: Profiles of children with 47,XXY, 47,XXX and 47,XYY karyotypes. In Robinson, A., Lubs, H. A. and Bregsma, D. (eds.) *Sex Chromosome Aneuploidy: Prospective Studies on Children*, pp. 261–266. (New York: A. R. Liss)
22  Shaffer, J. W. (1962). A specific cognitive deficit observed in gonadal aplasia (Turner's Syndrome). *J. Clin. Psychol.*, **18**, 403
23  Money, J. (1963). Cytogenetic and psychosexual incongruities with a note on space–form blindness. *Am. J. Psychiatry*, **119**, 820
24  Garrow, D. C. and Vander Stoep, L. P. (1969). Personality and intelleigence in Turner syndrome: A clinical review. *Arch. Gen. Psychiatry.*, **21**, 339
25  Garrow, D. (1977). Intelligence among persons with Turner's syndrome. *Behav. Genet.*, **7**, 105
26  Nielsen, J., Sørensen, A., Theilgaard, A., Froland, A. and Johnsen, S. G. (1969). A psychiatric–psychological study of 50 severely hypogonadal males, including 34 with Klinefelter's syndrome, 47,XXY. *Acta Jutlandia*, **41**
27  Theilgaard, A., Nielsen, J., Froland, A. and Johnsen, S. (1971). *A Psychiatric–Psychological Study of Patients with Klinefelter's syndrome.* (Copenhagen: Munksgaard)
28  Garvey, M. and Mutton, D. E. (1973). Sex chromosome aberrations and speech development. *Arch. Dis. Child.*, **48**, 937
29  Netley, C. and Rovet, J. (1980). Specific verbal deficits in children with abnormalities of sex chromosome complement. Presented at *International Neuropsychology Society Annual Meeting*, January 31–February 3, San Francisco
30  Tennes, K., Puck, M., Orfanakis, D. and Robinson, A. (1977). The early childhood development of 17 boys with sex chromosome abnormalities. *Pediatrics*, **59**, 574
31  Tennes, K., Puck, M., Bryant, K., Frankenburg, W. and Robinson, A. (1975). A developmental study of girls with trisomy X. *Am. J. Hum. Genet.*, **27**, 71
32  Haka-Ikse, K., Stewart, D. A. and Cripps, M. (1978). Early development of children with sex chromosome aberrations. *Pediatrics*, **62**, 761
33  Masica, D. N., Money, J., Ehrhardt, A. A. and Lewis, V. G. (1969). I.Q. fetal sex hormones and cognitive patterns: studies in the testicular feminizing syndrome of androgen sensitivity. *Johns Hopkins Med. J.*, **123**, 34
34  Mitwoch, U. (1973). *Genetics of Sex Differentiation.* (New York: Academic Press)
35  Money, J. and Ehrhardt, A. (1972). *Man and Woman, Boy and Girl: The Differentiation and Dimorphism of Gender Identity from Conception to Maturity.* (Baltimore: Johns Hopkins UP)
36  Money, J. (1968). Turner's syndrome and parietal lobe functions. *Cortex*, **9**, 313
37  Waber, D. P. (1979). Neuropsychological aspects of Turner's syndrome. *Devel. Med. Child Neurol.*, **21**, 58
38  Kolb, J. E. and Heaton, R. K. (1975). Lateralized neurologic deficits and psychopathology in a Turner syndrome patient. *Arch. Gen. Psychiatry*, **32**, 1198
39  Silbert, A., Wolff, P. H. and Lilienthal, J. (1977). Spatial and temporal processing in patients with Turner's syndrome. *Behav. Genet.*, **7**, 11
40  Barlow, P. W. (1973). X-chromosomes and human development. *Devel. Med. Child Neurol.*, **15**, 205
41  Levy, J. (1969). Possible basis for the evolution of lateral specialization of the human brain. *Nature (London)*, **224**, 614
42  Netley, C. (1977). Dichotic listening of callosal agencies and Turner's syndrome patients. In Segalowitz, S. and Gruber, F. (eds.) *Language Development and Neurological Theory*, pp. 133–143. (New York, San Francisco, London: Academic Press)
43  Almqvist, S., Linsten, J. and Lindvall, N. (1963). Linear growth, sulfation factor activity and chromosome constitution in 22 subjects with Turner's syndrome. *Acta. Endocrinol.*, **42**, 168
44  Rovet, J. and Netley, C. (1980). Dentition rates of children with sex chromosome abnormalities. (In preparation)
45  Hreczko, T. and Sigmon, B. (1980). The dermoglyphics of a Toronto sample of children with XXY, XYY and XXX aneuploidies. *Am. J. Phys. Anthropol.* (In press)
46  Rovet, J. and Netley, C. (1980). Processing deficits in Turner's syndrome. *Dev. Psychol.* (In press)

47 Ratcliffe, S. G. and Evans, H. J. (1975). Sex chromosome abnormalities and social class. *Lancet*, **1**, 1144
48 Walzer, S. and Gerald, P. S. (1975). Social class and frequency of XYY and XXY. *Science*, **190**, 1219
49 Daly, R. F. (1969). The XYY condition in childhood – clinical observation. *Pediatrics*, **44**, 621
50 Jacobs, P. A. (1979). The incidence and aetiology of sex chromosome abnormalities in man. In Robinson, A., Lubs, H. A. and Bergsma, D. (eds.) *Sex Chromosome Aneuploidy: Prospective Studies on Children*, pp. 3–14. (New York: A. R. Liss)
51 Money, J. and Alexander, D. (1966). Turner's syndrome: Further demonstration of the presence of specific cognitional deficiencies. *J. Med. Genet.*, **3**, 47
52 Rovet, J. and Netley, C. (1980). The mental rotation task performance of Turner syndrome subjects. *Behav. Genet.*, **10**, 437
53 Netley, C. and Rovet, J. (1980). Dichotic listening performance of Turner syndrome subjects. (In preparation)
54 Barlow, P. (1973). The influence of inactive chromosomes on human development. *Hum. Genet.*, **17**, 105
55 Holt, S. B. (1968). *The Genetics of Dermal Ridges*. (Springfield: Thomas)
56 Waber, D. P. (1976). Neurophysiological analysis of spatial ability in Turner's syndrome: Genetic implications. Presented at *Meeting of the American Psychological Association*, Washington, D.C.
57 Hamerton, J. L. (1979). Ethical considerations in newborn chromosome screening programs. In Robinson, A., Lubs, H. A. and Bergsma, D. (eds.) *Sex Chromosome Aneuploidy: Prospective Studies on Children*, pp. 267–278. (New York: A. R. Liss)
58 Caldwell, P. D. and Smith, D. W. (1972). The XXY (Klinefelter's) syndrome in childhood: Detection and treatment. *J. Pediatr.*, **80**, 250

# 8
# Syndrome delineation and its implications for the study of pathogenetic mechanisms*

## M. M. COHEN JR.

---

The process of syndrome delineation can be divided into the following stages:

(1) *Unknown-genesis syndromes*

    (a) Provisionally unique-pattern syndrome
    (b) Recurrent-pattern syndrome.

(2) *Known-genesis syndromes*

    (a) Pedigree syndrome
    (b) Chromosomal syndrome
    (c) Biochemical-defect syndrome
    (d) Environmentally-induced syndrome.

Let us consider each of these stages in turn.

## UNKNOWN-GENESIS SYNDROMES

*In an unknown-genesis syndrome, the cause is simply not known.*

### Provisionally unique-pattern syndrome

*In a provisionally unique-pattern syndrome, several anomalies are observed in the same patient such that the clinician does not recognize the overall pattern of defects from his own experience, nor from searching the literature, nor from consultation with the most learned colleagues in the field*[1]. The patient shown in Figures 8.1–8.3 has a provisionally unique-pattern syndrome consisting of a cloverleaf skull malformation with craniosynostosis, preaxial polydactyly, micropenis, and bifid scrotum[1]. Most likely these anomalies have a common

---

* Reprinted from *The Patient With Multiple Anomalies*, by M. M. Cohen Jr., (Raven Press, 1981), with permission.

103

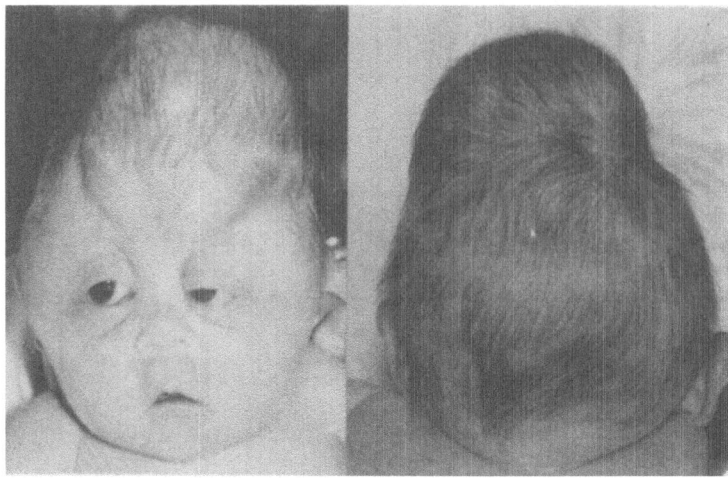

**Figure 8.1** An example of a provisionally unique-pattern syndrome. Cloverleaf skull malformation. See also Figures 8.2 and 8.3. (*From* Cohen[1])

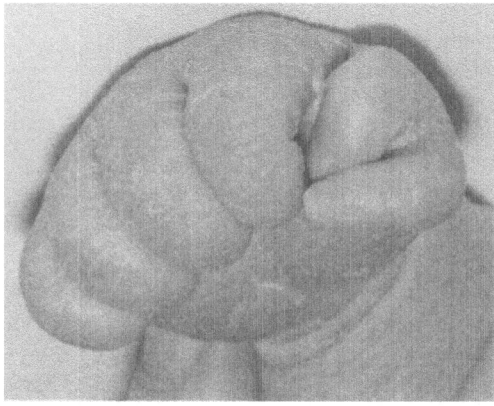

**Figure 8.2** An example of a provisionally unique-pattern syndrome. Preaxial polydactyly. See also Figures 8.1 and 8.3. (*From* Cohen[1])

cause, even though unknown, rather than having different causes acting independently. The probability that such anomalies occur in the same patient by chance becomes less likely the more anomalies the patient has and the rarer these anomalies are individually in the general population.

Obviously if a second example comes to light, the condition is no longer unique. A provisionally unique-pattern syndrome is a one-of-a-kind syndrome to a particular observer at a particular point in time. There may be a nineteenth century description of a similar instance that escapes his attention.

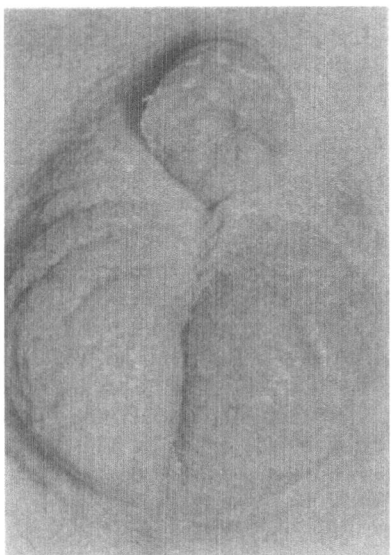

**Figure 8.3** An example of a provisionally unique-pattern syndrome. Micropenis, bifid
scrotum. See also Figures 8.1 and 8.2. (*From* Cohen[1])

There may also be some instances of the syndrome in different parts of the
world that remain as yet unrecognized. Thus, many syndromes appear to be
unique at the time the initial patient is discovered, but are no longer unique
when two or more examples become known. On the other hand, some
syndromes may be truly unique, and this possibility is considered later in this
chapter.

There is a widely held view that a recurrent pattern of anomalies in two or
more individuals constitutes a syndrome, but that a provisionally unique
pattern of anomalies in a single patient cannot be accorded syndrome status.
Clinicians who subscribe to this notion belong to the 'it-takes-two-or-more-
to-make-a-syndrome' school. Their reasoning in fallacious. 'A syndrome is a
syndrome is a syndrome.' To take an instructive example, paleontologists
sometimes define a whole species on the basis of a single fossil specimen. One
wants a definition that states clearly, 'Item X belongs to Class Y or it does not',
irrespective of how many X's there are and, in formal logic, a Class may have
as few as one member – or even none! Thus, the term syndrome can apply to a
one-of-a-kind condition as well as to a many-of-a-kind condition.

There is a definite need for a term such as 'provisionally unique-pattern
syndrome'. Today, perhaps half of all dysmorphic syndromes seen by
clinicians may be provisionally unique-pattern syndromes. Such a large and
important class of patients merits its own special designation. Even those
clinicians who belong to the 'it-takes-two-or-more-to-make-a-syndrome'
school recognize provisionally unique-pattern syndromes in practice if not
philosophically. For example, such a clinician may say to a parent, 'your child
has a pattern of anomalies that we do not recognize'.

Ideally, all provisionally unique-pattern syndromes should be published. In practice they are usually filed away and hardly ever published because their significance is a mystery to clinicians and to journal editors alike.* However, the publication of a provisionally unique-pattern syndrome is like an advertisement with a red flag (Figures 8.1–8.3); it reaches a large audience, which may allow one or more clinicians to react by publishing similar cases. When this happens, the process of syndrome delineation is under way.

### Recurrent-pattern syndrome

The next stage in syndrome delineation is the recurrent-pattern syndrome.

*A recurrent-pattern syndrome can be defined as a similar or identical set of anomalies in two or more unrelated† patients*[1,2]. A recurrent-pattern syndrome is illustrated in Figures 8.4–8.6. These two patients from different families share in common a wide bifrontal diameter, ocular hypertelorism, large ears, micrognathia, finger contractures at the proximal interphalangeal joints, deeply set fingernails, umbilical hernia, excessive growth, and a variety of other abnormalities[3]. The features of both patients are compared in Table 8.1, in which they are congruent for 22 of the 26 listed abnormalities. The same abnormalities in two or more patients suggest, but do not prove, that the pathogeneis in both cases may be the same. At the recurrent-pattern stage of syndrome delineation, the aetiology is still not known. In general, the validity of a recurrent-pattern syndrome increases the more abnormalities there are in the condition and the more patients that are known to have the syndrome. A better-known recurrent-pattern syndrome with over 120 reported cases is the Rubinstein–Taybi syndrome. Features include microcephaly, mental retardation, downslanting palpebral fissures, aquiline nose with the nasal septum extending below the alae, growth deficiency, and broad thumbs and great toes. Almost all instances are sporadic, although affected sibs have been observed on two occasions[4].

At the recurrent-pattern stage of syndrome delineation, the number of findings is usually expanded as the number of patients increases. However, because the aetiology remains unknown at this point in time, other examples of the syndrome tend to be selected because they most closely resemble the first case. This results in an artificial homogeneity of cases, which emphasizes the most severe aspects of the syndrome. Thus, we should be wary of estimated frequencies given in review articles and textbooks for various anomalies that occur in a recurrent-pattern syndrome; they tend to be overestimates.

### KNOWN-GENESIS SYNDROMES

*A known-genesis syndrome can be defined as several anomalies causally related on the basis of (1) occurrence in the same family or, less conclusively, the same*

---

* One of the missions of the journal *Syndrome Identification* is to publish provisionally unique-pattern syndromes to allow clinicians to react by publishing similar cases if they exist. This journal is one of the few outlets for such activities.

† An occasional patient may be related within a large syndrome sample, yet the delineation status remains at the level of a recurrent-pattern syndrome, as in the Rubinstein–Taybi syndrome.

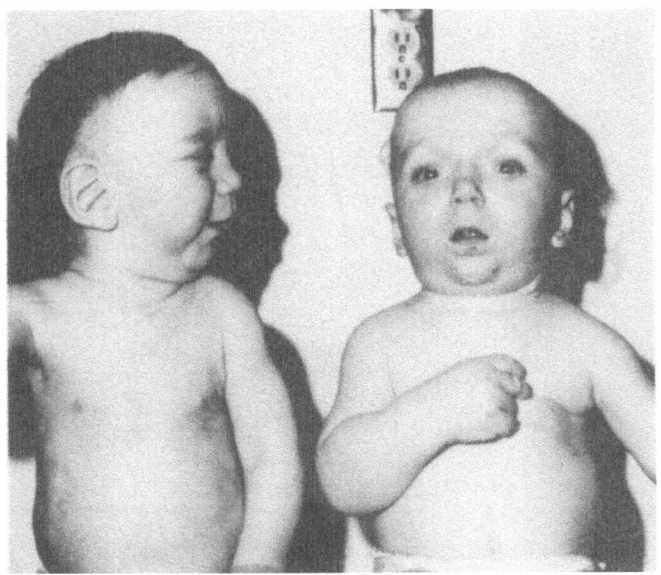

**Figure 8.4** An example of a recurrent-pattern syndrome in two patients. Note similarity. See also Figures 8.5 and 8.6. (*From* Weaver *et al.*[3])

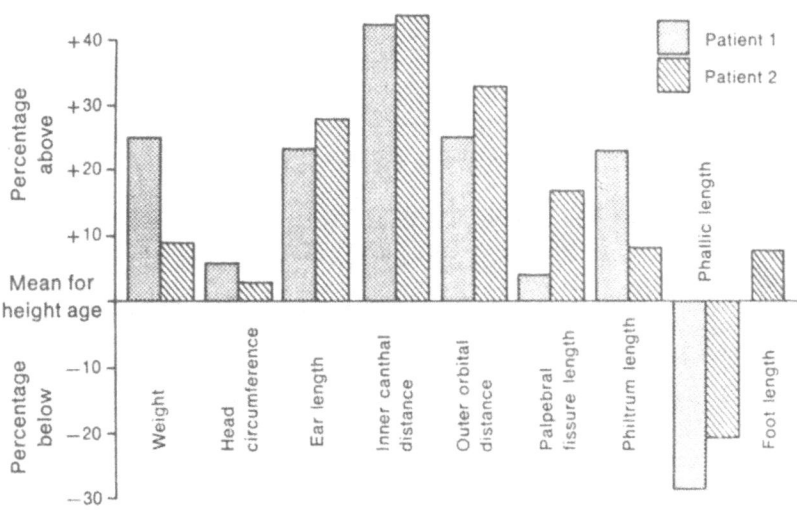

**Figure 8.5** An example of a recurrent-pattern syndrome in two patients. Note similarity of measurement patterns. See also Figures 8.4 and 8.6. (*From* Weaver *et al.*[3])

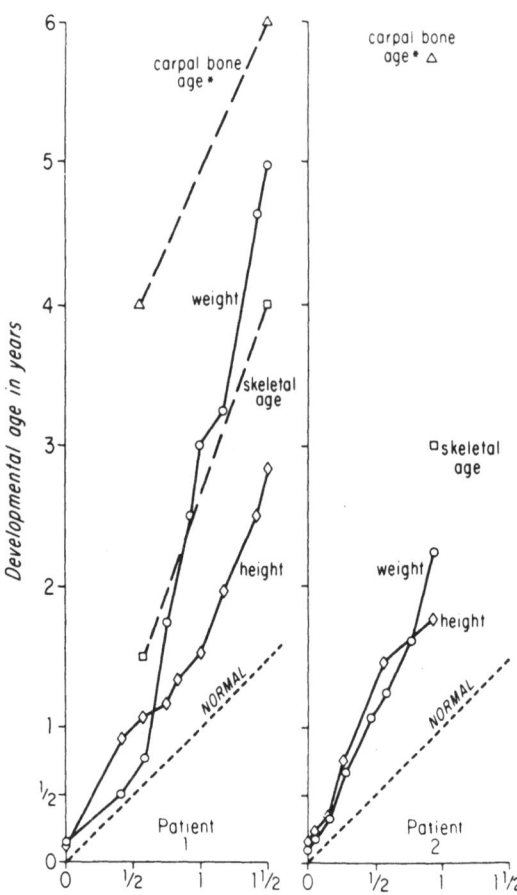

**Figure 8.6** An example of a recurrent-pattern syndrome in two patients. Note similarity in accelerated osseous maturation pattern. See also Figures 8.4 and 8.5. (*From* Weaver et al.[3])

mode of inheritance in different families, (2) a chromosomal defect, (3) a specific defect in an enzyme or structural protein or (4) an environmental factor[1,2].

## Pedigree syndrome

The term 'pedigree syndrome' refers to known genesis on the basis of pedigree evidence alone; the basic defect itself remains undefined, although the condition is known to represent a monogenic disorder.* A good example is the autosomal recessive Meckel syndrome characterized by features such as occipital

---

* The term pedigree syndrome could be applied to an inherited translocation, an enzymatic defect or an abnormal structural protein. The term could also be applied to a familial occurrence of a non-genetic syndrome. However, I have chosen to restrict the use of the term to undefined monogenic syndromes as a stage in syndrome delineation.

Table 8.1 Comparison of features of two patients with recurrent-pattern syndrome

|  | Patient 1 | Patient 2 |
|---|---|---|
| *Growth* | | |
| Excessive growth of prenatal onset | + + + | + + + |
| Accelerated osseous maturation | + + + + | + + + + |
| *Performance* | | |
| Hypertonia | + + | + |
| Hoarse, low-pitched cry | + + | + + |
| Developmental delay | ? | ? |
| Excessive appetite | + + | + + |
| *Craniofacial* | | |
| Wide bifrontal diameter | + + + | + + + |
| Flat occiput | + | + |
| Large ears | + + + | + + + |
| Ocular hypertelorism | + + | + + |
| Long philtrum | + + | + |
| Relative micrognathia | + | + |
| *Limbs* | | |
| Hands: | | |
|     Prominent finger pads | + + | + + |
|     Simian crease | + | − |
|     Camptodactyly | + + | + |
|     Broad thumbs | + + | + |
|     Thin, deep-set nails | + + | + + |
| Feet: | | |
|     Clinodactyly, toes | + | + |
|     Talipes equinovarus | + + | − |
|     Short fourth metatarsals | + | − |
| Limited early elbow and knee extension | + | + |
| Widened distal femurs and ulnas | + + | + + |
| *Skin* | | |
| Excessive, loose skin | + + | + + |
| Inverted nipples | + | + |
| Thin hair | + | + |
| *Other* | | |
| Umbilical hernia | + + | + |
| Inguinal hernias | + + | − |

+ = present, in varying degrees of severity; − = absent; ? = uncertain. (*From* Weaver *et al.*[3]).

encephalocoele, polydactyly, polycystic kidneys and other anomalies[4] (Figures 8.7 and 8.8).

## Chromosomal syndrome

*A chromosomal syndrome is cytogenetically defined*, such as the trisomy 13 syndrome[4] (Figure 8.9).

## Biochemical-defect syndrome

*In a biochemical-defect syndrome, specific enzymatic defects are known in recessive syndromes. The term is also meant to include specific defects in*

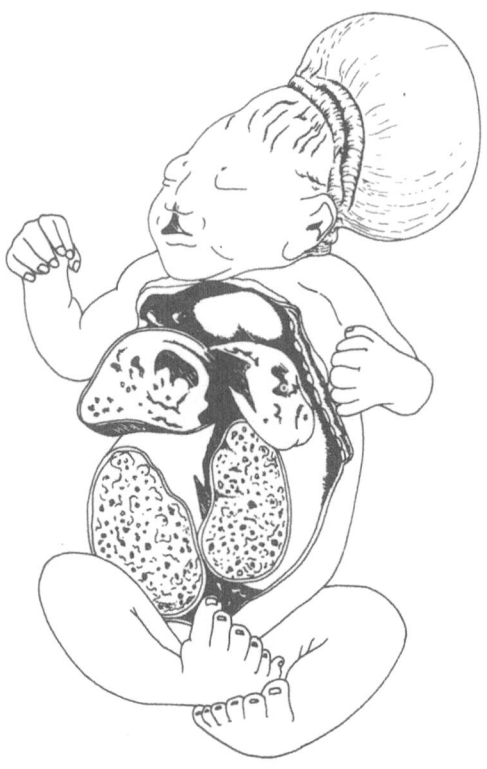

**Figure 8.7** An example of a known-genesis syndrome of the pedigree type. Autosomal recessively inherited Meckel syndrome. Note encephalocoele, cleft lip, polydactyly, polycystic kidneys and cysts of liver

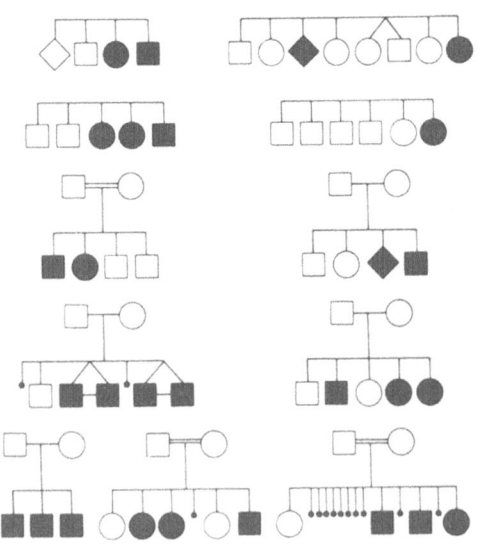

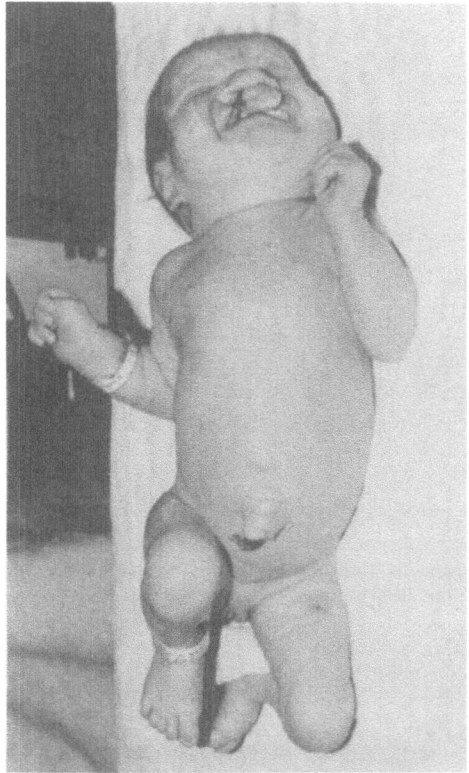

**Figures 8.9** Known-genesis syndrome of the chromosomal type. Trisomy 13 syndrome

*structural proteins as these become known in some of the dominant disorders.* The Lesch–Nyhan syndrome, characterized by hypoxanthine-guanine-phosphoribosyl transferase deficiency, is an X-linked recessive biochemical defect syndrome[5] (Figure 8.10).

## Environmentally-induced syndrome

*An environmentally-induced syndrome is defined in terms of the environmental factor or causative teratogen.* Infants born to mothers who are chronic alcoholics during their pregnancies have an increased risk of having growth deficiency of prenatal onset persisting into postnatal life, microcephaly, mental deficiency, narrow palpebral fissures, mild maxillary hypoplasia, long philtrum, joint deformities, cardiac malformations, and other anomalies[6] (Figure 8.11).

---

**Figure 8.8** Pedigrees of sibships with recurrent cases of the Meckel syndrome, a known-genesis syndrome of the pedigree type. Affected sibships and consanguinity are consistent with autosomal recessive inheritance. (Pedigrees are summarized from Hsia, Y. E. *et al.* (1971). *Pediatrics*, **48**, 237; Simopoulos, A.P. *et al.* (1967). *Pediatrics*, **39**, 931; Chemke, J. *et al.* (1977). *Clin. Genet.*, **11**, 285)

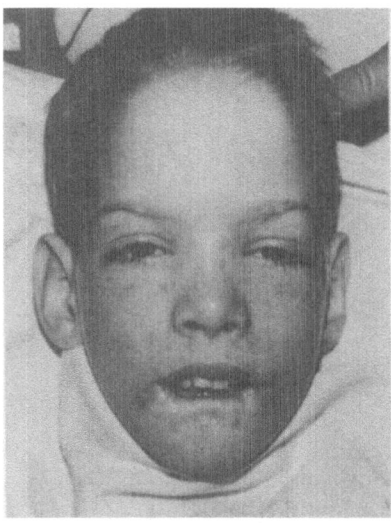

**Figure 8.10** Known-genesis syndrome of the biochemical-defect type. Lesch–Nyhan syndrome with enzymatic defect in purine metabolism that leads to self-mutilation of lips

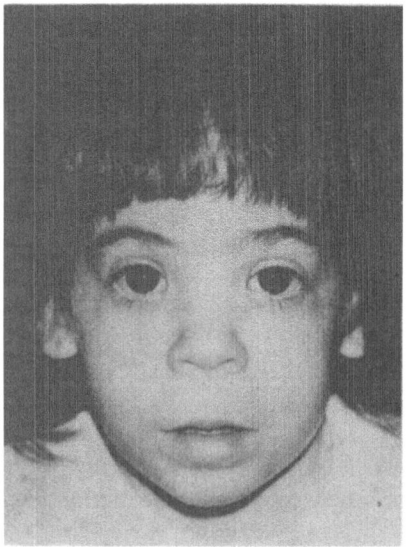

**Figure 8.11** Known-genesis syndrome of the environmentally-induced type. Fetal alcohol syndrome. Microcephaly, narrow palpebral fissures, mild maxillary hypoplasia, and short nose. (*From* Clarren and Smith[6])

## COMMENTS ON THE SYNDROME DELINEATION PROCESS

The process of syndrome delineation is summarized in Figure 8.12. Generally, a syndrome can be placed into one of the categories discussed above. Special categories such as chance patterns, associations, and variant additive patterns are treated later in this section and in the next section. Occasionally, a syndrome may be delineated in a one-step delineation, thus bypassing several of the stages mentioned earlier. For example, if a new chromosomal abnormality is discovered during the laboratory investigation of a patient clinically defined as having a provisionally unique-pattern syndrome, the patient represents a known-genesis syndrome of the chromosomal type in a one-step delineation. However, the variability of the clinical expression must await the discovery of more patients. In other instances, such as a large dominant pedigree with many affected individuals, a known-genesis syndrome of the pedigree type and much of its phenotypic variability can be determined in one step.

Provisionally unique-pattern syndromes occur with some frequency. Further delineation will often occur, given sufficient time. A truly unique-pattern syndrome may occur with a chromosomal anomaly involving two or more breaks. The condition may be sporadic or segregate within a family. Since the chance of an identical duplication-deficiency syndrome occurring in another family is very slight, the syndrome may be considered unique to an affected individual or an affected family[7].

The multiple anomalies that make up any syndrome are thought to be pathogenetically related[8]. However, anomalies may concur either by chance or as a statistically related association in the same patient. Some associations are well-stocked pools for future syndrome delineation. For example, in a large series of patients with the VATER association (vertebral anomalies, anal atresia, tracheo-oesophageal fistula, radial defects and renal anomalies)[9], there may be some recurrent-pattern syndromes and even some provisionally unique-pattern syndromes.

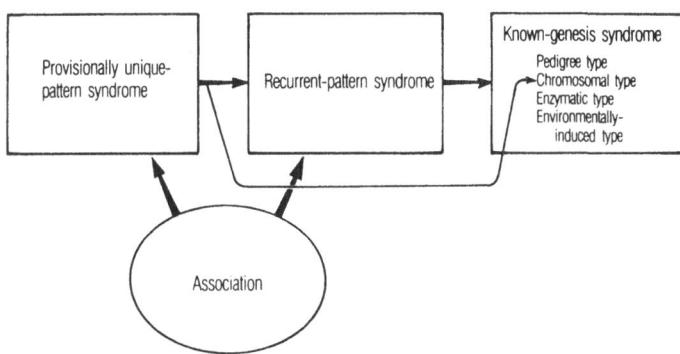

**Figure 8.12** Diagrammatic summary of the process of syndrome delineation. See text

113

## VARIANT ADDITIVE PATTERNS

At times, it may be difficult to distinguish between a true multiple anomaly syndrome, which is a form of discontinuous variability with respect to normal first degree relatives, and a variant additive syndrome, which is a form of continuous variability with respect to such relatives. *A variant additive pattern usually consists of several minor anomalies and often a major anomaly in the same individual such that the pattern of morphological findings in that individual is statistically unusual for the general population but biologically normal for the individual's family with the possible exception of the major anomaly. A variant additive pattern is an unusual chance pattern that can be identified in the population because the minor anomalies that make up the pattern in the proband may be observed separately in various relatives.* Thus, the pattern is not caused by a monogenic or chromosomal abnormality, but by a variety of different genes acting independently. In this context, the minor anomalies in the proband and his or her family should be regarded as normal morphological variants[1,7,10,11].

Often the parents will explain that their child is not at all abnormal but simply resembles other members of the family. Facial comparison may be particularly telling. Many variant additive patterns do not have a major anomaly. When a major anomaly occurs in addition by chance, the likelihood of referral to a syndromologist increases. Care must be taken not to overdiagnose a variant additive pattern as a true multiple anomaly syndrome of presumed unitary aetiology[7,10,11].

A typical example of a variant additive pattern is a proband with downslanting palpebral fissures, ear pits, mandibular prognathism, clinodactyly, cubitus valgus and ventricular septal defect. In examining the proband's relatives (Figure 8.13), the father is noted to have downslanting palpebral fissures and cubitus valgus, the mother is found to have ear pits and clinodactyly, and both the brother and the maternal grandfather are observed to have mandibular prognathism. Thus, all the features except ventricular septal defect are dispersed in various members of the family as minor anomalies. They all happened to come together in the proband by chance in addition to one major anomaly – a ventricular septal defect. The probability of all these anomalies coming together in future offspring in this family is slight, although various individual anomalies or combinations of anomalies may recur. For example, of three future offspring, a first might have ear pits; a second, ventricular septal defect, clinodactyly and downslanting palpebral fissures; a third, mandibular prognathism and cubitus valgus. Thus, a variant additive pattern may be considered unique to an affected individual or to an affected family. In this particular family, genetic counselling should include a multifactorial recurrence risk for ventricular septal defect[1].

## FURTHER COMMENTS ON THE SYNDROME DELINEATION PROCESS

Syndrome delineation and the use of the delineation terms proposed should be thought of as a dynamic, flexible, and continually changing framework in which to view various syndromes. These categories should never be thought of

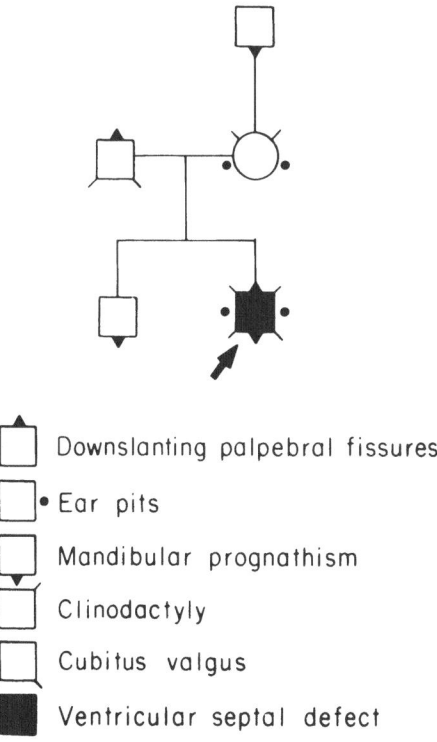

| | |
|---|---|
| Downslanting palpebral fissures | |
| Ear pits | |
| Mandibular prognathism | |
| Clinodactyly | |
| Cubitus valgus | |
| Ventricular septal defect | |

**Figure 8.13** Variant additive pattern. Note that all features of the pattern observed in the proband except ventricular septal defect are dispersed in various members of the family. See text. (*From* Cohen[1])

as static or immutable, even at the higher stages of syndrome delineation. Aetiological and clinical heterogeneity is common and should be expected to occur even when not readily apparent. A recurrent-pattern syndrome such as the Noonan syndrome is undoubtedly aetiologically heterogeneous[4]. In a known-genesis syndrome of the pedigree type, heterogeneity may also be present, as in the autosomal dominant and autosomal recessive forms of the Larsen syndrome[4]. Even in a known-genesis syndrome of the biochemical-defect type, aetiological heterogeneity may occur, as in the hydroxylysine-deficient and procollagen peptidase-deficient forms of the Ehlers–Danlos syndrome[4].

Finally, we should not confuse syndrome delineation with our understanding or lack of understanding of the syndrome's pathogenesis, even at the higher stages of delineation. In a pedigree syndrome, such as the autosomal recessively inherited Meckel syndrome, we know nothing about how the homozygous state of the Meckel gene produces such diverse features as encephalocele, polydactyly, and polycystic kidneys. We understand nothing about the pathogenesis of the trisomy 13 syndrome, although the aetiology is known. How an aetiological agent such as ethyl alcohol produces the abnormalities found in the fetal alcohol syndrome is still a mystery. Syndrome delineation does, however, aid in the preliminary sorting of disorders for the proper study of pathogenesis. This subject is treated more fully, in a later section.

## COMMENTS ON NEW SYNDROMES

Syndromologists frequently postulate, discuss and write about 'new syndromes'. The term has various shades of meaning which will be considered in this section. A common meaning for a 'new syndrome' is a condition that has always existed but has become newly recognized. Thus, the syndromologist calls attention to the entity. Sometimes the condition is remarkably well described in the older literature unbeknownst to the syndromologist who proposed the 'new syndrome'. When this happens, the 'new syndrome' simply represents a 'rediscovered syndrome'.*

In some instances, the syndromologist who proposed a 'new syndrome' is aware that the condition has been described previously. However, justification for new syndrome status is based upon more definitive delineation of the condition. For example, more complete description of the spectrum of anomalies and establishing autosomal recessive inheritance might be enough to consider it a 'new syndrome'. The earlier description might simply have been a one paragraph thumbnail sketch of a sporadic case.

Finally, some new syndromes are genuinely new, such as the thalidomide syndrome at the time of its discovery. The syndrome *per se* did not exist prior to the marketing of the drug.

## SPORADIC OCCURRENCE AND SYNDROME DELINEATION

It should be emphasized that sporadicity *per se* does not necessarily mean non-genetic. In the early stages of syndrome delineation, a number of sporadic instances of a new recurrent-pattern syndrome might represent a monogenic disorder. Because of the small size of the human family, more than half of all families in which both parents are heterozygous carriers for a given autosomal recessive disorder have only one occurrence of the disorder per family. Sporadic occurrences of a syndrome can also represent fresh mutations for an autosomal dominant disorder in which the genetic fitness is dramatically reduced. Almost all instances of the Apert syndrome (a known dominant disorder) are sporadic because the severe physical malformations and the presence of mental deficiency (in some cases) diminish their desirability as mates (Figure 8.14). In the few known instances in which mating has occurred, dominant transmission of the disorder has been observed[12] (Figure 8.15).

---

* Some provisionally unique and recurrent-pattern syndromes are remarkably well described in the older literature. Many workers today (1) do not have access to much of the old literature before the turn of the century, (2) do not have the linguistic prowess to read many languages or (3) do not have time to carry out such a literature search. Thus, it sometimes happens that after a 'new' syndrome becomes well delineated, a complete description of one or more affected patients is discovered in some eighteenth or nineteenth century reference. We owe a great deal to the early investigators who were limited only as prisoners of history in not being able to understand pedigree analysis, chromosomal aberrations or enzymatic defects, and in not having large numbers of colleagues to communicate with regularly at national meetings, through journals and by telephone.

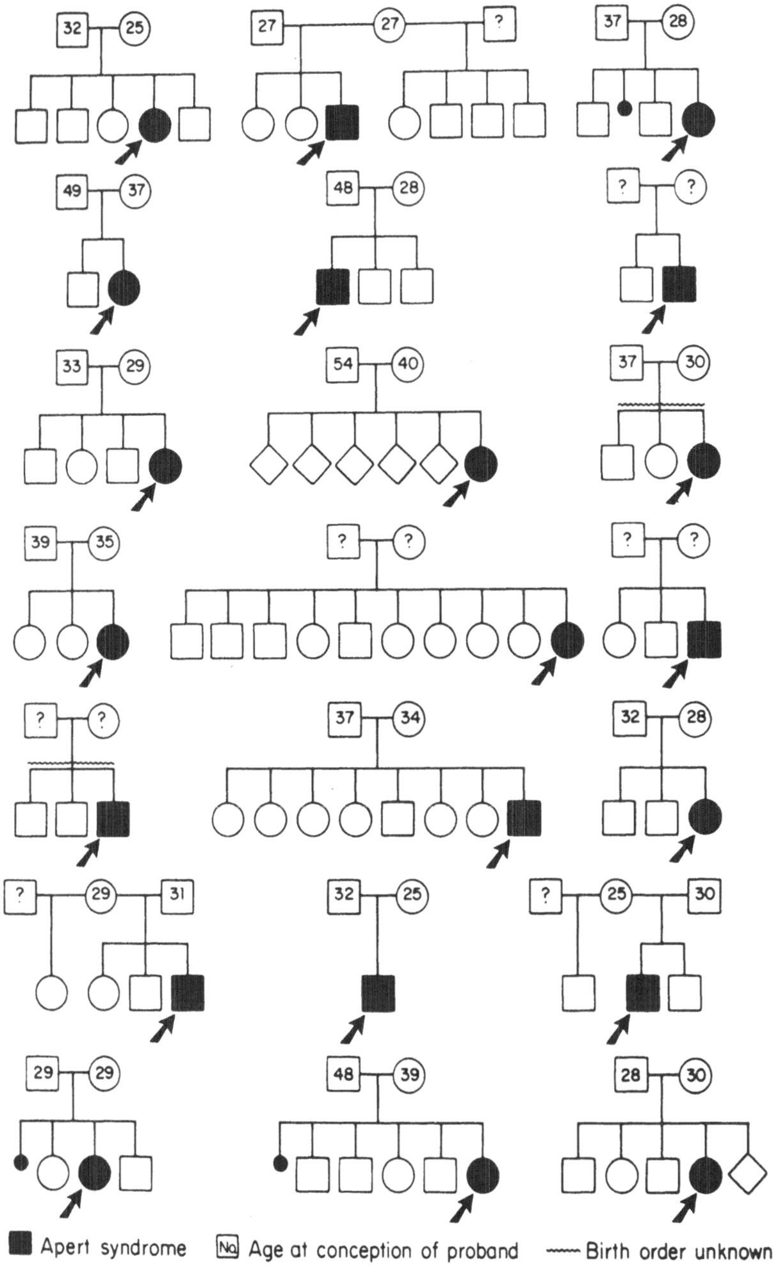

**Figure 8.14** Sporadic occurrences of the Apert syndrome represent fresh mutations for an autosomal dominant disorder in which the genetic fitness is dramatically reduced. (*From* Cohen, M. M. Jr. (1975). *Birth Defects*, **11**(2), 137)

117

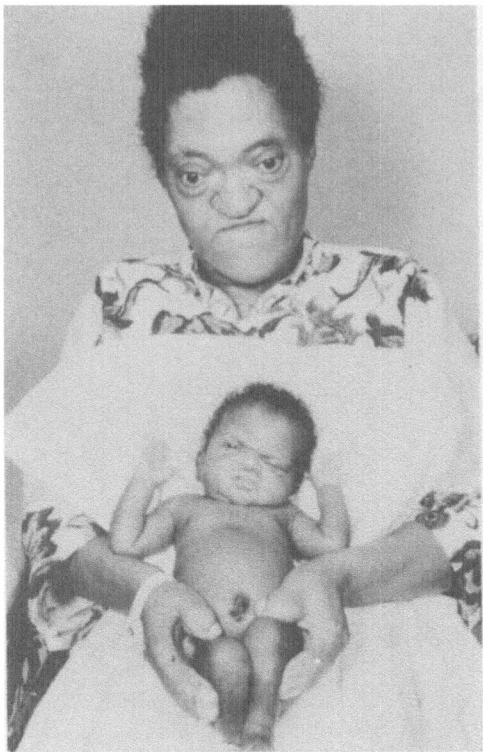

**Figure 8.15** Dominant inheritance of the Apert syndrome. Affected mother and child. (*From* Roberts, K. B. and Hall, J. G. (1971). *Birth Defects*, 7(7), 262)

## DIFFERENCES BETWEEN SYNDROMOLOGY AND CLASSICAL MEDICINE

The similarities and differences between syndromology and classical medicine are illustrated in Figure 8.16. In classical medicine, there are two general categories to consider. In the first category, clinical presentations become all-important and the challenge is in proper work-ups, from which diagnosis and treatment follow. In the second category, research is conducted on various disease entities to elucidate mechanisms of aetiology and pathogenesis. Although some disorders are well understood and others are poorly understood, classical medicine assumes that almost all disease categories at least have been described. Only rarely does a new disease entity, such as Legionnaires' disease, come to light.

Syndromology, like classical medicine, deals with the first two general categories. However, a third category becomes apparent since many patients examined by the syndromologist represent provisionally unique-pattern syndromes. Thus, the syndromologist is always asking, 'does the patient

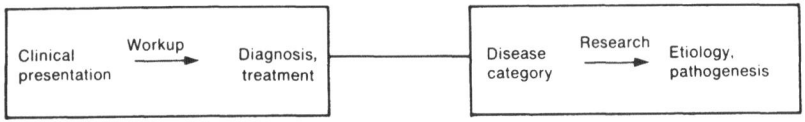

SYNDROME DELINEATION

CLASSICAL MEDICINE

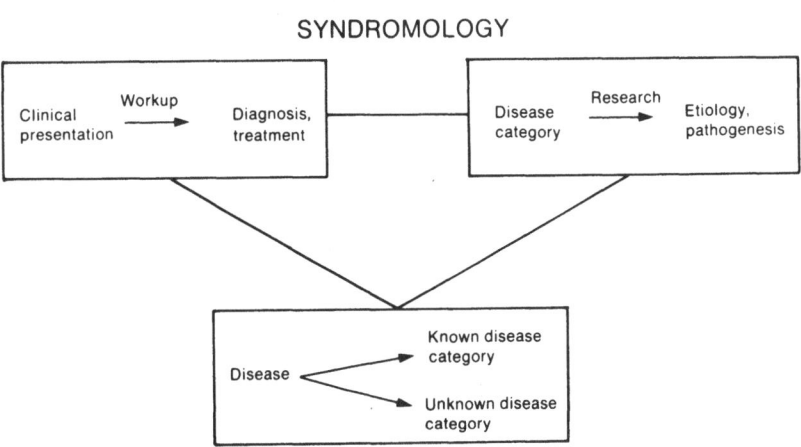

SYNDROMOLOGY

**Figure 8.16** Similarities and differences between syndromology and classical medicine. See text

represent a known disease category or an unknown disease category', and frequently concludes by postulating a 'new disease'. Clinicians not trained in syndromology or medical genetics usually become uncomfortable with this notion since they are not used to dealing with 'new diseases'.

## THE PACE OF SYNDROME DELINEATION

Syndrome delineation is proceeding at a very rapid rate. In 1971, for example, 72 syndromes were known in which orofacial clefting was one feature[13]. By 1978, however, 154 syndromes with orofacial clefting were known[14] – more than twice as many (Table 8.2). Thus, the discovery of new syndromes is taking place at a very rapid rate. Since various epidemiological surveys have shown that other anomalies may accompany as many as 13–50 % of all cases of cleft palate[4], the delineation of many more new syndromes from this group can be anticipated in the future.

## SIGNIFICANCE OF SYNDROME DELINEATION

The significance of syndrome delineation cannot be overestimated. As an unknown-genesis syndrome becomes delineated, its phenotypic spectrum, its natural history, and its inheritance pattern or risk of recurrence become known, allowing for better patient care and family counselling. If the

E                                    119

Table 8.2  **Syndrome delineation involving clefts from 1971 to 1978**

| Aetiology | Gorlin et al., 1971* | Cohen, 1978† |
|---|---|---|
| Monogenic | 39 | 79 |
| Autosomal dominant | (17) | (35) |
| Autosomal recessive | (18) | (39) |
| X-linked | (4) | (5) |
| Environmentally-induced | 0 | 6 |
| Chromosomal | 15 | 29 |
| Unknown genesis | 18 | 40 |
| Total | 72 | 154 |

* From Gorlin et al.[13]. † From Cohen[14].
The criteria for inclusion or exclusion of a syndrome with orofacial clefting differs slightly in these two compilations. Thus, a small proportion of the increase from 1971 to 1978 is attributable to this factor rather than to syndrome delineation *per se*

phenotypic spectrum is known, the clinician can search for suspected defects that may not be immediately apparent but which may produce clinical problems at a later date, such as a hemivertebra in the Goldenhar syndrome. If a certain complication can occur in a given syndrome, such as a Wilms tumour in the Beckwith – Wiedemann syndrome, the clinician is forewarned to monitor the patient for the possible development of neoplasia. Finally, if the recurrence risk is known, the parents can be counselled properly about future pregnancies. This is especially important if the risk is high and the disorder is severely handicapping or disfiguring, has mental deficiency as one component, or has a dramatically shortened life span. For example cleft palate or Robin sequence is a common feature of the Stickler syndrome, an autosomal dominant disorder with a 50 % recurrence risk when one parent is affected. In this condition, retinal detachment occurs in 20 % of reported cases and blindness occurs in 15 %[15]. Genetic counselling is of great importance because the risk of developing serious ocular problems is high. This relatively common condition also illustrates the importance of syndrome delineation because the entity was unknown and unrecognized before 1965, although surely it existed before that time. Thus, syndrome delineation fosters good patient care; the overall treatment programme gains rationality. In contrast, with a provisionally unique-pattern syndrome, the treatment programme and overall management frequently leave something to be desired.

## SYNDROME DELINEATION AND THE STUDY OF PATHOGENETIC MECHANISMS

Syndrome delineation aids in the preliminary sorting of disorders for the proper study of pathogenesis. Achondroplasia is a classic example. Conditions listed on the right hand side of Figure 8.17 are known to be separate entities today. However, a decade ago, they were thought to be examples of one and the same disorder – achondroplasia. Imagine the confusion that used to prevail when these conditions were lumped together as one entity. The disorder (1) was thought to be frequently fatal during the neonatal period

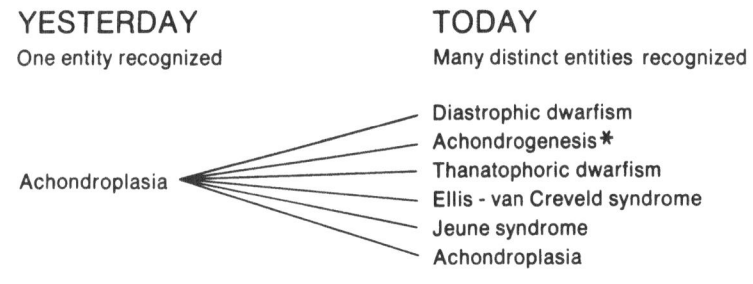

YESTERDAY
One entity recognized

TODAY
Many distinct entities recognized

Achondroplasia

- Diastrophic dwarfism
- Achondrogenesis*
- Thanatophoric dwarfism
- Ellis - van Creveld syndrome
- Jeune syndrome
- Achondroplasia

## IMPLICATIONS

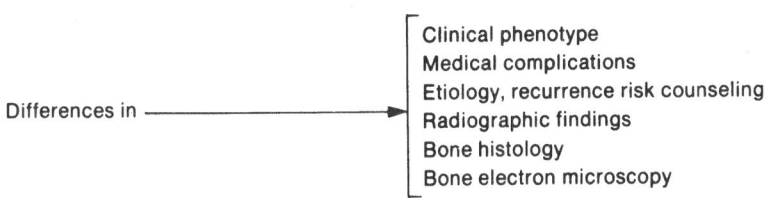

Differences in →

- Clinical phenotype
- Medical complications
- Etiology, recurrence risk counseling
- Radiographic findings
- Bone histology
- Bone electron microscopy

**Figure 8.17**  Significance of syndrome delineation for the study of pathogenetic mechanisms. Asterisk indicates that achondrogenesis itself is aetiologically heterogeneous. See text

(confusion with thanatophoric dysplasia and achondrogenesis),* (2) might have clubfeet that were extremely resistant to correction (confusion with diastrophic dysplasia) and (3) might have polydactyly (confusion with Ellis–van Creveld syndrome and Jeune syndrome). The medical complications and recurrence risks differ in the conditions listed on the right hand side of Figure 8.17. Since histological and electron microscopic findings also differ[16], imagine the results of a study that failed to separate them in analysis. It is somewhat like studying fruit instead of analysing apples, oranges, peaches and pears separately.

The early delineation of these disorders was based upon physical examination, radiographic findings, and genetic history. These differences suggested heterogeneity *before* histological and electron microscopic studies were carried out. Thus, syndrome delineation aids in the sorting of disorders for the proper study of pathogenesis.

## HETEROGENEITY AND PLEIOTROPY

In syndromology, a decision must be made about whether similar patients have an identical disorder with slightly different manifestations or aetiologically separate disorders with somewhat similar manifestations. Two basic principles of genetic nosology are heterogeneity and pleiotropy. Heterogeneity refers to multiple causes resulting in the same effect. Pleiotropy means multiple effects from a single cause. 'Splitting' occurs with genetic heterogeneity and 'lumping' occurs with pleiotropy[17]. Clinical, genetic and bio-

* Achondrogenesis itself is aetiologically heterogeneous.

121

chemical methods are used to recognize genetic heterogeneity, and these have been discussed elsewhere[17]. The process of syndrome delineation can lead to the establishment of separate genetic entities that were once thought to constitute a single phenotype. Such a fate has befallen the mucopolysaccharidoses, 'achondroplasia' and the Ehlers–Danlos syndromes[18].

Examples of heterogeneity and pleiotropy are illustrated in Figure 8.18. At one time, the Marfan syndrome and homocystinuria were thought to be one and the same disorder, but they are now known to be separate genetic entities. A Marfanoid habitus is non-specific. It is indicative of the Marfan syndrome, but may also be observed in homocystinuria[18], congenital contractural arachnodactyly[19], the Marfanoid hypermobility syndrome[20] and, occasionally, in the basal cell naevus syndrome[21].

Syndromologists are lumpers to the extent that they pull together pleiotropic effects of a single genetic disorder. If one patient has arachnodactyly and dolichostenomelia with ectopia lentis and another patient has a Marfanoid habitus with aortic regurgitation, both patients have the Marfan syndrome because the different manifestations are pleiotropic effects of the same gene.

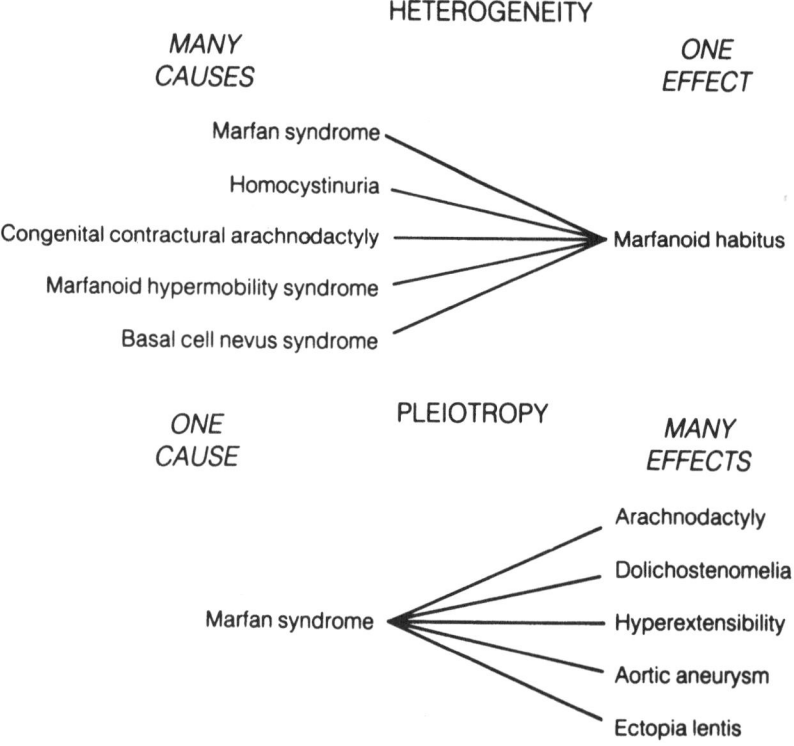

**Figure 8.18** With genetic heterogeneity, aetiologically separate disorders may have somewhat similar manifestations. With pleiotropy, different manifestations have a common cause. On the left hand side of the diagram, the causes are actually the various genes responsible for the disorders listed; the phenotype is listed simply for convenience

Many clinicians prefer the 'spectrum thinking' of classical medicine to the 'discontinuous thinking' of medical geneticists. The former emphasize relationships and similarities between various disorders; the latter emphasize the differences and discontinuities of the same disorders. These seemingly different perspectives are actually very compatible. Disorders that are aetiologically heterogeneous (discontinuous) may have similar or identical pathogenetic pathways.

## PATHOGENETIC HETEROGENEITY

The aetiology and pathogenesis of a disorder should be considered separately. Figure 8.19 diagrams some of the possible relationships between aetiology, pathogenesis and the phenotype. Model I shows aetiologically heterogeneous disorders with a common pathogenetic mechanism producing a single phenotype. The oligohydramnios sequence serves as an example. It results in intrauterine compression, producing characteristic facial deformities and limb positioning defects. Oligohydramnios has many causes, including amniotic tears, bilateral renal agenesis (which itself is aetiologically heterogeneous), and sirenomelia, a more extensive malformation sequence with fused lower extremities and absent kidneys and genitalia[4].

Model II shows aetiologically heterogeneous disorders with similar but not identical pathogenetic mechanisms and a single phenotype. Gaucher disease, type 2, and Niemann–Pick disease, type A, serve as examples. Both disorders have profound central nervous system disturbances and hepatosplenomegaly of early onset. They are difficult to distinguish clinically in the early stages.

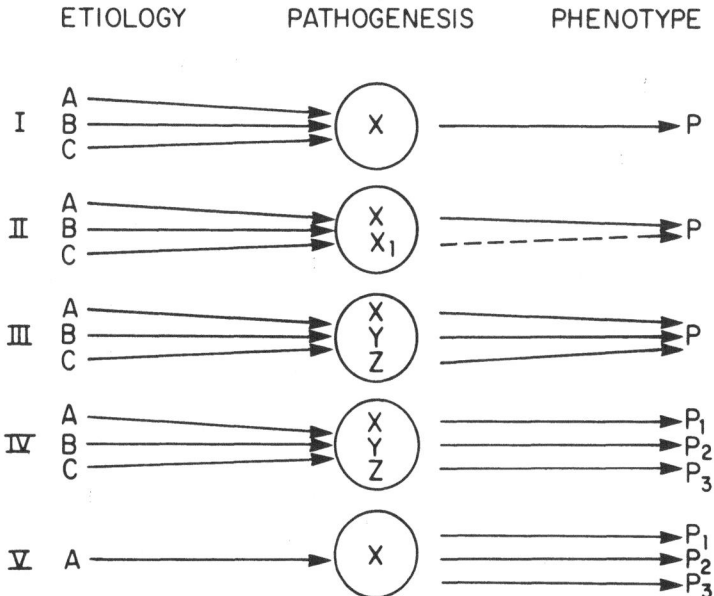

**Figure 8.19** Possible relationships between aetiology, pathogenesis and the phenotype

Both are autosomal recessive lipid storage diseases, Gaucher disease being caused by $\beta$-glucosidase deficiency and Niemann–Pick disease being caused by sphingomyelinase deficiency. The pathogenesis is similar. In Gaucher disease, glucosyl ceramide accumulates in the cells of the reticuloendothelial system; in Niemann–Pick disease, an increased phosphorylcholine ceramide is found[22]. Tay–Sachs disease and Sandhoff disease also fit Model II, as do Sanfilippo syndrome A and Sanfilippo syndrome B.

Model III shows aetiologically heterogeneous disorders with pathogenetically heterogeneous mechanisms resulting in one and the same phenotype. Because the major focus of this chapter is the patient with multiple anomalies, a detailed discussion of three anomalies – the Robin sequence, hemifacial microsomia, and craniosynostosis – will be presented to illustrate Model III. However, this analysis will be deferred until Models IV and V have been considered.

In Model IV, aetiologically heterogeneous disorders with different pathogenetic mechanisms produce slightly different phenotypes. The Marfan syndrome, homocystinuria and congenital contractural arachnodactyly, discussed earlier (Figure 8.18), serve as examples. Despite their Marfanoid habitus, they all have slightly different phenotypes that are distinguishable clinically, although formerly they were viewed as a single disorder. Variably in the autosomal dominantly inherited Marfan syndrome, patients may have ectopia lentis with superior displacement occurring most commonly, mitral and aortic regurgitation, progressive dilatation of the ascending aorta and dissecting aortic aneurysm. Homocystinuria has autosomal recessive inheritance, deficiency of the enzyme cystathione $\beta$-synthase and increased urinary excretion of homocystine. Patients may have ectopia lentis with inferior displacement occurring most commonly, thromboembolic episodes involving medium sized arteries and veins, malar flushing, osteoporosis and mental deficiency. Congenital contractural arachnodactyly has autosomal dominant inheritance, joint contractures that tend to improve gradually and crumpled ears[4,18,19,22]. Obviously the pathogenetic mechanisms in these three connective tissue disorders (although not well understood in the Marfan syndrome or in congenital contractural arachnodactyly) are different.

Finally, Model V shows a single disorder with a single pathogenetic mechanism and variably expressed phenotypes. Maroteaux–Lamy syndromes A and B serve as examples. Autosomal recessive inheritance, enzymatic deficiency of arylsulphatase B and excessive urinary excretion of dermatan sulphate are characteristic of both. Type A has a mild phenotype which becomes apparent at approximately 6 years of age with short stature and spinal deformity. Type B has severe clinical dysmorphism in early childhood with strikingly short stature, marked facial and skeletal alterations, severely impaired vision and hearing and prominent cardiac defects which frequently lead to death in adolescence. The two types may result from allelic mutations[4,18,22].

We now turn to a detailed analysis of the Robin sequence, hemifacial microsomia, and craniosynostosis[23,24] to illustrate Model III.

## The Robin sequence

The Robin sequence consists of micrognathia, cleft palate and glossoptosis (Figure 8.20). Pathogenesis is usually thought to be based upon a small mandible which prevents the normal descent of the tongue. Thus, the tongue interferes with palatal fusion. The mandible is said to exhibit significant catch-up growth in time (Figure 8.21).

The syndromology perspective on the Robin sequence adds considerably to our understanding of the pathogenesis. Table 8.3 lists some syndromes in which the Robin sequence is one feature. The aetiology of each condition is different. Thus, the Robin sequence is probably pathogenetically heterogeneous. The autosomal dominantly inherited Stickler syndrome[25]

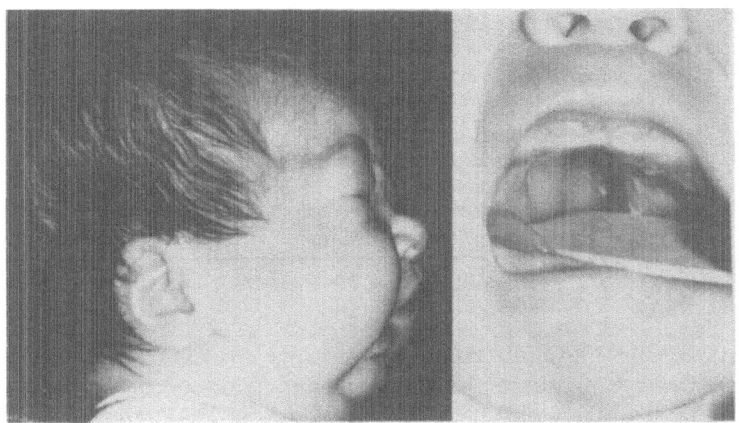

**Figure 8.20** Patient with Robin sequence showing micrognathia and cleft palate. (*From* Cohen[23])

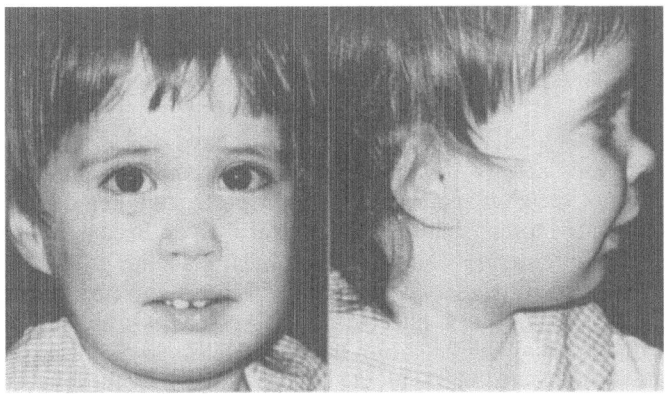

**Figure 8.21** Patient with Robin sequence showing mandibular catch-up growth. Same patient as in Figure 8.20. (*From* Cohen[23])

125

**Table 8.3  Conditions associated with the Robin sequence\***

*Monogenic syndromes*
  Beckwith–Wiedemann syndrome
  Campomelic syndrome
  Cerebrocostomandibular syndrome
  Diastrophic dwarfism
  Donlan syndrome
  Myotonic dystrophy
  Persistent left superior vena cava syndrome
  Radiohumeral synostosis syndrome
  Spondyloepiphyseal dysplasia congenita
  Stickler syndrome

*Chromosomal syndromes*
  Partial trisomy 11q syndrome

*Teratogenically-induced syndromes*
  Fetal alcohol syndrome
  Fetal hydantoin syndrome
  Fetal trimethadione syndrome

*Unknown-genesis syndromes*
  Femoral dysgenesis–unusual facies syndrome
  Martsolf syndrome
  Robin-accessory metacarpal syndrome
  Robin-amelia syndrome

\* Modified and updated from M. M. Cohen Jr. (1976). *J. Oral. Surg.*, 34, 587

(Figure 8.22) is the most common syndrome associated with the Robin sequence. Since abnormalities of bones and joints occur in the Stickler syndrome, the major pleiotropic effect appears to be on connective tissue. Thus, in this condition, the Robin sequence may result from intrinsic mandibular hypoplasia and failure of connective tissue penetration across the palate. Another condition that may have the Robin sequence is the partial trisomy 11q syndrome[26]. With the growth deficiency that accompanies most chromosomal syndromes, there may not be significant mandibular catch-up growth in patients with the partial trisomy 11q syndrome who survive. Therefore, to include such patients in a mandibular growth study of the Robin sequence would be a study of 'fruit' since 'oranges' are being confounded with 'apples'.

Some instances of the Robin sequence have been associated with oligohydramnios[7]. It is thought that reduced amniotic fluid results in compression of the chin against the sternum, restricting mandibular growth and impacting the tongue between the palatal shelves. Because micrognathia is based upon intrauterine moulding, mandibular catch-up growth is expected after birth when intrauterine deforming forces are no longer acting. Poswillo[27] has produced a phenocopy of the Robin sequence in rats by puncturing the amniotic sac prior to palatal closure. Some experimental animals also had deformities of the limbs, ranging from clubfeet to ring constrictions and intrauterine amputations. Such limb abnormalities have also been associated with the Robin sequence in humans[28,29].

Finally, the Robin sequence has been associated with congenital

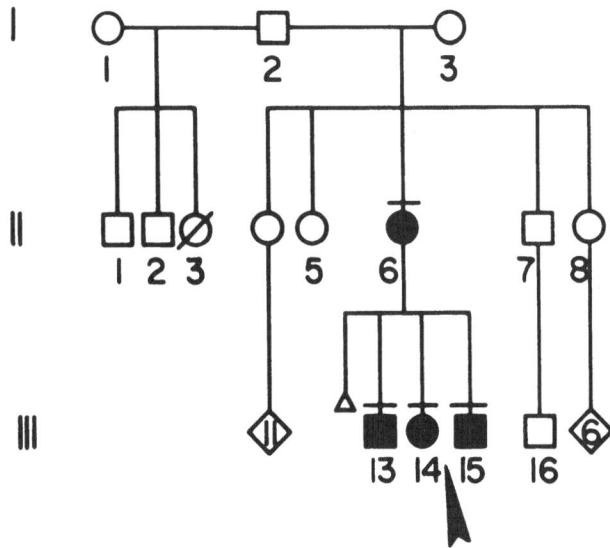

**Figure 8.22** Stickler syndrome exhibiting dominant inheritance. One affected patient has Robin sequence. (*From* Herrmann, J. *et al.* (1975). *Birth Defects*, **11**(2), 76)

hypotonia[7]. If neurogenic hypotonia occurred prior to complete closure of the palate, it is conceivable that the Robin sequence might result from the lack of mandibular exercise. Different aetiologies and pathogenetic possibilities are summarized diagrammatically in Figure 8.23.

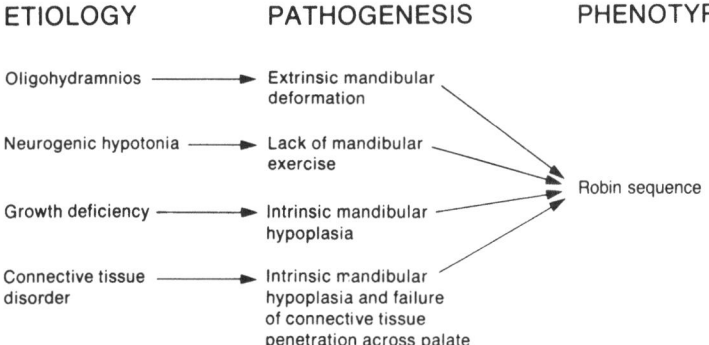

**Figure 8.23** Aetiological heterogeneity suggests pathogenetic heterogeneity in the Robin sequence. The following pathogenetic possibilities should be considered. (I) Oligohydramnios results in decreased amniotic fluid, compressing the chin against the sternum and thus restricting mandibular growth. (II) If hypotonia restricts mouth opening during early fetal life prior to complete palatal closure, the Robin sequence might result from lack of mandibular exercise. (III) Growth deficiency, as observed in chromosomal syndromes such as the partial trisomy 11q syndrome, may produce the Robin sequence by intrinsic mandibular hypoplasia. (IV) In a connective tissue disorder, such as the Stickler syndrome, the Robin sequence may result from intrinsic hypoplasia and failure of connective tissue penetration across the palate. (*From* Cohen, M. M.[23])

## Hemifacial microsomia

Hemifacial microsomia is a well-known condition affecting aural, oral and mandibular growth (Figure 8.24). The disorder may be mild or severe, and involvement is limited to one side in most cases, but bilateral involvement is also known to occur with more severe expression on one side[4]. Poswillo[30] reported a phenocopy of hemifacial microsomia in mice following maternal administration of triazene and in monkeys following maternal ingestion of thalidomide. He proposed that the pathogenesis was based upon embryonic haematoma formation arising from the anastomosis that precedes formation of the stapedial artery stem. Variation in the severity of hemifacial microsomia was found to depend upon the size and extent of haematoma formation, large haematomas interfering more severely with branchial arch growth by taking longer to resolve than small haematomas.

It seems most probable that hemifacial microsomia is both aetiologically and pathogenetically heterogeneous. Haematoma formation has heterogeneous causes, including hypoxia, hypertension, pressor agents, salicylates and anticoagulants[30]. Although embryonic haematoma formation may explain some human cases of hemifacial microsomia, it probably doesn't explain all cases. For example, familial instances are known[4] and, in some cases, affected relatives may have only preauricular tags (Figures 8.25 and 8.26). It is difficult to conceive of any basic mechanism causing haematoma formation to explain these cases. In minimally affected individuals, the ear and mandible are well formed; the preauricular tag seems to represent an accessory auricular hillock – an example of embryonic redundant morphogenesis. To postulate separate pathogenetic mechanisms for instances of hemifacial microsomia and accessory ear tags in the same family (Figure 8.25) seems unnecessarily complicated. The most frugal hypothesis should take into account that the

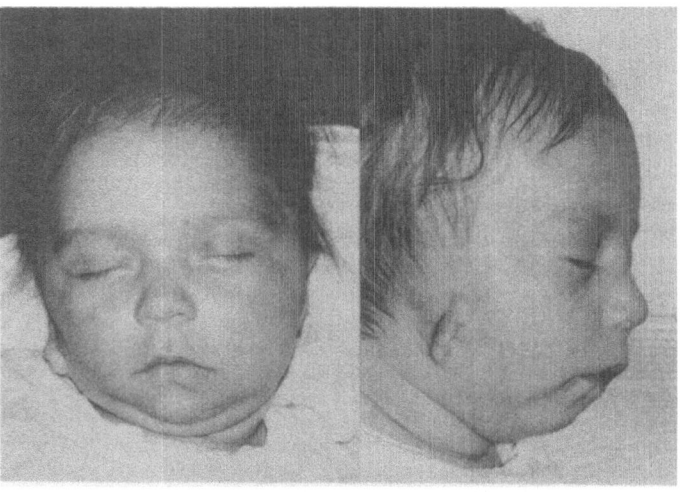

**Figure 8.24** Hemifacial microsomia. (*From* Cohen[23])

128

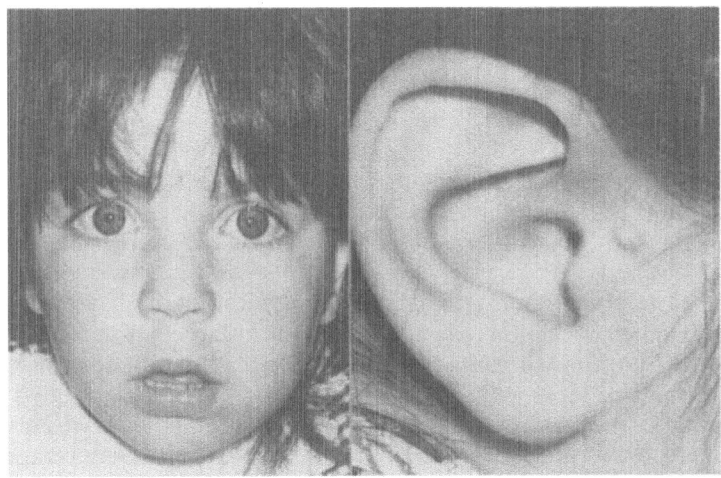

**Figure 8.25** Girl has preauricular tag, representing an accessory auricular hillock. The ear is otherwise normal. Sister of patient shown in Figure 8.24. (*From* Cohen[23])

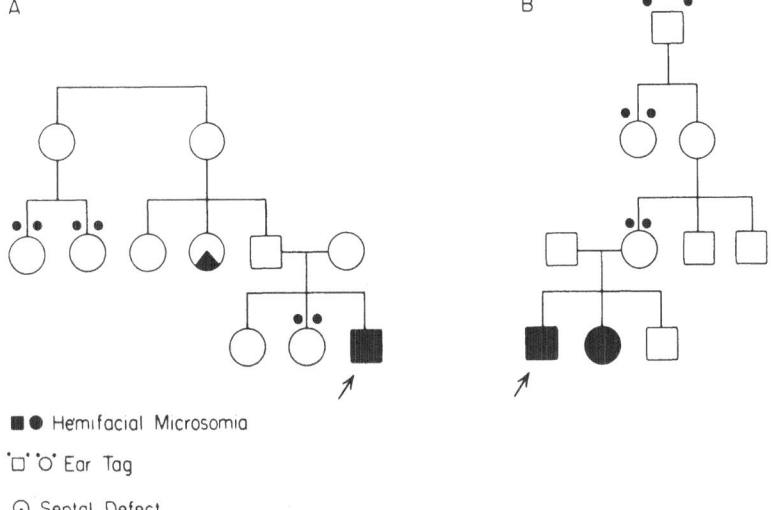

■ ● Hemifacial Microsomia

˙□˙ ˙○˙ Ear Tag

◕ Septal Defect

**Figure 8.26** Pedigrees with hemifacial microsomia in which minimally affected relatives have only preauricular tags. A, Family of proband and sister illustrated in Figures 8.24 and 8.25. B, Family from Pruzansky, S. (1973). *ASHA Rep.*, **8**, 62

two pedigrees in Figure 8.26 represent a single entity which is variably expressed and which is genetically transmitted.

Hemifacial microsomia is known to occur as an isolated defect or together with a variety of other anomalies[4]. A recurrent-pattern syndrome has been described consisting of hemifacial microsomia, occipital encephalocoele, renal agenesis, hypoplastic lung and vertebral anomalies[31] (Figures 8.27–8.29). These malformations most likely have a common cause (even though we don't know what it is) rather than being caused by different factors acting independently. Whatever mechanism is responsible for one malformation should be responsible for the other malformations as well. To date, there is no experimental evidence that haematoma formation can cause encephalocoeles or renal agenesis. Therefore, it seems unlikely that haematoma formation has anything to do with the pathogenesis of this recurrent-pattern syndrome.

Some cases of the amniotic band spectrum of disruptions have been noted to simulate hemifacial microsomia[32]. In such instances, hemifacial microsomia may possibly be caused by intrauterine compression secondary to oligohydramnios. Kennedy and Persaud[33] extracted amniotic fluid from pregnant rats at 16 days of gestation and studied the embryos at various times thereafter. On histological examination, they found haemorrhage and oedema followed by tissue necrosis in the cartilage and mesenchymal preskeleton of the developing limbs. Thus, the observed reduction defects and amputations of the limbs resulted from the venous stasis, hypervolaemia and embryonic oxygen deficiency caused by intrauterine compression. Kennedy and Persaud[33] did not give a detailed histological evaluation of the branchial arch region, but they did note that micrognathia was observed in addition to subcutaneous haemorrhages in the head region. Thus, intrauterine compression might be construed as a possible mechanism for producing haematoma formation in the branchial arch region, resulting in hemifacial microsomia. Since human hemifacial microsomia may be observed occasionally with limb reduction defects[4], the association appears to have a common pathogenesis compatible with Poswillo's hypothesis. If hemifacial micro-

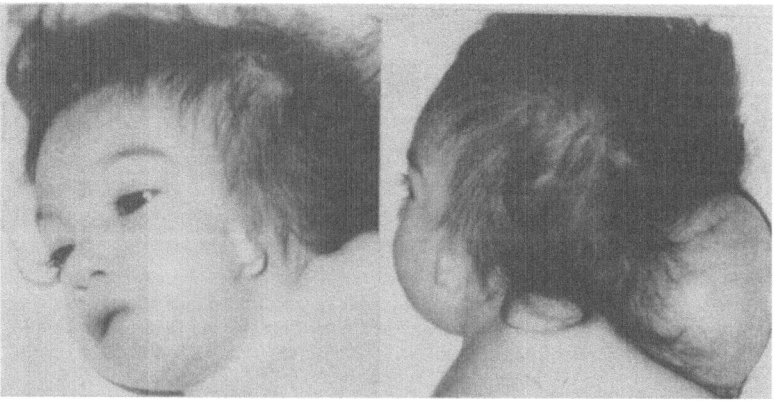

**Figure 8.27** Patient with recurrent-pattern syndrome consisting of hemifacial microsomia, occipital encephalocoele, renal agenesis, hypoplastic left lung and vertebral anomalies. See Figures 8.28 and 8.29. (*From* Cohen, M. M.[23])

130

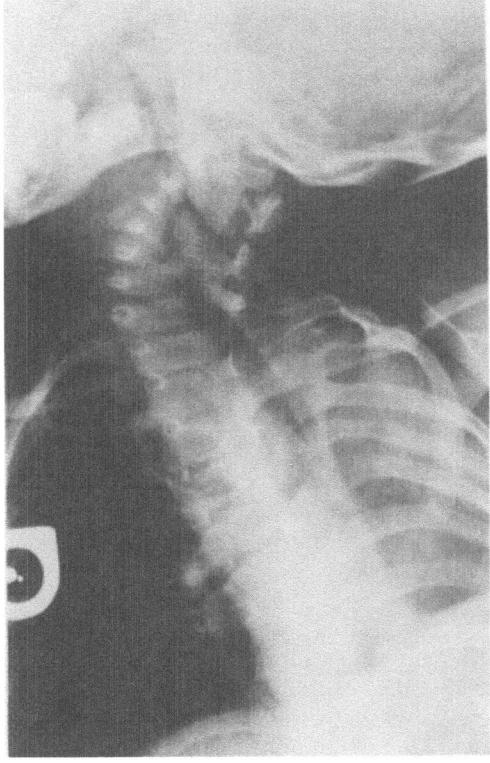

**Figure 8.28**   Hypoplastic left lung and vertebral anomalies. See Figures 8.27 and 8.29. (*From* Cohen, M. M.[23])

somia is ever observed with frank amniotic band-related limb anomalies,* this too would be compatible with Poswillo's hypothesis.

## Craniosynostosis

Three classic theories have been advanced to explain craniosynostosis. Virchow[34] believed that craniosynostosis was a primary malformation and that the associated cranial base deformity was secondary to craniosynostosis. The converse was postulated by Moss[35]; the cranial base malformation was the primary anomaly, resulting in secondary premature fusion of the cranial sutures. In speculating on the pathogenesis of the Apert syndrome, Park and Powers[36] postulated a primary defect in the mesenchymal blastema that led to both craniosynostosis and an abnormal cranial base.

Currently, Moss's theory is the most popular of the three. Unfortunately, most discussions of craniosynostosis assume that there is a single pathogenetic mechanism that remains to be elucidated, and that once this is done,

* To our knowledge, this has not been reported to date.

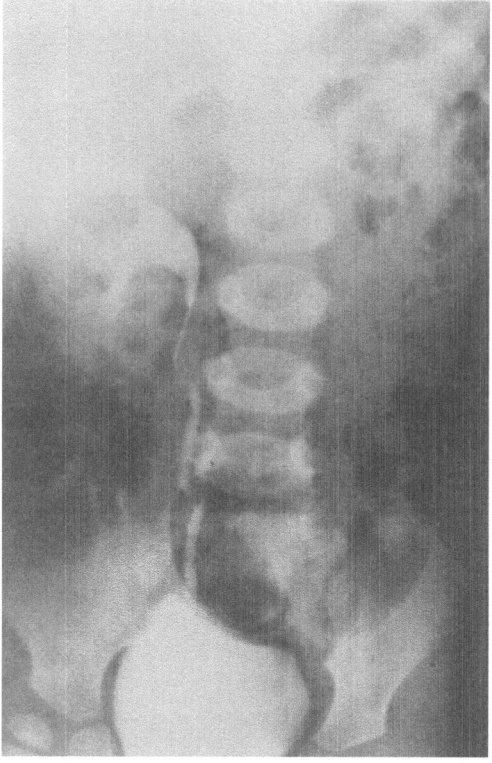

**Figure 8.29** Intravenous pyelogram showing absence of one kidney. See Figures 8.27 and 8.28. (*From* Cohen[23])

alternative hypothesis will be shown to be incorrect. Syndrome delineation and clinical evidence to date strongly suggest that craniosynostosis is pathogenetically heterogeneous. Some 57 different syndromes have been recognized in which craniosynostosis is a feature (Table 8.4). Some are known to be, and others are presumed to be, aetiologically heterogeneous. Such aetiological heterogeneity, of course, suggests the possibility of pathogenetic heterogeneity. Thus, all three theories are probably correct; each may be implicated in some, but not all, cases of craniosynostosis.

According to Moss's theory, spatially malformed lesser sphenoidal wings in coronal synostosis, and spatially malformed cribriform plate and crista galli in sagittal synostosis are viewed as primary abnormalities which, at the points of dural attachment, transmit aberrant tensile forces upward through the dura, leading to premature fusion of the overlying sutural tissues.

Many familial instances of isolated (non-syndromic) craniosynostosis have been observed[37]. Most instances are compatible with autosomal dominant transmission. In some families, involvement is variable – some family members having fusion of the sagittal suture, some having fusion of the coronal suture, and still others having synostosis of both coronal and sagittal sutures

Table 8.4 Summary of syndromes with craniosynostosis*

| Types of syndromes | | No. of syndromes |
|---|---|---|
| Chromosomal syndromes | | 11 |
| Monogenic syndromes | | 26 |
|     Autosomal dominant | (12) | |
|     Autosomal recessive | (12) | |
|     X-linked | ( 2) | |
| Teratogenic syndromes | | 2 |
| Total syndromes with known genesis | | |
|     (all categories above) | | 39 |
| Unknown genesis syndromes | | 18 |
| | | — |
| *Total* | | 57 |

* *From* Cohen[37]

(Figures 8.30 and 8.31). It is difficult to conceive of dramatically different primary abnormalities of the cranial base occurring in the same family as a dominant trait. Thus, a pathogenetic mechanism different from the one proposed by Moss[35] may be operative in such families.

A primary abnormality of the cranial base as the cause of craniosynostosis in the Apert syndrome has been proposed by Moss[35] and supported by Stewart[38]. However, clinical evidence to date indicates that the pathogenetic mechanism for craniosynostosis in this syndrome may possibly be unique. Features of the Apert syndrome include progressive calcification and fusion, with time, of the bones of the hands, feet, and cervical spine[39]. Progressive generalized bony dysplasia with ankylosis of joints as well as progressive limitation of motion at these joints have been documented[40]. Finally, progressive calcification of the cartilaginous nasal septum and stylohyoid ligament has also been observed (S. Pruzansky, personal communication,

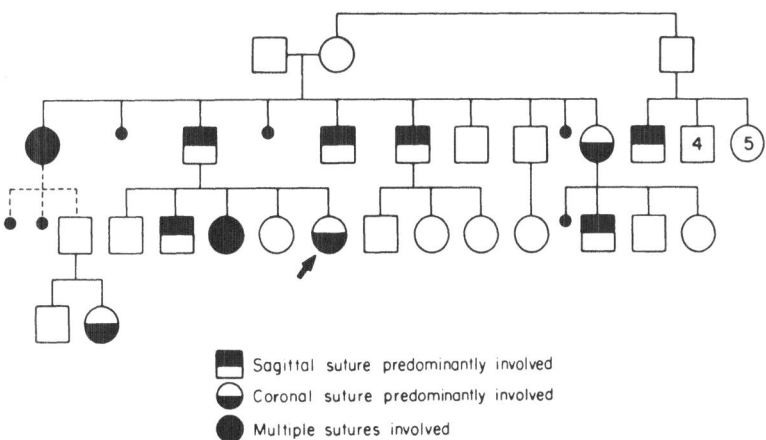

■ Sagittal suture predominantly involved
◒ Coronal suture predominantly involved
● Multiple sutures involved

**Figure 8.30** Autosomal dominant inheritance of simple craniosynostosis illustrating variability of sutural fusion. (Modified from Herrmann, J. *et al.* (1969). *Rocky Mt. Med. J.*, **76**, 45)

**Figure 8.31** Dominantly inherited craniosynostosis. Father has sagittal synostosis. Infant has unilateral coronal synostosis. (*From* Anderson, F. M. and Geiger, L. (1965). *J. Neurosurg.*, **22**, 229)

1977). The most frugal hypothesis would be that whatever mechanism is responsible for progressive calcification throughout the body is also responsible for premature craniosynostosis in the Apert syndrome. Because neither the cranial base nor the points of dural attachment can be invoked to explain progressive calcification elsewhere, it is probable that they have nothing to do with craniosynostosis in the Apert syndrome either.

Recently, Persson and his co-workers[41] showed that craniosynostosis occurred after the experimental application of methyl cyanoacrylate adhesive to the coronal suture in 9-day-old rabbits. The resultant immobilization produced constraint of growth stretch across the sutural area which, presumably, caused craniosynostosis to occur. Graham and colleagues[42] hypothesized that, in similar fashion, human prenatal head constraint may be responsible for some cases of craniosynostosis. They noted that some mothers of infants with isolated sagittal synostosis gave a history of early descent of their abdominal silhouette and severe pelvic pressure during the last 1–3 months of gestation. These symptoms were interpreted as a sign of early descent of the fetal head into the lower pelvis.

Lack of growth stretch at the sutures may also be implicated in three malformations in which premature craniosynostosis may occur as a complicating feature. First, sutural fusion may accompany some cases of microcephaly[43]. Lack of central nervous system growth may result in lack of growth stretch across the sutural areas, producing secondary craniosynostosis. Second, several reports have linked shunted hydrocephaly to craniosynostosis[44,45]. Low pressure systems may be implicated, in which growth stretch at the sutural areas suddenly becomes totally deficient. Finally, some cases of encephalocoele have been associated with craniosynostosis[46]. Such 'blow-out' lesions may sometimes result in lack of growth stretch across the sutures.

Premature craniosynostosis may result from several dysmetabolic states. It has been observed to accompany hyperthyroidism during childhood[47-49]. Sutural fusion may be caused by primary thyroid hyperplasia, but more commonly results from excessive thyroxine treatment for congenital hypothyroidism (Figure 8.32).

Premature sutural fusion may also occur in the Hurler syndrome, which is characterized by $\alpha$-L-iduronidase deficiency. Craniosynostosis involves the sagittal and lambdoidal sutures[4].

Craniosynostosis is observed in various aetiologically distinct forms of rickets (Table 8.5). Premature sutural fusion was found in a third of 59 rachitic

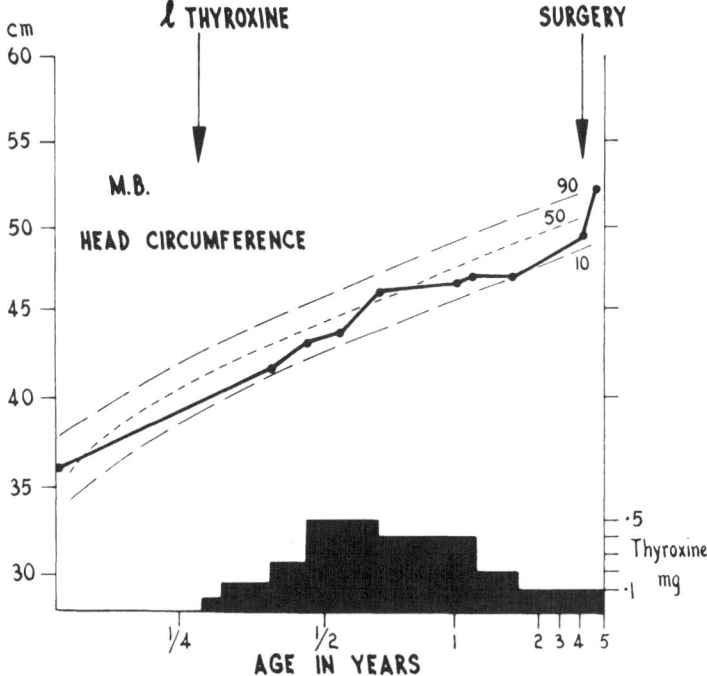

**Figure 8.32** Craniosynostosis from excessive thyroxine treatment. Note acceleration of head circumference during period of intensive thyroxine replacement therapy and resumption of growth following neurosurgical intervention to relieve sutures. (*From* Penfold and Simpson[49])

Table 8.5  Craniosynostosis and rickets*

| Type of rickets | Patients studied | Craniosynostosis | | Approximate proportion affected |
| --- | --- | --- | --- | --- |
| | | Present | Absent | |
| Vitamin D deficiency | 16† | 4 | 12 | 1/4 |
| *Vitamin D-refractory* | | | | |
| Simple hypophosphatemic | 14 | 6 | 8 | 1/3 |
| Hypophosphatemic, hypocalcemic amino-aciduric | 5 | 1 | 4 | 1/5 |
| Cystine storage disease | 4 | 0 | 4 | 0 |
| Hypophosphatemic, amino-aciduric, cirrhotic | 2 | 1 | 1 | 1/2 |
| Renal tubular acidosis | 1 | 0 | 1 | 0 |
| Azotemic osteodystrophy | 4 | 1 | 3 | 1/4 |
| Hepatic rickets | 3 | 0 | 3 | 0 |
| Hypophosphatasia | 10 | 3(+3‡) | 4 | 2/3 |
| Total (all types of rickets) | 59 | 16(+3‡) | 40 | 1/3 |

\* From Reilly, B. J. *et al.* (1964) *J. Pediatr.*, 64, 369
† An additional 7-month-old male infant with severe vitamin D deficiency rickets had unilateral coronal synostosis which was probably present at birth. It is assumed that the occurrence of rickets and craniosynostosis were coincidental and the case was omitted from the study
‡ Three very severely affected infants with almost completely uncalcified calvaria died in early infancy

children under 9 years of age in the study of Reilly and associates[50]. The severity of the rachitic process was not related to the type of rickets. However, the severity was related to the extent of synostosis and the more severe cases tended to have more severe degrees of craniosynostosis. The age at which rickets first occurred was also related to the extent of synostosis, the earlier diagnosed cases tending to be more severe. Alkaline phosphatase, a reflection of the severity of the rachitic process (with the obvious exception of hypophosphatasia), was significantly related, and higher values were associated with a higher incidence of craniosynostosis. No demonstrable relationship was found between synostosis and serum calcium levels or vitamin D intake.

Craniosynostosis may occur in various haematological disorders. Hyperplasia of the marrow with compensatory bony overgrowth of the calvaria can 'lock' the sutures. Conditions known to result in premature fusion of sutures include the thalassaemias, sickle cell anaemia, congenital haemolytic icterus, and polycythaemia vera[43,51].

Premature fusion of sutures may occasionally accompany a variety of miscellaneous disorders as an abnormality. These conditions include, among others, ataxia telangiectasia[52], the epidermal naevus syndrome[53], and Job syndrome[54]. The pathogenesis in such instances is completely obscure.

## References

1  Cohen, M. M. Jr. (1977). On the nature of syndrome delineation. *Acta Genet. Med. Gemellol. (Roma)*, **26**, 103
2  Opitz, J. M., Herrmann, J. and Dieker, H. (1969). The study of malformation syndromes in man. *Birth Defects*, **5**(2), 1

3 Weaver, D. D., Graham, C. B., Thomas, I. T. and Smith, D. W. (1974). A new overgrowth syndrome with accelerated skeletal maturation, unusual facies and camptodactyly. *J. Pediatr.*, **84**, 547

4 Gorlin, R. J., Pindborg, J. J. and Cohen, M. M. Jr. (1976). *Syndromes of the Head and Neck.* 2nd Edn. (New York: McGraw-Hill)

5 Nyhan, W. L., James, J. A., Teberg, A. J., Sweetman, L. and Nelson, L. G. (1969). A new disorder of purine metabolism with behavioral manifestations. *J. Pediatr.*, **74**, 20

6 Clarren, S. K. and Smith, D. W. (1978). The fetal alcohol syndrome. *N. Engl. J. Med.*, **298**, 1063

7 Herrmann, J. and Opitz, J. M. (1974). Naming and nomenclature. *Birth Defects*, **10**(7), 69

8 Benirschke, K., Cohen, M. M. Jr., Hall, J. G., Lenz, W., Lowry, R. B., Opitz, J. M., Pinsky, L., Schwarzacher, H. G., Smith, D. W. and Spranger, J. (1980). Terms pertaining to morphogenesis and malformations. Presented at the *Birth Defects Meeting*, June 8–11, New York

9 Quan, L. and Smith, D. W. (1973). The VATER association, Vertebral defects, Anal atresia, T–E fistula with esophageal atresia, Radial and Renal dysplasia. A spectrum of associated defects. *J. Pediatr.*, **82**, 104

10 Neuhäuser, G. and Opitz, J. M. (1975). Studies of malformation syndrome in man. XXXX. Multiple congenital anomalies/mental retardation syndrome or variant familial developmental pattern; differential diagnosis and description of the McDonough syndrome (with XXY son from XY/XXY father). *Z. Kinderheilk.*, **120**, 231

11 Opitz, J. M. (1979). Terminological and epistemological considerations of human malformations. In Harris, H. and Hirschhorn, K. (eds.) *Advances in Human Genetics*, pp. 71–107 (Chap. 2, Part 1). (New York: Plenum Publishing Corporation)

12 Erickson, J. D. and Cohen, M. M. Jr. (1974). A study of parental age effects on the occurrence of fresh mutations for the Apert syndrome. *Ann. Hum. Genet.*, **38**, 89

13 Gorlin, R. J., Cervenka, J. and Pruzansky, S. (1971). Facial clefting and its syndromes. *Birth Defects*, **7**(7), 3

14 Cohen, M. M. Jr. (1978). Syndromes with cleft lip and cleft palate. *Cleft Palate J.*, **15**, 306

15 Herrmann, J., France, T. D. and Opitz, J. M. (1975). The Stickler syndrome. *Birth Defects*, **11**(6), 203

16 Rimoin, D. L. (1975). The chondrodystrophies. In Harris, H. and Hirschhorn, K. (eds.) *Advances in Human Genetics*, pp. 1–118 (Vol. 5, Chap. 1). (New York: Plenum Publishing Corporation)

17 McKusick, V. A. (1969). On lumpers and splitters, or the nosology of genetic disease. *Birth Defects*, **5**(1), 23

18 McKusick, V. A. and Rimoin, D. L. (1980). *Heritable Disorders of Connective Tissue.* 5th Edn. (St Louis: Mosby)

19 Beals, R. K. and Hecht, F. (1971). Congenital contractural arachnodactyly. *J. Bone Jt. Surg.*, **53A**, 987

20 Walker, B. A., Beighton, P. H. and Murdock, J. L. (1969). The Marfanoid hypermobility syndrome. *Ann. Intern. Med.*, **71**, 349

21 Cohen, M. M. Jr. (1976). Human dysmorphic syndromes with craniofacial anomalies. In Stewart, R. and Prescott, G. (eds.) *Oral–Facial Genetics*, p. 604. (St Louis: Mosby)

22 Stanbury, J. B., Wyngaarden, J. B. and Frederickson, D. S. (eds.) (1978). *The Metabolic Basis of Inherited Disease.* 4th Edn. (New York: McGraw-Hill)

23 Cohen, M. M. Jr. (1979). Syndromology's message for craniofacial biology. *J. Maxillofacial Surg.*, **7**, 89

24 Cohen, M. M. Jr. (1980). Perspectives on craniosynostosis. *West. J. Med.*, **132**, 507

25 Herrmann, J., France, T. D., Spranger, J. W., Opitz, J. M. and Wiffler, C. (1975). The Stickler syndrome (hereditary arthroophthalmology). *Birth Defects*, **11**(2), 76

26 Aurias, A. and La'urent, C. (1975). *Trisomie IIq.* Individualisation d'un nouveau syndrome. *Ann. Génét.*, **18**, 189

27 Poswillo, D. (1966). Observations of fetal posture and causal mechanisms of congenital deformity of palate, mandible, and limbs. *J. Dent. Res.*, **45**, 584

28 Routledge, R. T. (1960). The Pierre Robin syndrome. A surgical emergency in the neonatal period. *Br. J. Plast. Surg.*, **13**, 204

29 Smith, J. L. and Stowe, F. R. (1961). The Pierre Robin syndrome. *Pediatrics*, **27**, 128

30  Poswillo, D. (1973). The pathogenesis of the first and second branchial arch syndrome. *Oral Surg.*, **35**, 302
31  Cohen, M. M. Jr., Zellweger, H., Waziri, M., Hanson, J. W. and Jones, K. L. (1980). A recurrent-pattern syndrome of branchial arch anomalies, encephalocele and renal anomalies. (In preparation)
32  Hall, B. D. (1979). Syndromes and situations simulated by amniotic bands. Presented at the *Birth Defects Conference*, June 24–27, Chicago
33  Kennedy, L. A. and Persaud, T. V. N. (1977). Pathogenesis of developmental defects induced in the rat by amniotic sac rupture. *Acta Anat.*, **97**, 23
34  Virchow, R. (1851). Über den Cretinismus, namentlich in Franken, and über pathologische Schädelformen. *Verh. Phys. Med. Ges. Wurzburg*, **2**, 231
35  Moss, M. L. (1959). The pathogenesis of premature cranial synostosis in man. *Acta Anat.*, **37**, 351
36  Park, E. A. and Powers, G. F. (1920). Acrocephaly and scaphocephaly with symmetrically distributed malformations of the extremities. *Am. J. Dis. Child.*, **20**, 235
37  Cohen, M. M. Jr. (1979). Craniosynostosis and syndromes with craniosynostosis. Incidence, genetics, penetrance, variability, and new syndrome updating. *Birth Defects*, **15**(5B), 13
38  Stewart, R. E., Dixon, G. and Cohen, A. (1977). The pathogenesis of premature craniosynostosis in acrocephalosyndactyly (Apert's syndrome). A reconsideration. *Plast. Reconstr. Surg.*, **59**, 669
39  Schauerte, E. W. and St.-Aubin, P. M. (1966). Progressive synosteosis in Apert's syndrome (Acrocephalosyndactyly) with a description of roentgenographic changes in the feet. *Am. J. Roentgenol.*, **97**, 67
40  Harris, V., Beligere, N. and Pruzansky, S. (1977). Progressive generalized bony dysplasia in Apert syndrome. *Birth Defects*, **14**(6B), 175
41  Persson, K. M., Roy, W. A., Persing, J. A., Rodeheaver, G. T. and Winn, H. R. (1979). Craniofacial growth following experimental craniosynostosis and craniectomy in rabbits. *J. Neurosurg.*, **50**, 187
42  Graham, J. M., deSaxe, M. and Smith, D. W. (1979). *Sagittal craniosynostosis*. Fetal head constraint as one possible cause. *J. Pediatr.*, **95**, 747
43  Duggan, C. A., Keener, E. B. and Gay, B. B. (1970). Secondary craniosynostosis. *Am. J. Roentgenol.*, **109**, 277
44  Andersson, H. (1966). Craniosynostosis as a complication after operation for hydrocephalus. *Acta Paediatr. Scand.*, **55**, 192
45  Kloss, J. L. (1968). Craniosynostosis secondary to ventriculoatrial shunt. *Am. J. Dis. Child.*, **116**, 315
46  Lorber, J. (1967). The prognosis of occipital encephalocele. *Devel. Med. Child. Neurol.*, **13** (Suppl.), 75
47  Johnsonbaugh, R. E., Bryan, R. N., Hierlwimmer, U. R. and Georges, L. P. (1978). Premature craniosynostosis. A common complication of juvenile thyrotoxicosis. *J. Pediatr.*, **93**, 181
48  Menking, M., Wiebel, J., Schimdt, W. U., Schmidt, W. T., Ebel, K. D. and Ritter, R. (1972). Premature craniosynostosis associated with hyperthyroidism in 4 children with reference to 5 further cases in the literature. *Monatsschr. Kinderheilk.*, **120**, 106
49  Penfold, J. L. A. and Simpson, D. A. (1975). *Premature craniosynostosis*. A complication of thyroid replacement therapy. *J. Pediatr.*, **86**, 360
50  Reilly, B. J., Lemming, J. M. and Fraser, D. (1964). Craniosynostosis in the rachitic spectrum. *J. Pediatr.*, **64**, 396
51  Dykstra, O. H. and Halbertsma, T. (1940). Polycythaemia vera in childhood. *Am. J. Dis. Child.*, **60**, 907
52  Robinson, A. (1962). Ataxia-telangiectasia presenting with craniostenosis. *Arch. Dis. Child.*, **37**, 652
53  Moynahan, E. J. and Wolff, O. H. (1967). A new neurocutaneous syndrome (skin, eye, brain) consisting of linear nevus, bilateral lipodermoids of the conjunctiva, cranial thickening, cerebral cortical atrophy, and mental retardation. *Br. J. Dermatol.*, **79**, 651
54  Smithwick, E. M., Finelt, M., Pahwa, S. and Good, R. A. (1978). Cranial synostosis in Job's syndrome. *Lancet*, **1**, 826

# 9
# Numerical taxonomy in the study of birth defects

JANE A. EVANS

## INTRODUCTION

The idea of classification is intrinsic to man, stemming from the time when it was vital to recognize plants as food or non-food, animals as fierce or non-fierce, but increasing knowledge especially of biological systems made the development of more complex classifications methods a necessity. Taxonomy has been defined as 'the theoretical study of classification including its bases, principles, procedures and rules'[1] and numerical taxonomy refers to the use of mathematical and statistical techniques in classification. Such methods were developed by workers in the fields of botany and zoology, to deal with the similarities and differences both between individuals and between populations. It is only in the last few years that researchers in the medical sciences have started to adapt and use numerical taxonomy techniques. However, there is now increasing awareness of their value as an objective tool in studying patients, syndromes and disease entities.

The principles and methods of numerical taxonomy are a science in their own right and cannot be covered adequately here. Detailed treatment of the topic is found in several excellent texts, notably those of Sneath and Sokal[2] and Clifford and Stephenson[3]. The purpose of this paper is rather to introduce very briefly some of the concepts involved in numerical taxonomy, to review some of the applications of the techniques to various aspects of the study of birth defects and to apply certain methods to a set of data concerning children with renal and/or limb malformations to show how such studies may further our knowledge of the interrelationship between birth defects.

## METHODOLOGY

The basic principle involved in numerical taxonomy, as in any other method of classification, is to place a large group of varied objects into meaningful subunits for further consideration. To introduce some taxonomy nomenclature, the objects to be classified are referred to as 'operational taxonomic units', or OTUs, and they are represented by a series of descriptors or 'character states'. The units into which they are grouped are known as 'taxa' (singular: *taxon*). These words however, are unfamiliar to most workers in medicine and for our purposes can be replaced by the more commonly understood terms, 'individual', 'character' and 'subgroup', respectively.

One of the problems facing any attempt at realistic classification is that of biological continuity. In many situations, individuals are not truly individual in their characteristics but represent a continuum of variation. An example of such a continuum would be the array of phenotypic findings seen in patients with Down syndrome. A classification system has the difficulty therefore of producing meaningful subgroups, from what may appear to be a continuous or almost continuous spectrum of variation. Numerical taxonomy techniques approach this problem in one of two ways. First, either they attempt to divide the large heterogeneous population into smaller homogeneous subgroups (divisive techniques) or they build individuals up one by one into larger units (agglomerative techniques). Second, they can try to maintain the continuous nature of the data by defining it by certain multifactorial parameters and seeing where particular individuals fit in the spectrum of variation (ordination techniques). The method chosen for any particular study will depend on the form of the data and on the size of the array and computing methods available, but all will attempt to discover meaningful similarities between individuals rather than maintain arbitrary and probably artificial boundaries. A second choice in taxonomic method comes about in the decision whether to use simple key characters for classification (monothetic classification) or to set up subgroups on the basis of several characters shared in common (polythetic classification).

There are several distinct methods of numerical taxonomy, therefore, and these can be summarized as monothetic agglomerative, monothetic divisive, polythetic agglomerative, polythetic divisive and ordination methods. Only three are in actual use. Monothetic agglomerative methods are discarded as trivial, since they would involve the grouping of individuals together on the basis of a single character, e.g. placing a banana and a lemon in the same group because they are both yellow. The formulae and computations involved in polythetic divisive methods are too complex to be readily available yet. The remaining methods are all in use and will be discussed in more detail later.

The application of a numerical taxonomy method to a data set involves several steps – selection of the characters to be used and initial formulation of the individual-character matrix, selection of measures of similarity or difference, selection of classification techniques and clustering method, and final interpretation of results. The choice made at any step may restrict or influence the choices possible in other parts of the process.

## Selection of data and formulation of the matrix

In numerical taxonomy the data will consist of a population of individuals (or OTUs) represented by observations of several characteristics. It is not valid to make preconceived judgements as to which characters will prove to be important for further classification of the population. All observable characters may be useful and as many as possible should be taken into account. Almost all characters which are amenable to accurate and unambiguous recording can be utilized. However, they may take a variety of forms. The commonest forms are binary, disordered or ordered multistate, meristic or continuous data. Binary data is two state data often, especially in the study of birth defects, involving presence or absence of a certain character. Multistate data can involve several states either in a random or disordered fashion, e.g. colour – red, white, blue, or in an ordered way, e.g. short, medium, tall or absent, unilateral, bilateral. Meristic data involve whole numbers and, though commonly used in plant or animal taxonomy (e.g. petal number), are less useful in the study of birth defects. Continuous data are of a metrical nature, e.g. birth weight, head circumference, where variation is truly continuous and subdivisions, where they exist, are due to lack of finer measurement. In some taxonomic methods where data must be in a specific form, e.g. binary state, the data may be transformed as necessary. For example, head circumference data could be transformed into two state characters based on percentiles; less than 5th percentile (microcephaly): yes/no, or greater than 95th percentile (macrocephaly): yes/no, and so on.

Selection of data will vary to a certain extent with the technique to be applied. One constraint on the choice of characters is therefore their ability to be measured and recorded in a suitable form for the analysis chosen. A second restriction involves character number. As each individual usually has to be compared in pairs with all others the number of characters to be studied will depend on the availability of computerized methods of analysis. A matrix of more than 20 characters in 20 individuals is difficult to handle manually both in terms of time and user's sanity! Computer programs allowing much larger data arrays are fortunately available for many numerical taxonomy techniques. Once the characters have been chosen and recorded a character-individual matrix is formulated for use in further analysis.

## Measures of similarity and differences

The next step in numerical taxonomy, once the data have been collected and transformed as necessary, is to apply some criterion to measure the similarity or dissimilarity between pairs of individuals, and to produce a matrix of similarity/dissimilarity scores for the total population. Similarity and dissimilarity are obviously closely interrelated and can generally be considered as different ends of a spectrum. The term 'similarity' will therefore be used as a general expression to cover both meanings. The similarity of any two individuals will depend to a great extent on the total variation within the population and the purpose of numerical classification is to use such measures to identify more homogeneous groups of individuals within the heterogeneous sample of individuals available for study. The choice of similarity measure is influenced by the type of analysis to be performed. Common measures involve coefficients of similarity and coefficients of association. Other measures involving study of the total variation within groups (information measures), similarity based on probability estimates or measures of distance between entities may also be used.

Many coefficients of similarity have been devised and not all are in current use. They can be divided into those used to compare individuals on the basis of binary state characters and those used for meristic or continuous data. Basically they compare the number of characters that occur concordantly or discordantly in any pair of individuals. Most binary state measures are constrained between 0 and 1, with 0 implying that two individuals have no characters in common and are totally dissimilar and 1 indicating that all characters are shared and the individuals are indistinguishable. Common measures are simple matching and Jaccard's coefficient. Different measures such as that of Bray and Curtis[4] are used for meristic or continuous data and these may vary from 0 to infinity.

Coefficients of association, again by studying the numbers of characters occurring together and separately, determine the degree of association between pairs of individuals. The commonest coefficient used in numerical taxonomy for binary data is chi-square ($\chi^2$). For meristic and continuous data correlation coefficients can be used. Details of these coefficients of similarity and association are given in Sneath and Sokal[2] or Clifford and Stephenson[3]. Once the similarity measure has been chosen and calculated, a second matrix can be drawn up showing the values for each pair of individuals. This matrix will then be used in the classification itself.

## Selection of classification techniques

As previously mentioned, the choice of classification techniques lies between monothetic divisive, polythetic agglomerative and ordination techniques. The first two methods are probably those most often used. In either method the classification may be of individuals or attributes. The first analysis, grouping individuals by reference to their characters, is known as normal or $Q$ analysis. This is most commonly used in taxonomy. Inverse or $R$ analysis which groups characters by their sites of occurrences (individuals) has seen more use in ecological studies but may also be valuable in birth defect studies as a method of studying malformation associations. Both $Q$ and $R$ analyses in monothetic divisive and polythetic agglomerative techniques seek to produce non-overlapping groups or clusters of individuals. It is difficult to state which method is 'best'. The usefulness of the techniques depends both on the nature of the data and the purpose of the analysis. Each method has its advantages and drawbacks.

### Monothetic divisive techniques

The principle of monothetic divisive methods is to split the sample of individuals into subgroups and further divide these subgroups in a hier-archical fashion into smaller units until all individuals within a group are homogeneous or at least until no further significant associations of characters within the individuals can be generated. The splitting at each stage of the analysis is done on the basis of the possession or non-possession of specific single key characters, which may obviously be different in different parts of the hierarchy. Monothetic divisive techniques therefore stress constancy in subgroups. Constancy implies that all members of a given taxonomic class or

142

group possess a specific constant character. To use a genetic example, all members of the class 'Down syndrome' are trisomic for all or part of chromosome 21. This trisomy is a constant character. The groups generated by such an analysis are therefore non-ambiguous and stable. The most important point in this method is therefore the choice of the key characters. This choice must be made by recourse to some measure of similarity or dissimilarity between individuals. By far the most common measure used is chi-square ($\chi^2$), and the key character at each stage of the division is that which generates the greatest degree of both negative and positive association with other characters. The methodology is therefore simple. All characters are compared in pairs using the $\chi^2$ coefficient and then the sum of chi-squares obtained for each character. The character with the highest $\chi^2$ becomes the key character for division and the population is split into two groups, one possessing the character and one in which it is absent. The analysis then proceeds on each subgroup independently and new key characters are identified and the population further split until a level is reached when no further significant associations are generated.

This analysis is usually termed 'normal association analysis' and was developed by Williams and Lambert[5] for studying plant communities. It has several advantages. Firstly, the method is simple and easy to use and is fast as it concentrates on producing large clusters first. Computer programs exist to facilitate the analysis, so it is relatively inexpensive and well used for preliminary analysis of data, especially to reduce matrices to smaller size for use in agglomerative techniques. On the negative side, it requires binary state data and therefore transformation of metrical characters. Also, missing data are difficult to allow for. This is less of a drawback in birth defect studies where, in some cases, a defect may be presumed to be absent if it is not recorded. The main disadvantage to this method, however, is an intrinsic one, in that it is liable to produce occasional misclassifications due to the emphasis on single key characters. Careful review and reclassification of individuals may become necessary.

A second use of normal association analysis is also the study of actual character associations and this has been taken further by Williams and Lambert in inverse association analysis[6] and nodal analysis[7]. The former is just the $R$ analysis of the same data matrix, i.e. producing subgroups of characters rather than individuals. Nodal analysis involves the amalgamation of both normal and inverse analysis to produce 'individual-character clusters'. This analysis shows promise for the investigation of birth defects and an example will be given later.

## Polythetic agglomerative techniques

These methods have found greatest favour in numerical taxonomy, primarily because they attempt to resolve the problems of misclassification inherent in monothetic methods. Polythetic methods using many characters attempt to classify subgroups by an overall 'gestalt', or sense of belonging, rather than by possession of a single key character or even combination of characters. Polythetic classifications are therefore less rigorous than monothetic ones and less easy to define, but are also more flexible.

In general terms, polythetic methods involve the clustering of single individuals using a matrix of measures of similarity between individuals. The two individuals which are most alike are grouped into a cluster, then other individuals join this cluster or make new clusters based on further perusal of the measurement matrix. The further methodology in polythetic agglomerative techniques therefore involves the choice of clustering strategy. Three common clustering methods, each with different properties, are nearest neighbour, furthest neighbour and group average clustering, though several others are in use. Nearest neighbour clustering begins, as do all methods, with the joining of the two most similar individuals, A and B. Then if a third individual, C, is closer by similarity measurement to either A or B than to a fourth, D, it joins A and B's cluster. If it is less similar it forms a new cluster with D. The disadvantage to nearest neighbour clustering is that it forms 'chains' of individuals rather than discrete clusters. As a classification system it is rather loose and seldom used. Furthest neighbour clustering is the opposite of nearest neighbour. Here the joining of an individual to a cluster is based on the similarity of that individual to the most distant member of that cluster or, for two clusters, on the degree of similarity between the most dissimilar individuals in each group. As opposed to the looseness of nearest neighbour, it forms very tight clusters. Perhaps preferable to either method is group average clustering, where the mean value of the similarity measures of an individual to all members of a cluster is used to determine if that individual joins that group or another. Group average clustering is less prone to misclassification than either nearest or furthest neighbour and it is intermediate in the tightness of its clustering, and for these reasons it has become widely used.

The disadvantages of polythetic classifications are that they are relatively time consuming as they start from individuals and only slowly develop clusters. This also restricts to some extent the size of the data matrix that can be studied. Computer methods of clustering do, however, resolve these difficulties and have led to the increasing popularity of these methods.

## Ordination

Both monothetic divisive techniques and polythetic agglomerative methods are used primarily to produce hierarchical clusters. One drawback of such techniques is, of course, that they display multidimensional systems as two-dimensional forms. The third major method of classification – ordination – attempts to avoid such two-dimensional representations, and the loss of complexity they involve. Obviously, if the data could be looked at in terms of multiple dimensions, such diversity could be retained. Multivariate methods of numerical taxonomy, including ordination, attempt to view relationships between individuals in multidimensional space or 'hyperspace' and then display them in two-dimensional form. Multivariate analyses include ordination, which starts with an ungrouped population, and canonical and factor analysis, which generally use grouped data. Only ordination will be discussed further here.

As mentioned previously, what distinguishes ordination from techniques producing hierarchical clusters is that it visualizes individuals as representing points on a continuous spectrum of variation. The aims of the 'classification' therefore are not to produce discontinuities or boundaries unless they are inherent in the data but rather to arrange the individuals in a meaningful sequence. The ordination technique most widely applied is that of principal components analysis. In this analysis individuals are defined in terms of a series of 'n' axes where 'n' is the number of characters. The individual can then be thought of as existing in n-dimensional hyperspace. The analysis then chooses those axes in multidimensional space which most appropriately display the relationships between the individuals. These axes will be the ones which achieve maximum separation of the individuals in one dimension. To get this one-dimensional separation individuals must be projected onto a single line. This line is unlikely to be either axis, as this will represent a single variable, but will lie somewhere between them. This is known as the first principal component. At right angles to this will be a second principal component. If the individuals are projected onto a plane defined by the first and second principal components then the individuals will be separated but the relationships between them preserved. The distribution of the individuals and any obvious clustering produced can then be studied. Principal components analysis is also available in a simplified form using a technique developed by Bray and Curtis[4]. This analysis takes as a principal component the axis joining the two most similar sites and then projects other individuals onto that axis to show the relationship between sites.

## Interpretation of results

The final step in any classification, once the analysis has been carried out and the final hierarchy or plot obtained, must be the interpretation of the results. The purpose of the classification should be to discover the nature and extent of the underlying relationships between the entities classified. One test of the strength of these underlying relationships or the taxonomic structure of the population will lie in comparison of the results of different classification techniques. If very different groups are produced by different methods then the underlying relationships are probably weak. If the use of almost any technique produces the same results then the intrinsic structure of the population as revealed by the analysis is solid and probably meaningful. Once a strong classification has been obtained then numerical taxonomy can be said to have fulfilled its objective of improving the classification of variation among objects. The classification can then be studied in detail for clues to better understanding of the cause of such variability.

Two steps are necessary before this can be done profitably. Firstly, the classification should be checked for obvious misclassification. Misclassifications may occur in divisive techniques if otherwise similar individuals are separated by key character differences. In agglomerative methods they may occur because clustering starts at a point far removed from the final groupings and inclusion of a individual in a group may be based on chance similarities. Such misclassified individuals can then be assigned to more appropriate groups. Secondly, the groups or clusters produced should be defined. This may be done either by reference to the actual members of that group or by a general description of the characteristics of the group members. The second approach is often more useful, as the classification can then be tested using other individuals not in the original data set. If the classification is not generally applicable to the general population then the group definition may be poor and not reflect the underlying relationships between members or the sample is not representative[8].

Once the classification is acceptable and the groups well defined, it can be examined more closely for both biological sense and usefulness in interpreting underlying biological relationships. This stage will involve some hypothesizing as attempts are made to seek the reasons for these relationships. Obviously in the study of birth defects the clustering of two or three individuals into a group does not imply genetic closeness as it might in a study of animal or plant taxonomy, but rather is analogous to ecological classifications where the occurrence of species together in the same habitat may well be due to similar environmental needs. The grouping of individuals together on the basis of physical attributes including malformations and dysmorphic features may imply common cause. Numerical classification, by identifying such groups, can play a role in the understanding of causation of birth defects, by complementing clinical observation and intuitive reasoning.

## APPLICATIONS OF NUMERICAL TAXONOMY TO THE STUDY OF BIRTH DEFECTS

Numerical taxonomy techniques have been utilized by several workers in birth defects, especially in the areas of syndrome identification and delineation. Pinsky[9,10] has made use of these concepts in his development of the idea of syndrome communities. He believes that numerical classification may be useful both in recognizing new syndromes or malformation associations (analogous to the recognition of new species in taxonomy) and in recognizing communities of syndromes or associations (more analogous to ecological classification of plant communities).

Pinsky considers that polythetic classifications are probably most valuable in studies of syndrome communities because of the inconstant nature of many characters. For example, he uses a simple polythetic classification to classify manually the hand–foot–ectodermal dysplasias into two classes: those with alar–lip anomalies including EEC syndrome, the orofacialdigital and Mohr syndromes and the SC pseudothalidomide phocomelia syndrome among others and those without alar–lip anomalies[10]. Alar–lip anomalies are not the only findings separating these groups. Sensory neural deafness is much more common in those syndromes lacking alar–lip anomalies but is rare in syndromes with alar–lip anomalies that lack or have minimal hand–foot findings.

The recognition of such groupings may be useful both for syndrome identification and for seeking cause. Pinsky has been criticized for using this approach on the basis that phenotypic clustering of syndromes into communities may confuse understanding of the underlying dysmorphogenetic mechanisms[11]. He rightly points out, however, that classification of disease entities into groups is a habit that predates numerical taxonomy and frequently helps rather than hinders our knowledge of genetic heterogeneity. Moreover, heterogeneous disorders that have been grouped into phenotypic communities have in some cases been shown to possess similar dysmorphogenetic mechanisms by later research. One example Pinsky quotes is that of the reduction limb defects–haematopoietic anomalies community, including such syndromes as Fanconi's anaemia and thrombocytopenia–absent radius syndrome where underlying susceptibility to chromosome breakage or DNA damage has been noted.

Other workers, notably Preus and MacGibbon in Montreal, have used numerical classification for detailed analysis of syndromes and for syndrome identification rather than community analysis. They recognize that problems in syndrome identification occur because of phenotypic overlap between syndromes and because, in any one syndrome, phenotypic characters are not constant even when the cause is known. These problems are compounded when the aetiology of the syndrome is not known. Numerical classification using a large sample of cases and a large number of unbiased characters is perhaps a better approach to syndrome delineation than the previously used intuitive methods of deciding what characters constituted a syndrome and then 'trying to fit new patients into old moulds'[12]. They have used a polythetic agglomerative technique to study syndrome delineation[12], between four well recognized syndromes; Down, Cornelia de Lange, Rubinstein–Taybi and Williams syndromes. They use Gower's coefficient as a similarity measure[2] but have developed a new clustering method. This distorted shell method utilizes the concept of multidimensional space in a similar fashion to ordination techniques. Using this method patients with these syndromes formed well defined and non-overlapping clusters. In a further application to a larger sample of children with and without Down syndrome[13], the distorted shell method was more efficient than nearest neighbour and furthest neighbour and as efficient as group average in clustering children with and without Down syndrome into distinct groups. In addition, distorted shell clustering led to fewer misclassifications than the other methods. This technique, and polythetic agglomerative methods in general, seem to be well suited to syndrome delineation especially in well recognized syndromes even when the aetiology is not known. These authors now intend to apply these techniques to syndromes with greater phenotypic variability and overlap.

A different area of birth defects research that has been studied by numerical taxonomy techniques is that of mental retardation. Morton and colleagues in Hawaii used principal components ordination to divide a cohort of mentally retarded individuals into medical, biological and sociofamilial clusters[14]. They found that low socioeconomic status as measured by parental education, occupation, residence and ethnic origin tended to be associated with high performance of retarded patients on the basis of their intelligence and social quotients, degree of retardation and educability. Patients' performance, symptomatology and changes in performance secondary to institutionalization were all closely interrelated but were not as closely associated with medical history or social class. Path analysis of these five variables further indicated that the better performance of institutionalized patients from poorer backgrounds was not due to a greater probability of institutionalization but rather due to a higher incidence of mild retardation of probable polygenic and environmental cause among the children from low socioeconomic groups.

They found that principal components analysis was relatively effective in distinguishing genetic or biological retardation from sociofamilial retardation (cases of known environmental aetiology were excluded prior to the analysis). For example, 88 % of Down syndrome patients were correctly assigned to the biological group. Having developed the groupings they went on to define the classes by various parameters including performance, sex ratio, mortality rates and age at admission as well as social class, symptoms and history and found striking differences in these parameters. For example mean IQ was low and mortality rates were high in the genetic and environmental groups but the reverse was seen in the socioeconomic group. Detailed examination of other factors also helped in the detection of underlying potential misclassifications. For example, the coefficient of inbreeding was high for the genetic groups, as would be expected, but was also high for the socioeconomic group, indicating possible unrecognized autosomal recessive conditions. As this work shows, even in a disorder as heterogeneous as mental retardation, numerical taxonomy and careful interpretation of results can be a useful tool.

One additional heterogeneous area that has been studied using numerical classification is that of patient referrals for clinical genetics evaluation. Hermann[15] used an inverse association analysis (or $Q$ analysis) to study the relationships between characters in 75 patients or families referred for genetic evaluation. Characters included factors in the family, pregnancy, delivery and neonatal period, the physical appearance of the proband, intellectual and motor function, medical history including laboratory findings and the course of the condition. He used the comparison of a character's rank and frequency in any particular association with its overall rank and position to determine the significance of associations. Clusters of characters in unexpectedly frequent associations and their associated characters were formed. Intra- and inter-cluster heterogeneity was not estimated quantitatively. In interpretation of these clusters and associations, he found that some were 'nonsense' associations or were non-informative, e.g. absence of seizures–permanent nature of condition. Others were redundant, e.g. severe impact of condition–severe mental retardation, or obvious, e.g. decreased weight–decreased length. In other cases, however, there could well be a biological basis for the associations or clusters seen, e.g. severe mental retardation–microcephaly–hypotonicity; microcephaly–severe mental retardation–abnormal karyotype–low pregnancy weight gain; or multiple radiographic lesions–markedly abnormal upper and lower limbs–slowly progressive course–occurrence in sibs. Hermann considers that this approach has value both for management and counselling of family even when precise aetiologies are not known and for planning an appropriate diagnostic work-up for the patient. Also individual clusters can obviously be studied in detail to test aetiological, prognostic or other hypotheses.

## APPLICATION OF NUMERICAL TAXONOMY TO STUDY OF RENAL AND LIMB ANOMALIES

During epidemiological investigations of limb defects and renal agenesis in Manitoba, several children with multiple congenital anomalies were ascertained. In many cases a precise diagnosis was available but others represented unknown patterns of malformations. Numerical classification techniques were applied to cases for two purposes. Firstly, they were used to determine if numerical taxonomy techniques were efficient in separating individuals with distinct phenotypes into clusters with little ambiguity and to compare polythetic agglomerative and monothetic divisive techniques in the classification of the same individuals. Secondly, it was intended to apply these techniques to a more heterogeneous group of children with limb defects and/or renal malformations and other components of the VACTERL association[16] to see if more homogeneous subgroups within this variable group could be defined.

### Population A

*Methods*
This population consisted of 24 patients ascertained through records of the Genetics clinic, Pathology Department or Medical Records of Children's Hospital, Winnipeg, Manitoba or the Manitoba Congenital Anomalies Registry. The patients included 4 with Poland syndrome with unilateral hypoplasia of the hand, 4 with trisomy 18 with reduction defects of the limb, 4 with classic Potter syndrome of renal agenesis, 8 with three or more components of the VACTERL association including 4 with and 4 without renal involvement and finally four cases of children with severe hypoplasia of one or both lower limbs and other components of the caudal regression association.

Fifty-three malformations occurring in two or more cases were chosen as the characters of the study. All data was in binary form as presence or absence. Data was considered complete on all patients, as a malformation was considered absent if its occurrence had not been noted. All four cases with Poland syndrome were still alive as were four children with the VACTERL association. All but one of the 16 deaths had had postmortem examinations.

The two-way matrix of 24 individuals and 53 characters was subjected to a monothetic divisive technique (normal association analysis) using $\chi^2$ and to a polythetic agglomerative technique using Jaccard's similarity coefficient[3] with group average clustering. The hierarchies produced by each method are shown in Figures 9.1 and 9.2 respectively.
/

*Results*
Association analysis produced nine groups, three of which consisted of a single individual. Two groups were clustered without ambiguity: the children with Poland syndrome (H) and those with Potter syndrome (I). Three cases of trisomy E were clustered at the first stage of the hierarchy using the key characters 'camptodactyly'. Failure to record this malformation, if present, in

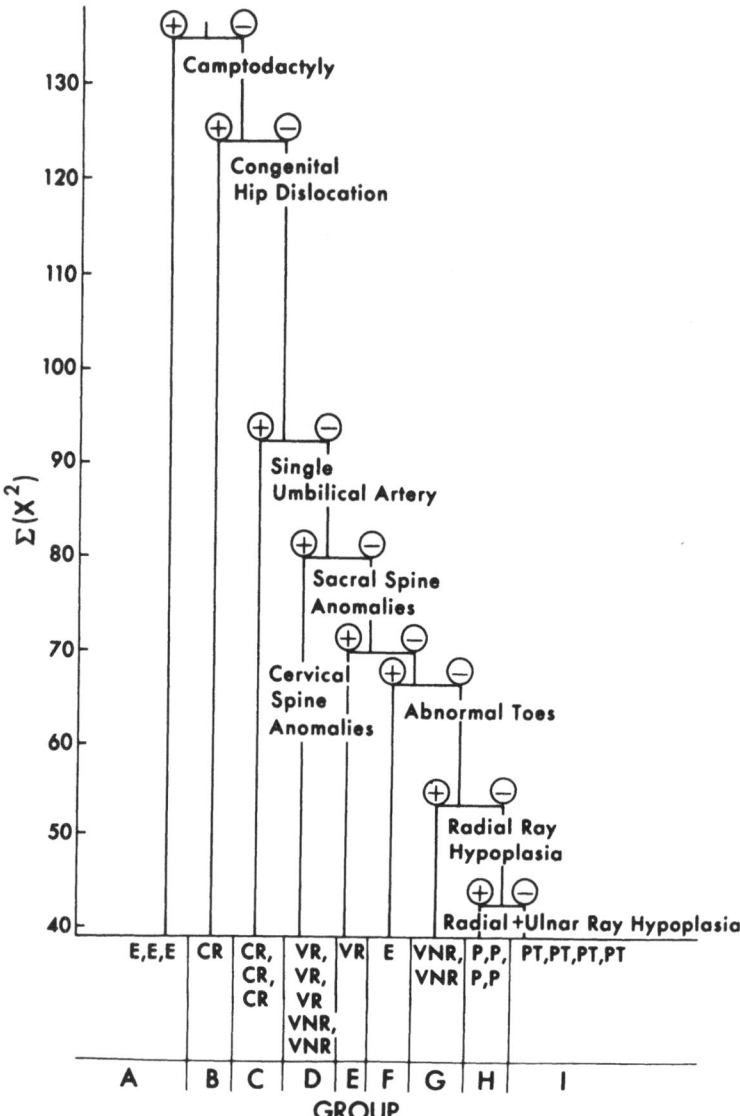

**Figure 9.1** Normal association analysis of Population A (24 individuals with limb hypoplasia and/or renal agenesis). E = trisomy 18; CR = caudal regression; VR = VACTERL association with renal involvement; VNR = VACTERL association without renal involvement; P = Poland syndrome; PT = Potter syndrome

the fourth case may have led to its misclassification. The four cases of 'caudal regression' formed groups B and C. The cases of VACTERL association were obviously more heterogeneous and formed three groups, D, E and G. The key characters splitting these groups were sacral spine anomalies, cervical spine anomalies and radial ray anomalies respectively.

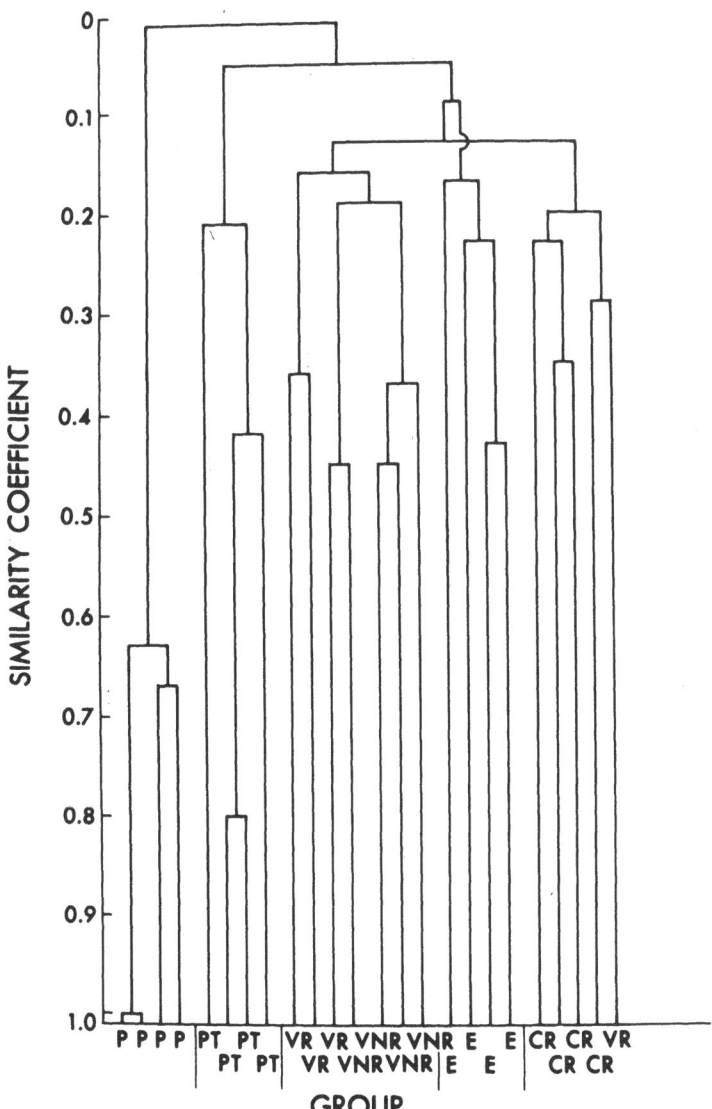

**Figure 9.2** Group average clustering of Population A (24 individuals with limb hypoplasia and/ or renal agenesis). E = trisomy 18; CR = caudal regression; VR = VACTERL association with renal involvement; VNR = VACTERL association without renal involvement; P = Poland syndrome; PT = Potter syndrome

Similarity coefficient analysis produced five clusters below a similarity value (*s*) of 0.15. Again the cases with Poland syndrome (1) and Potter syndrome (2) were clearly and unambiguously separated from the other groups. All other groups joined at an *s* value of 0.13. Group 3 contained all the trisomy 18 cases; group 4 consisted of seven patients with VACTERL association and group 5 included all cases of caudal regression plus the final VACTERL patient, one with renal involvement.

## Discussion

In both analyses Poland and Potter syndromes were well defined, as was trisomy 18 to a large extent. The greatest ambiguity involved the patients with VACTERL association and caudal regression. Both methods appeared equally effective in producing a realistic classification which closely followed clinical diagnosis. As Estabrook[8] points out, when the taxonomic structure of the population is high then almost any reasonable method of numerical classification will detect it. The variable nature of the classification for the VACTERL and caudal regression groups in both analyses implies a greater heterogeneity in those groups and possibly a relatively weak taxonomic structure. In order to study these malformation complexes in more detail it was decided to repeat the analyses on a larger sample of children with these disorders and to expand the association analysis to include inverse and nodal analyses to detect malformation associations and specific malformation-patient clusters.

## Population B

### Methods

This population consisted of the eight VACTERL and four caudal regression cases from population A, and 13 additional patients with three or more components of the VACTERL association, i.e. vertebral, anal, cardiovascular, tracheo-oesophageal, renal or limb anomalies. To simplify the analyses the malformations were restricted to 16. These characters and their overall frequency in the 25 cases are given in Table 9.1. The frequencies of limb and renal malformations are relatively high, owing to the biased ascertain-

Table 9.1    Malformation characters in Population B (25 cases)

| Character | Frequency (%) |
|---|---|
| 1  Hypoplasia upper limb | 52 |
| 2  Hypoplasia lower limb | 24 |
| 3  Cervical spine/thoracic spine/rib anomalies | 68 |
| 4  Lumbar/sacral spine anomalies | 52 |
| 5  Tracheo-oesophageal fistula/oesophageal atresia | 36 |
| 6  Imperforate anus/anal ectopia | 68 |
| 7  Ventricular septal defect | 20 |
| 8  Single umbilical artery | 36 |
| 9  Unilateral renal agenesis | 72 |
| 10  Bilateral renal agenesis | 12 |
| 11  Aberrant external genitalia | 40 |
| 12  Other musculoskeletal malformations | 52 |
| 13  Other respiratory system malformations | 28 |
| 14  Other alimentary system malformations | 52 |
| 15  Other cardiovascular system malformations | 52 |
| 16  Other genitourinary system malformations | 56 |

ment of limb/renal defect patients. The sex ratio among the 25 patients was 0.92. Nine children were alive, many having undergone extensive surgical procedures, as had the only one of 16 dead children who had not had postmortem examination.

The two-way matrix of 25 patients and 16 characters was subjected to similarity coefficient analysis using Jaccard's coefficient and group average clustering (Figure 9.3). In additional normal (Figure 9.4) and inverse association analysis using $\chi^2$ was carried out. Using the $Q$ and $R$ divisions obtained, nodal analysis was performed[7]. This technique conceives the population as consisting of 'character-individual' units doubly defined and doubly extracted from the data and thus combines normal and inverse analyses in order to establish the existence and relative importance of coincidences. In effect, each group established by normal analysis is subjected to inverse analysis and the key character identified but the group is not further subdivided. Similarly the inverse analysis groups are subjected to normal analysis. The total population of individuals and characters can be defined therefore as part of cells of a two-way matrix. The further key characters identified can be used to determine partial and total coincidences of occurrence of individuals and characters. Cells defined in both directions by

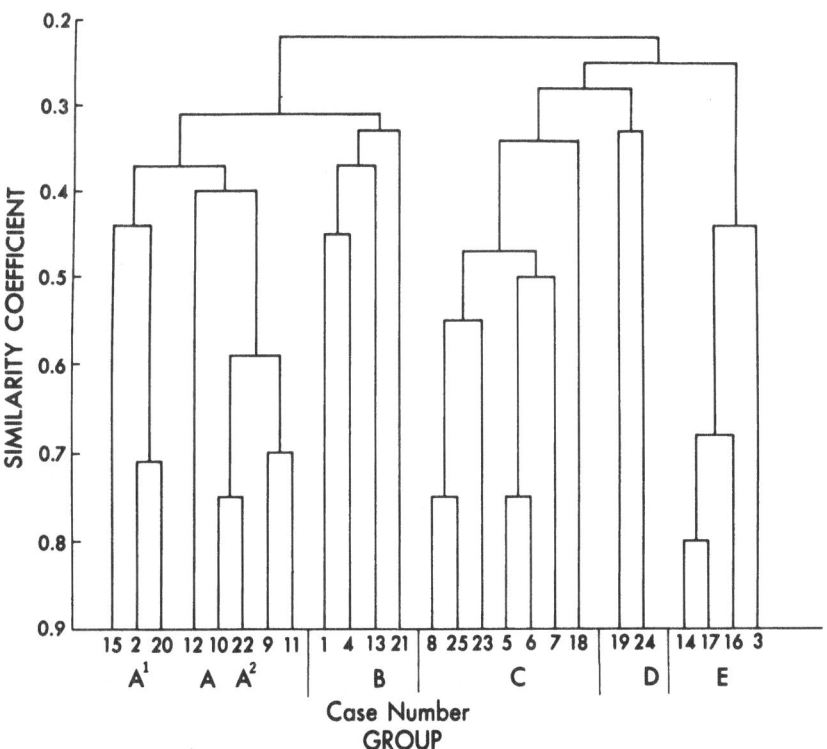

**Figure 9.3** Group average clustering of Population B (25 individuals with 3 or more components of the VACTERL association)

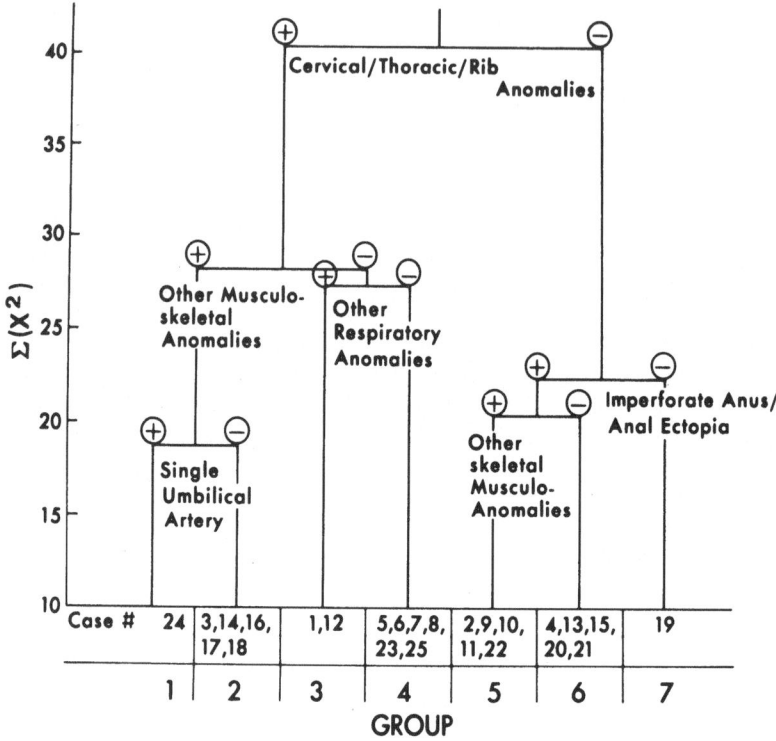

**Figure 9.4** Normal association analysis of Population B

characters and individuals are referred to as 'noda', those defined in one direction only become 'sub noda'. Many cells are defined in neither direction and the records of these cells are discarded. (For details of the method see Lambert and Williams[7].) Figure 9.5 shows the full nodal analysis of this population.

*Results*

Choosing an arbitrary *s* of 0.32, groups, A–E, are formed by polythetic agglomerative techniques. Group A has been broken into two subgroups, $A^1$ and $A^2$. If each group is defined by the malformations present in more than 70% of group members, the groups are characterized by different combinations of anomalies. Group $A^1$ (three cases) have anal anomalies, unilateral renal genesis and other alimentary malformations. Malformations in these children are restricted to the lower part of the body. Two of the three patients including 1 with sacral anomalies are known to be children of diabetic mothers. This group is closely associated with group $A^2$ (five cases) characterized by hypoplasia of the lower limb, anal anomalies, single umbilical artery, unilateral renal agenesis, absent external genitalia, other musculoskeletal anomalies and the other genitourinary anomalies. This group probably represents a more severe expression of caudal regression. The main difference between $A^1$ and $A^2$ is single umbilical artery (SUA) which is constant in $A^2$ but was not reported in $A^1$.

Group B (four cases) has individuals with tracheo-oesophageal anomalies, anal malformations and aberrant external genitalia. Limb and cardiovascular malformations are rare. Group C is the largest group (seven cases). It is characterized by upper limb defects, anomalies of both the upper and lower spine, tracheo-oesophageal defects and imperforate anus. Group E (four cases) shows anomalies of the cervical and thoracic spine and ribs, unilateral renal agenesis and other musculoskeletal anomalies. Tracheo-oesophageal defects and cardiovascular anomalies were absent. Group D consists of 2 children, No. 19 and No. 24 who join each other and the rest of the population at a low level of similarity. Case 19 had radial anomalies, ventricular septal defect and a horseshoe kidney. Vertebral anomalies were not reported at autopsy. Case 24 was the stillborn infant of a diabetic mother. He had an unusual combination of anomalies including bilateral renal agenesis, hypoplastic lungs, thoracic hemivertebrae, short sternum and rib anomalies, lumbar spine defects, single umbilical artery, cleft palate, brachydactyly and axial polydactyly of the feet. This may well represent a separate malformation complex. Karyotypes were not available on either of these infants.

Normal association analysis (Figure 9.4) produced seven groups (1–7). Group 1 was the single case, No. 24, described above which was again separated from all other patients. Group 2 (five cases) was very similar to group E from the polythetic analysis, while group 4 closely resembled group C (see Figure 9.4). Group 7 was the single patient, No. 19, referred to above and again separated from all others by this analysis. The main difference in the classifications was the precise grouping of patients in groups 5 and 6 and A and B. Both analyses split off 'unique' patients from the population and made distinct groups of either those children with the combination of upper limb defects, spinal defects, tracheo-oesophageal anomalies and anal anomalies or those with severe caudal regression. Children with other combinations of the VACTERL association where limb defects were uncommon but renal defects common were less clearly distinguished.

The results of the inverse analysis showing malformation groupings are given in Table 9.2. Groups i–iv appear to have some morphogenetic basis. Group v includes three very different and unrelated malformations. This is due to the nature of the monothetic analysis. These characters in the last group in the hierarchy are these which are not found in any of the key individuals in the division and so are not grouped on the basis of significant positive association.

Nodal analysis, combining the inverse and normal association analysis produces a two-way matrix with 35 cells (Figure 9.5). However, the records cluster into fewer than half these cells and only ten cells are complete (i.e. have records in all columns or rows) in at least one direction. All other cells can be discarded. The coincidences of malformations and individuals in the ten remaining cells can now be determined using the key individual or attribute determined by nodal analysis in each group.

The subdivision parameters of the nodal analysis (based on $Q$ division of $R$ groups and $R$ division of $Q$ groups) are shown as solid arrows in the margins.

**Table 9.2  Malformation groupings by inverse association analysis**

*Group i*
Lumbar/sacral anomalies
Imperforate anus/anal ectopia
Other alimentary anomalies

*Group ii*
Unilateral renal agenesis
Other musculoskeletal defects (usually clubfeet)

*Group iii*
Single umbilical artery
Aberrant external genitalia
Other respiratory anomalies
Other cardiovascular anomalies
Other genitourinary anomalies

*Group iv*
Hypoplasia radial ray
Anomalies cervical/thoracic spine/ribs
Tracheo-oesophageal fistula/oesophageal atresia

*Group v*
Hypoplasia lower limb
Ventricular septal defect
Bilateral renal agenesis

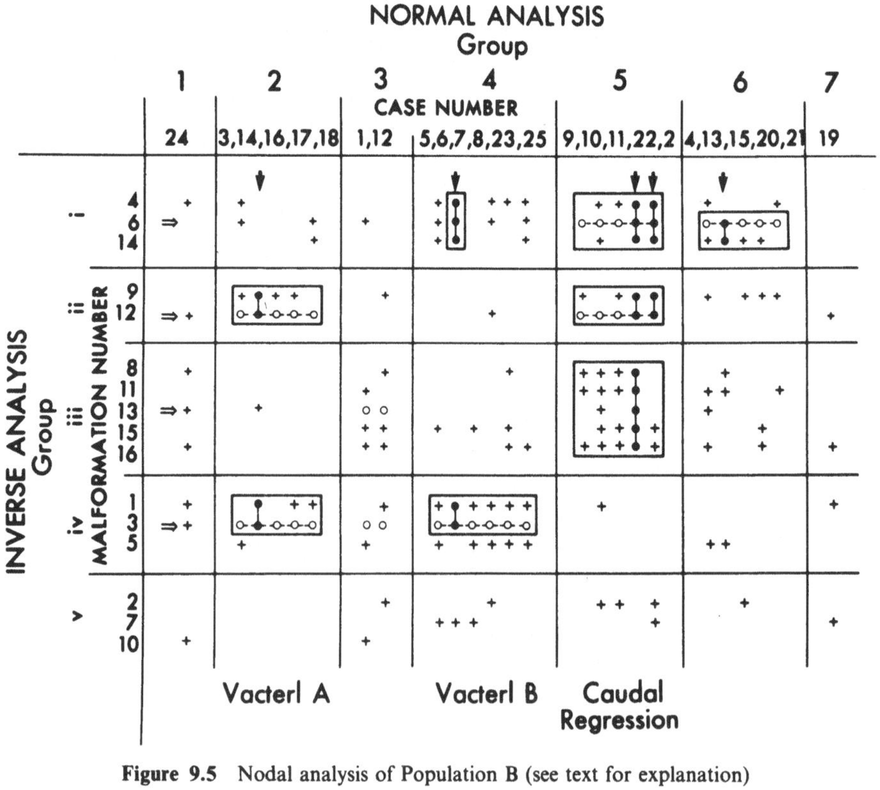

**Figure 9.5**  Nodal analysis of Population B (see text for explanation)

156

Not all groups including all inverse groups have enough residual variation for a statistically significant parameter to be generated. The parameters responsible for establishment of the original groups are shown as open arrows. Six cells (2ii, 4iv, 5i–iii and 6i) are complete in both directions. As coincidence parameters are lacking in all inverse groups no noda of high rank are seen. However, weaker definition is produced by the subdivision parameters in 2ii, 5i and 5ii. These groups may be considered noda of low rank. In cells 4iv and 6i there are similar parameters in both directions but they are incomplete in one direction. These may be considered major subnoda of low rank as may 2iv. The other cell complete in both directions (5iii) has no complete subdivision parameter and so is considered a minor subnoda, as is 4i. Other cells can be discarded as having no coincidence parameters of value. In all therefore, eight cells or 'malformation-individual' clusters of greater or less significance separate in this analysis.

In total, 21 of 25 patients and 12 of 16 malformations are involved in these clusters. Different groups of patients are defined by different malformation associations. Patients 3, 14, 16, 17, 18 are characterized by two associations: 'unilateral renal agenesis–other musculoskeletal anomalies' and 'hypoplasia of the upper limb–cervical thoracic anomalies'. This is considered VACTERL group A. Patients 5, 6, 7, 8, 23 and 25 have two associations: 'lumbar/sacral anomalies–imperforate anus/anal ectopia–other alimentary anomalies' and 'hypoplasia upper limb–cervical/thoracic/rib anomalies'. They are considered VACTERL group B. Patients 2, 9, 10, 11 and 22 have three associations: 'lumbar/sacral anomalies–anal anomalies–other alimentary anomalies' and 'unilateral renal agenesis–other musculoskeletal anomalies' and 'single umbilical artery–aberrant external genitalia–other respiratory anomalies–other cardiovascular anomalies–other genitourinary anomalies'. This is the Caudal Regression group. Finally, patients 4, 13, 15, 20 and 21 are weakly defined by one association: 'imperforate anus/anal ectopia–other musculoskeletal anomalies.'

## Discussion

The three strongly defined groups VACTERL A and B and Caudal Regression separate out distinctly in all three analyses used – polythetic agglomerative, simple association analysis (monothetic divisive) and nodal analysis – although there is phenotypic overlap between them. There are no apparent differences in sex ratio in any group. Mortality in the two VACTERL groups is approximately 50% despite surgical intervention. Children in the Caudal Regression group have a greater mortality rate and most have lethal malformations. There is apparently normal intelligence in survivors in all groups. Both VACTERL groups may have vertebral, limb and anal components to their malformation patterns. The differences lie in the distribution of renal, tracheo-oesophageal and cardiovascular anomalies. Renal agenesis is constant in VACTERL A but absent in VACTERL B though one child had an ectopic kidney. There are no cases of tracheo-oesophageal fistula (TEF) and only one of oesophageal atresia in VACTERL A but 5 of 6 children in B have TEF. No children in VACTERL A have cardiovascular anomalies but these are reported in four of six in the

VACTERL B group. In addition lumbar/sacral vertebral anomalies are rare in VACTERL A despite the presence of renal agenesis but common in VACTERL B. The chief differences between the VACTERL groups and Caudal Regression are the frequency of single umbilical artery (one of 11 in VACTERL, four of five in Caudal Regression), the high incidence of lower limb hypoplasia in Caudal Regression (three out of five compared to one out of 11 in VACTERL) and the type of anal anomaly. In Caudal Regression four of five children had a blind ending large bowel with absent rectum as well as imperforate anus. In VACTERL, of the six children with anal anomalies, three had anal ectopia and the three with imperforate anus had rectourinary or rectovaginal fistula.

The numerical taxonomy techniques described do therefore appear to separate three relatively homogeneous groups from the larger sample of children with components of the VACTERL association. It is not yet known if these groups reflect underlying morphogenetic and/or aetiological differences between the patients or are the result of some ascertainment bias. Different combinations and frequencies of malformations in VACTERL have been reported by other workers. Temtamy and Miller[16] for example report a high incidence of cardiovascular anomalies and a low incidence of renal agenesis in their patients, all of whom were ascertained by tracheo-oesophageal fistula. This agrees with the differences seen in the VACTERL A and B groups distinguished in this analysis. In the present study renal anomalies were present in only 56 % of patients with TEF but in all the patients lacking this malformation. Of the two patients described in detail by Temtamy and McKusick[17], patient 1 fits into VACTERL group A as defined in the present study while patient 2 has more features of Caudal Regression, as the authors point out. Both the original cases described by Quan and Smith[18] fall into VACTERL B which appears to be the more common picture seen in children with VACTERL findings. The rarer association of cervical anomalies and renal agenesis seen in VACTERL A has some similarity to the MURCS association (Mullerian duct aplasia, renal aplasia and cervicothoracic somite dysplasia) reported by Duncan et al.[19]. However, although four of the five VACTERL A patients are female none is known to have Mullerian duct aplasia and all have normal female external genitalia.

The aetiologies of the VACTERL and Caudal Regression associations are not known. Both are believed to be due to defects in mesoderm probably prior to the 35th day of morphogenesis, though whether the defect is a primary abnormality in mesodermal tissue or secondary to other external or intrinsic causes is not known. External causes including maternal intake of progesterones and maternal diabetes have been implicated in some cases. Other theories involving external stresses, such as overdistension of the neural tube[20] or overflexion of the cranial and caudal ends of the fetus[21] resulting in disruption of underlying mesodermal tissue, or haemodynamic imbalance secondary to single umbilical artery[22], have also been put forward. It is probable that these malformations associations may be the end result of any one of a number of different causes and that their heterogeneity and phenotypic overlap are a reflection of this varied aetiology.

The application of numerical taxonomy methods to this sample of children with limb, renal defects and/or VACTERL association has proved feasible and the most important value of the technique would appear to be the production of consistent subgroups for further study. Individual groups can be explored for further aetiological clues and awareness of specific malformation associations may aid in understanding of morphogenetic relationships. However, there may also be clinical applications of such research. Certainly the very strong association of cervicothoracic vertebral anomalies and renal agenesis in VACTERL A might lead to renal studies in babies noted to have short necks or other strong signs of spinal defects in this area. In addition it stresses the importance of very careful recording of anomalies and their site (e.g. the position of vertebral defects should be specified) if malformation associations are to be correctly defined. This work is preliminary and the findings reported reflect only a small sample of patients. Further analysis involving larger groups of patients is now anticipated, using cases ascertained by other malformations forming part of the VACTERL association to determine if unbiased ascertainment alters the groupings produced by numerical classification and to study further the malformation patterns in this fascinating malformation association.

## References

1 Simpson, G. G. (1961). *Principles of Animal Taxonomy*. (New York: Columbia UP)
2 Sneath, P. H. A. and Sokal, R. R. (1973). *Numerical taxonomy. The Principles and Practice of Numerical Classification*. (San Francisco: Freeman)
3 Clifford, H. T. and Stephenson, W. (1975). *An Introduction to Numerical Classification*. (New York: Academic Press.)
4 Bray, R. J. and Curtis, J. T. (1957). An ordination of the upland forest communities of southern Wisconsin. *Ecol. Monog.*, **27**, 325
5 Williams, W. T. and Lambert, J. M. (1959). Multivariate methods in plant ecology. I. Association analysis in plant communities. *J. Ecol.*, **47**, 83
6 Williams, W. T. and Lambert, J. M. (1961). Multivariate methods in plant ecology. III. Inverse association–analysis. *J. Ecol.*, **49**, 717
7 Lambert, J. M. and Williams, W. T. (1962). Multivariate methods in plant ecology. IV. Nodal Analysis. *J. Ecol.*, **50**, 775
8 Estabrook, G. F. (1978). Objective methods for classification and the study of birth defects. *Birth Defects: Orig. Art. Ser.*, **13**(3A), 5. (New York: National Foundation)
9 Pinsky, L. (1974). A community of human malformation syndromes involving the Mullerian ducts, distal extremities, urinary tract and ears. *Teratology*, **9**, 65
10 Pinsky, L. (1977). The polythetic (phenotypic community) system of classifying human malformation syndromes. *Birth Defects: Orig. Art. Ser.*, **13**(3A), 13. (New York: National Foundation)
11 Melnick, M. (1977). Letter. *J. Pediatr.*, **90**, 663
12 Preus, M. and MacGibbon, B. (1977). An application of numerical taxonomy to the classification of syndromes. *Birth Defects: Orig. Art. Ser.*, **13**(3A), 31. (New York: National Foundation)
13 MacGibbon, B. and Preus, M. (1979). The distorted shell method of clustering for syndrome identification. *Am. J. Hum. Genet.*, **31**, 498
14 Morton, N. E., Matsuura, J., Bart, R. and Lew, R. (1978). Genetic epidemiology of an institutionalized cohort of mental retardates. *Clin. Genet.*, **13**, 449
15 Hermann, J., (1977). Numerical taxonomy in clinical genetics. *Birth Defects: Orig. Art. Ser.*, **13**(3A), 39. (New York: National Foundation)
16 Temtamy, S. A. and Miller, J. D. (1974). Extending the scope of the VATER association. Definition of a VATER syndrome. *J. Pediatr.*, **85**, 345

17 Temtamy, S. A. and McKusick, V. A. (1978). The genetics of hand malformations. *Birth Defects: Orig. Art. Ser.*, **14**(3), 135. (New York: National Foundation)

18 Quan, L. and Smith, D. W. (1973). The Vater association, Vertebral defects, Anal atresia, *T-E* fistula with oesophageal atresia, Radial and Renal dysplasia: A spectrum of associated defects. *J. Pediatr.*, **82**, 104

19 Duncan, P. A., Shapiro, L. R., Stangel, J. J., Klein, R. M. and Addonizio, J. C. (1979). The MURCS Association: Mullerian duct aplasia, renal aplasia and cervicothoracic somite dysplasia. *J. Pediatr.*, **95**, 399

20 Gardner, W. J. and Breuer, A. C. (1980). Anomalies of heart, spleen, kidneys, gut and limbs may result from an overdistended neural tube: A hypothesis. *Pediatrics*, **65**, 508

21 Stephens, F. D. (1979). The association and embryology of esophageal and anorectal anomalies. Presented at a *Symposium on Associated Congenital Anomalies*, May 31, Toledo, Ohio

22 Chaurasia, B. D. (1974). Single umbilical artery with caudal defects in human fetuses. *Teratology*, **9**, 287

# 10
# A comparison of the anatomical variations found in trisomies 13, 18 and 21

J. C. PETTERSEN AND E. T. BERSU

## INTRODUCTION

Little or nothing is known about how genetic information translates into body form. However, it is recognized that small changes in the genome can result in serious malformations and that environmental factors can cause various defects depending on the timing of the interference. It is a difficult task to unravel the mechanisms involved in such cases and most attempts to explain the results are based on descriptions of the defects in question, a knowledge of how the body parts develop normally and inferences drawn from embryological experiments on animal models which provide clues that can be used to explain how the developing tissues interact.

It is now well known that unbalanced chromosomal constitutions result in multiple congenital anomaly syndromes. This has aroused considerable interest in determining the genotype–phenotype relationships which are associated with the observed malformations. Aneuploidy of a particular sort yields outcomes which appear to be quite specific. Cases of trisomy 13, 18 and 21 are now classic examples of such multiple congenital anomaly syndromes that can be diagnosed by physical examination at birth and confirmed later by cytogenetic analysis. Autopsy findings in these cases can provide additional confirmatory information.

Certain features have proven of value in the diagnosis of each of these trisomy syndromes. Examples of useful physical examination findings include dermatoglyphics, inner canthal distance, extra or missing digits, external ear shape and position, skull shape and size, facial characteristics etc. Autopsy findings such as particular heart, kidney, digestive tract and brain defects also group themselves in reasonably consistent ways. Such physical examination and autopsy data compiled for numerous cases have resulted in constellations of traits ('defects') which *in toto* can be said to define a given syndrome. However, no single feature can be said to be pathognomonic.

When our studies on gross anatomical variations associated with multiple congenital anomaly syndromes began, we were mainly interested in expanding the total constellation of defects for each syndrome to include those body parts not accessible to physical examination and not usually considered at autopsy. We felt such findings would, in addition to characterizing each syndrome more completely, contribute in some way to a better understanding of their pathogenesis.

Anatomical findings in eight cases of 13-trisomy, nine cases of 18-trisomy and five cases of 21-trisomy are presented on the following pages. A single case of trisomy for the distal four fifths of the long arm of chromosome 13 is also included. We present first those 'new' findings which we believe are reasonably unique to each of these syndromes. This is followed by findings which two or all three of the syndromes share to some degree. The differences and the similarities are discussed and pathogenetic hypotheses are presented.

This report, although it necessarily includes reference to specific and unusual variations, is a synthesis of several years of work which has been published in detail elsewhere[1-8].

## UNIQUE ANATOMICAL FEATURES OF TRISOMIES 13, 18 AND 21

Table 10.1 indicates those 'new' features we have found which characterize trisomies 13, 18 and 21. Features 2–4 indicated for trisomy 18 are not 'new' generalizations but the specific anatomical variations underlying these statements are new.

### Trisomy 13

In 1960 Patau et al.[9] identified an extra D-group chromosome in a female infant with cleft lip, cleft palate, apparent anophthalmia and polydactyly of the left foot. Subsequent cases suggested that a constellation of physical

Table 10.1 Unique, 'new' variations in trisomies 13, 18 and 21

---

*Trisomy 13*
1 Pectorodorsalis muscle, present
2 Peroneus tertius and plantaris muscles, absent
3 Extra muscle from central tendon of the diaphragm to the pericardium, present
4 Muscle variations not confined to discrete regions
5 Presence of cervical rib and absence of 12th rib

*Trisomy 18*
1 Musculocutaneous nerve, absent
2 Anomalous development of preaxial region of upper limb
3 Otomandibular developmental field defect
4 Improperly positioned extensor digitorum tendons over metacarpophalangeal joints

*Trisomy 21*
1 Anomalous distribution of inferior mesenteric artery
2 Failure of facial muscles to differentiate
3 Presence of first cervical nerve dorsal root ganglia associated with spinal accessory nerve
4 Multiple roots for vertebral artery

---

examination findings plus a constellation of autopsy findings defined the 13-trisomy syndrome[10-12]. We present below the gross anatomical findings which can be added to the phenotype.

The anomalies of the brain and midface comprise a developmental field defect which in many respects resembles the defect seen in cases of alobar holoprosencephaly. In other respects this defect manifests itself quite differently in the 13-trisomy condition and our ongoing anatomical study of cases of alobar holoprosencephaly continues to address these differences. The most striking examples we have noted to date are the broadened nasal bones in trisomy 13 versus their usual absence in alobar holoprosencephaly and the constant presence of the *crista galli* and an essentially normal ethmoid bone in trisomy 13 versus absence of the former and severe hypoplasia of the latter in alobar holoprosencephaly.

Mention is here made of cleft lip, cleft palate, microphthalmia, colobomata and arhinencephaly because these features appear to us to be the only clustered set of anomalies found in the 13-trisomy syndrome which are explainable using a developmental field defect hypothesis.

New features found in the 13-trisomy cases but not found in trisomies 18 and 21 include cervical ribs accompanied by absence of the 12th ribs in five of eight cases, pectorodorsalis muscles present bilaterally in seven of eight cases, extra muscles from the central tendon of the diaphragm to the pericardium in three of eight cases, bilateral absence of the peroneus tertius muscle in all 8 cases and absence of the plantaris muscle in 13 of 16 lower limbs. Figure 10.1 illustrates some of these features and includes a drawing of a typical trisomy 13 face.

In addition to these rather specific features, there were numerous variations in the musculature which were less specific, either because they occurred in less than half of the cases or because they were also found in one or both of the other trisomic conditions studied. Table 10.2 indicates these muscle variations. Other muscle variations were unique to single cases and are not included in the table. Because absence of tibialis anterior (bilateral in one subject) and the presence of a hepaticodiaphragmaticus muscle are unusual, they are included in the table even though they occurred in single cases.

Inspection of Table 10.2 suggests that deviations in the muscles are not preferentially confined to any particular body region or to any particular type. Within a given region, e.g. the forearm and hand, one can find extra muscles such as radiocarpus, muscle absences such as palmaris longus and palmaris brevis, and variations such as in extensor indicis, extensors carpi radialis longus and brevis and adductor pollicis. Similar ranges of variation occur in the suprahyoid region and the leg and foot and, in a less striking fashion, in the arm, thigh, pectoral region and gluteal region. We suggest that this widespread occurrence of muscle variations which includes absences, extra muscles, and deviations in structure and attachments is a characteristic of the 13-trisomy condition and reflects the influence of the extra number 13 chromosomal material on the mesoderm. Aside from the extra postaxial digits, rib cage variations and characteristic skull, the skeleton is otherwise grossly normal.

The one case of partial trisomy 13 studied resulted from a maternal

163

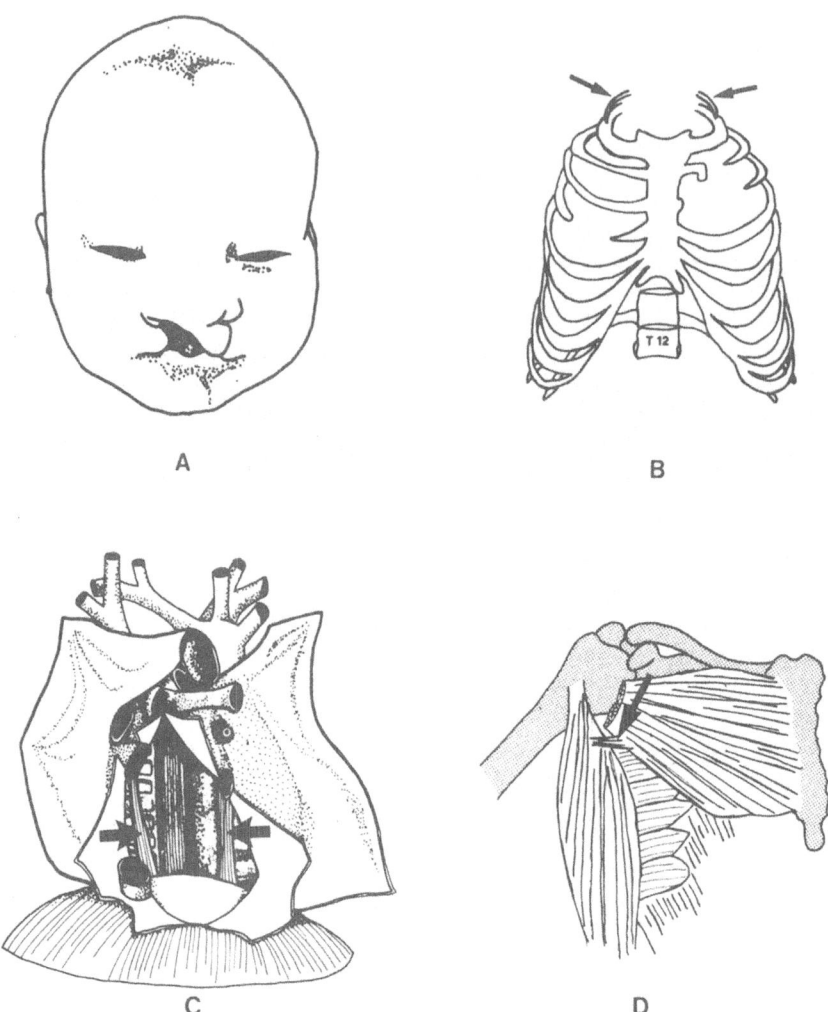

**Figure 10.1**   Features unique to trisomy 13. A, Typical facial features. B, Rib cage anomalies, including cervical ribs (arrows) and absence of 12th ribs. The anterior defects shown in ribs 2 and 3 on the left and 4 on the right were unique to this single case. C, Arrows indicate striated muscles extending from the central tendon of the diaphragm to the pericardium. D, Arrow indicates small muscle (pectorodorsalis) between latissimus dorsi and pectoralis major muscles

balanced translocation involving most of a number 22 chromosome which attached to the distal portion of the long arm of 13 at 13q12. The subject's karyotype was 46,XY, − 22,der(13),t(13;22)(13qter →13q12: :22p11 →22qter). The muscle variations in this case were unique in two respects: no muscle absences occurred; extra muscles and variations in muscles exceeded in number those found in any of the eight complete trisomy 13 cases. Many of these extra muscles and variations were among those listed in Table 10.2, although there were nine deviations in muscles which were unique to this case.

Table 10.2   Non-specific muscle variations found in eight cases of complete trisomy
13

| Variation | Incidence/side or limb |
|---|---|
| Palmaris longus, absent | 16/16 |
| Palmaris brevis, absent | 13/14 |
| Extensor indicis, variations | 14/16 |
| Extensors carpi radialis longus and brevis, variations | 10/16 |
| Biceps and coracobrachialis, variations | 10/16 |
| Posterior belly, digastricus, variations | 8/16 |
| Stylohyoid, variations | 8/16 |
| Anterior belly, digastricus and mylohyoid, variations | 6/16 |
| Occipital platysma, present | 6/16 |
| Flexor digitorum brevis, absent tendons | 7/16 |
| Trans. hd., adductor pollicis, variations | 5/16 |
| Subclavius posticus, present | 4/16 |
| Peroneus quartus, present | 4/16 |
| Tendon of flexor dig. sup. to fifth digit, absent | 5/16 |
| Trans. hd., adductor hallucis, variations | 4/16 |
| Radiocarpus, present | 3/16 |
| Tibialis anterior, absent | 2/16 |
| Hepaticodiaphragmaticus, present | 1/8 bodies |

These are presented elsewhere[7]. One case of complete trisomy 13 had eight unique muscle deviations but the average was about four such deviations for each of the remaining seven cases.

Escobar and Yunis[13] defined features which distinguish between proximal and distal 13q trisomy. Our findings in this single case of distal 13q trisomy suggest much overlap with those features found in complete trisomy 13 with the notable exception that palmaris longus, palmaris brevis, plantaris and peroneus tertius muscles were present. These muscles were almost uniformly absent in our eight cases of complete trisomy 13.

## Trisomy 18

Physical features which are used to distinguish the 18-trisomy syndrome from the 13- and 21-trisomy syndromes most often include a dolichocephalic skull, low-set, malformed ears and a small jaw, a characteristic positioning of the fingers (See Figure 10.5A), and a tendency towards a radius aplasia defect. With the exception of the skull shape, a majority of the anatomical variations observed in the nine bodies which have been dissected[1,2,4] are associated with the otomandibular region and the forearm and hand, thus defining more accurately the involvement of these developmental fields or regions in the syndrome.

Figure 10.2 shows the appearance of the otomandibular regions in one case of trisomy 18 which we dissected. These two sides more or less illustrate the spectrum of external ear malformations which are seen in the syndrome, ranging from absence, as on the right, to a low-set, 'pixie-shaped' ear on the left. Figure 10.3 shows the muscle variations observed in the otomandibular regions in this body. As might have been expected, there were a number of

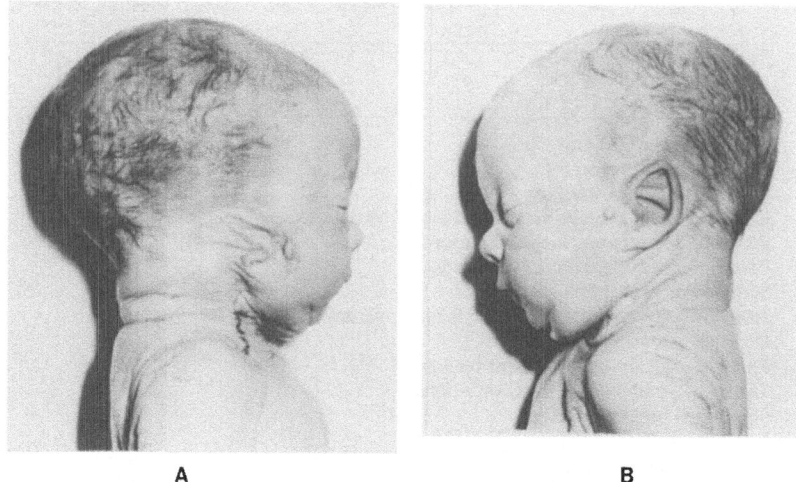

**Figure 10.2** Trisomy 18. A, Severely distorted otomandibular region on right side. B, Milder manifestation of the same defect on the left side

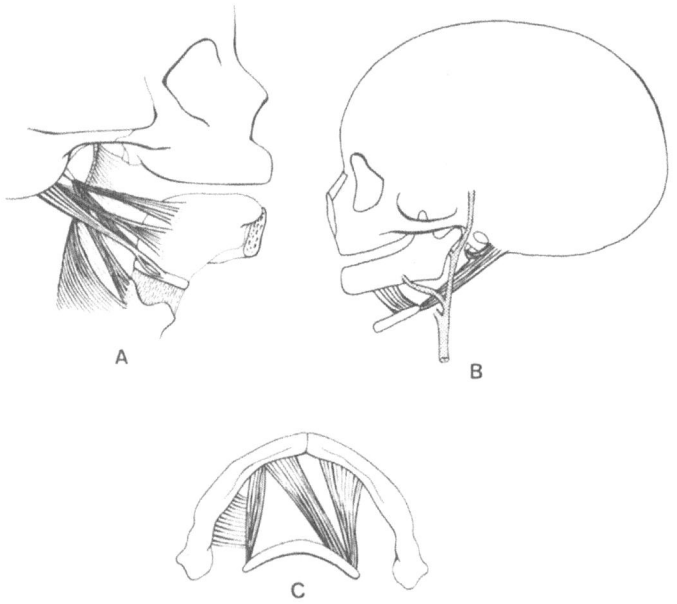

**Figure 10.3** Trisomy 18. A, Doubling of right styloid process muscles and absence of the tympanic membrane in the otomandibular region shown in Figure 10.2A. The posterior belly of digastricus has been removed. B, Absence of stylohyoideus, external carotid artery superficial to posterior belly of digastricus and slip from digastricus to styloid process in otomandibular region shown in Figure 10.2B. C, Bilateral variations in anterior belly of digastricus in subject shown in Figure 10.2

variations on the right side, including absence of the external meatus, a thin and small zygomatic arch, hypoplastic mandibular ramus, an 'extra' set of styloid process mucles and absence of the parotid and submandibular salivary glands (details are published elsewhere[2]). On the left side, where one would not have anticipated such defects, the stylohyoid muscle and the parotid and submandibular salivary glands were absent, the branching pattern of the external carotid artery varied from the more common pattern and there were variations in the styloid process muscles that were present. The anterior belly of digastricus varied bilaterally. Similar variations were found in these structures in the eight other bodies dissected, even though external features were not indicative of such variations. This consistent pattern of defects among the deeper structures in the otomandibular region, along with observations of defects of the middle and inner ears in trisomy 18[14,15], suggests that the characteristic mandibular hypoplasia and appearance of the ears are component manifestations of some extensive developmental disturbance which occurs in the developing first and second arch region in all 18-trisomies.

A similar statement can be made for the tendency towards the preaxial reduction defects in the upper limb in trisomy 18. Pfeiffer and Santelman[16] estimate that 5–10 % of all newborn 18-trisomies show preaxial skeletal deficiencies severe enough to be called radius hypoplasia or aplasia. The skeleton of the forearm and hand of one of the 18 limbs from our sample, diagnosed as 'radius hypoplasia', is shown in Figure 10.4C. In this case, the first metacarpal was absent and a rudimentary thumb was attached by soft tissue to the radial side of the second metacarpal, several of the carpal bones were hypoplastic and the tendons of the forearm muscles which normally attach to the bones of the thumb had anomalous attachments to the carpal bones. The forearm in Figure 10.4B appeared normal upon external examination but, as shown, there were doublings of tendons and abnormal attachments of the long tendons to the thumb and thenar muscles were absent. In addition to similar muscle variations which occurred in all of the other cases, there was a tendency for the radial artery to be replaced by other arterial branches and, in all cases, the definitive musculocutaneous nerve was absent. As shown in Figure 10.4A, one can assume that its fibres were conveyed by the median nerve.

Discussions concerning possible pathogenetic mechanisms responsible for these constellations or fields of defects are found in the original publications[2,4]. To summarize, one possibility is that each may have been caused by haemorrhaging in the respective region with subsequent tissue damage, followed by repair. This could be the result of an hypoxic condition caused by placental insufficiency in trisomy 18[17,18] which affects the rather labile stapedial and axial arteries which supply, respectively, the first and second branchial arch region and the upper limb bud during the fifth week of gestation.

Figure 10.5B shcws the characteristic flexion contractures of the fingers found in trisomy 18. One contributing factor to this positioning may be the radial or ulnar displacement of the membranous expansions of the extensor digitorum tendons over the metacarpophalangeal joints. There is no apparent

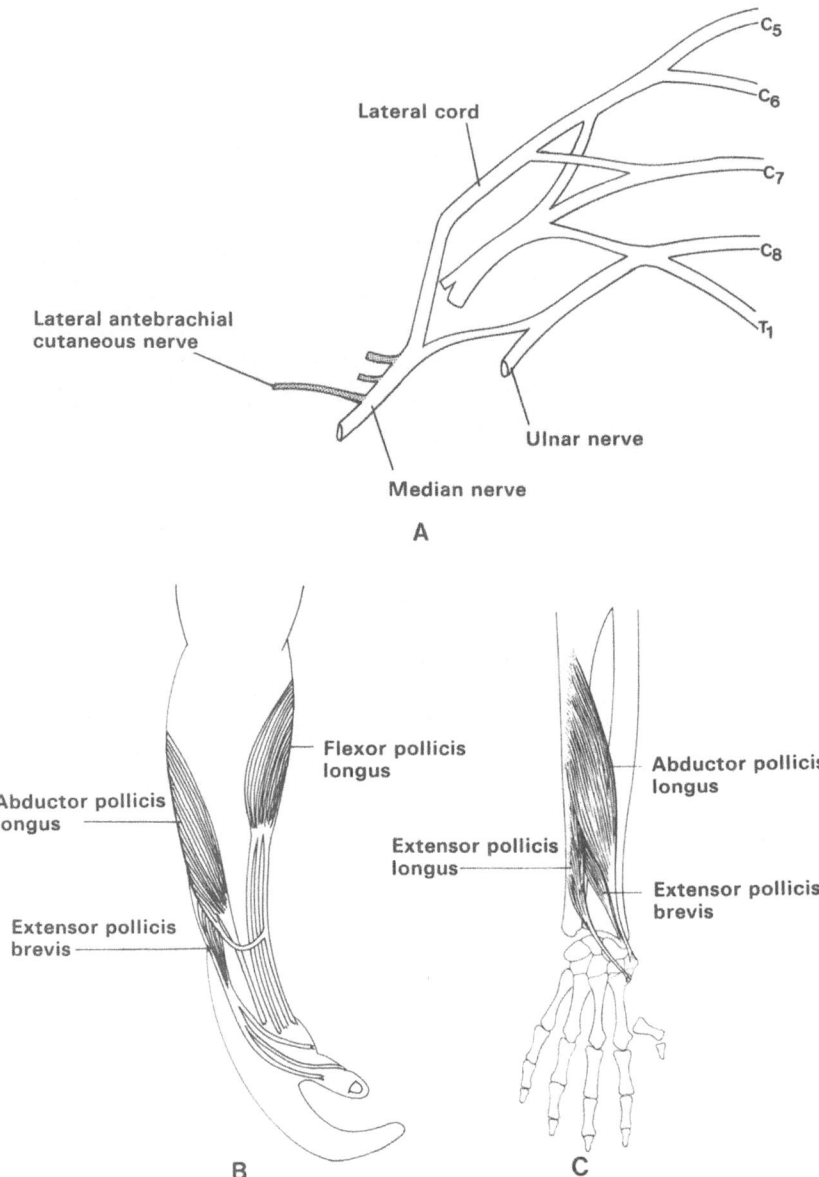

**Figure 10.4** Trisomy 18. A, Typical brachial plexus illustrating absence of a definitive musculocutaneous nerve. B, Aberrant forearm muscles in a case without obvious external indications. C, Aberrant muscle attachments in a case of 'radius hypoplasia'

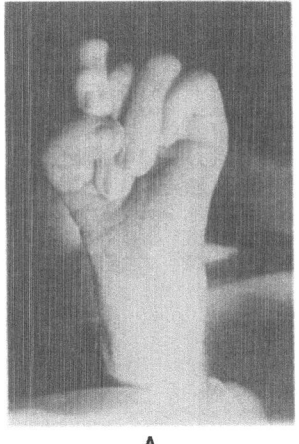
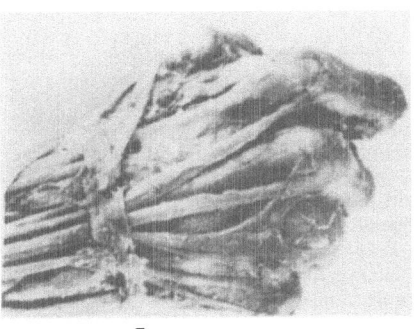

B

A

**Figure 10.5**  Trisomy 18. A, Typical position of digits. B, Aberrant position of extensor tendons over metacarpophalangeal joints

pathogenetic explanation for this variation, and Rehder[19] has observed the anomalous position of the fingers in trisomy 18 fetuses of 18–20 weeks gestation.

Several other patterns of muscle variations were found among the nine bodies which tended to be characteristic for the syndrome, but for which there was some overlap with the 13- and 21-trisomy syndromes (Table 10.3). These included extreme hypoplasia of the facial muscles which may contribute to the mask-like appearance of the face in trisomy 18, the presence of small occipital platysma muscles, found also in trisomies 13 and 21 (See Figure 10.6A), extra muscles associated with pectoralis major, and accessory muscles extending from the fibula to the tarsal bones, e.g. peroneocalcaneus medialis (see Figure 10.8B). Variations of this latter type were also found in the 13- and 21-trisomies and are discussed in a subsequent section.

## Trisomy 21

In his book, *Human Chromosomes*, Ford[20] suggests that, because of the rather mild constellation of phenotypic features associated with trisomy 21, this aneuploid condition represents a 'near miss' at a normal individual. This generalization is more striking when the Down syndrome phenotype is compared with that of the 13- and 18-trisomy syndromes, and is also consistent with the anatomical variations which we have observed[8]. For example, apart from the muscles of facial expression, described below, muscle variations were generally limited to absences of muscles which are not infrequently absent in the normal population: palmaris longus, psoas minor and flexor tendons to the fifth digits of the hand and foot (Table 10.3). These muscles are also characteristically absent in the 13- and 18-trisomy

169

Table 10.3   Muscle absences and extra muscles common to trisomies 13, 18 and 21

| | *Incidence/number of sides in* | | | |
| | *13* | *18* | *21* | *Normals** |
|---|---|---|---|---|
| *Muscle absences* | | | | |
| Palmaris longus | 16/16 | 18/18 | 8/10† | 12.6% |
| Tendon of flexor dig. sup. to 5th digit | 5/16 | 9/18 | 4/10 | not known |
| Tendon of flexor dig. brevis to 5th digit | 7/16 | 11/18 | 3/8 | 20% |
| Psoas minor | 15/16 | 18/18 | 5/19 | 50% |
| Palmaris brevis | 13/14 | 5/18 | 0/10 | 2% |
| Stylohyoid | 2/16 | 6/18 | 0/10 | 0.5% |
| *Extra muscles* | | | | |
| Platysma occipitalis | 6/16 | 16/16 | 10/10 | see text |
| Peroneocalcaneus medialis | 2/16 | 8/18 | 2/10 | not known |
| Peroneus quartus | 4/16 | 6/18 | 0/10 | not known |
| Radiocarpus | 3/16 | 3/18 | 0/10 | 5% |

* Sources for incidences were Macalister[31], LeDouble[32] and Mortensen and Pettersen[33]
† Found in a single case but varied structurally

syndromes, as shown in Table 10.3. Possible developmental explanations for these absences are considered below.

The four 'new' variations observed in the syndrome are shown in Table 10.1 and Figures 10.6 and 10.7. The most subtle of the variations, observed in all five bodies, was in the blood supply to the ascending and transverse segments of the colon, where these two segments were supplied by a single artery which arose as the first branch of the inferior mesenteric artery (Figure 10.7). Normally, these segments are supplied by separate middle and right colic arteries which arise from the superior mesenteric artery. This more normal pattern is, itself, subject to considerable variation[21] but a variation similar to the pattern observed in the Down syndrome individuals has not been described in the literature. There is no obvious developmental explanation for this pattern.

The three remaining variations (Figure 10.6) can all be explained tentatively as a failure of regression of usually transient embryonic structures. (1) In all five bodies, the midfacial muscles were represented by a continuous muscle sheet suggestive of an early stage in their morphogenesis, and there was an extra narrow muscle, which we have called occipital platysma, which extended from the origin of trapezius to the corner of the mouth. This muscle, also seen in cases of trisomy 13 and 18, is reminiscent of the 'platysma occipitalis' muscle described by Gasser[22] in human fetuses between 7 and 12 weeks gestation. The muscle is not observed in fetuses beyond 12 weeks, and it can be considered to regress. (2) Four of the 5 bodies had first cervical dorsal rootlets and small ganglia containing nerve cell bodies which were connected with, or surrounded, the spinal component of the spinal accessory nerve. Streeter[23] noted that such a relationship normally exists in human fetuses through the first trimester, but after this time the first cervical ganglia and associated fibres appear to reach a stage where they fail to grow further and disappear at the macroscopic level, possibly due to the continued growth of the fibre and cellular elements of the surrounding neural structures. (3) Three of the five

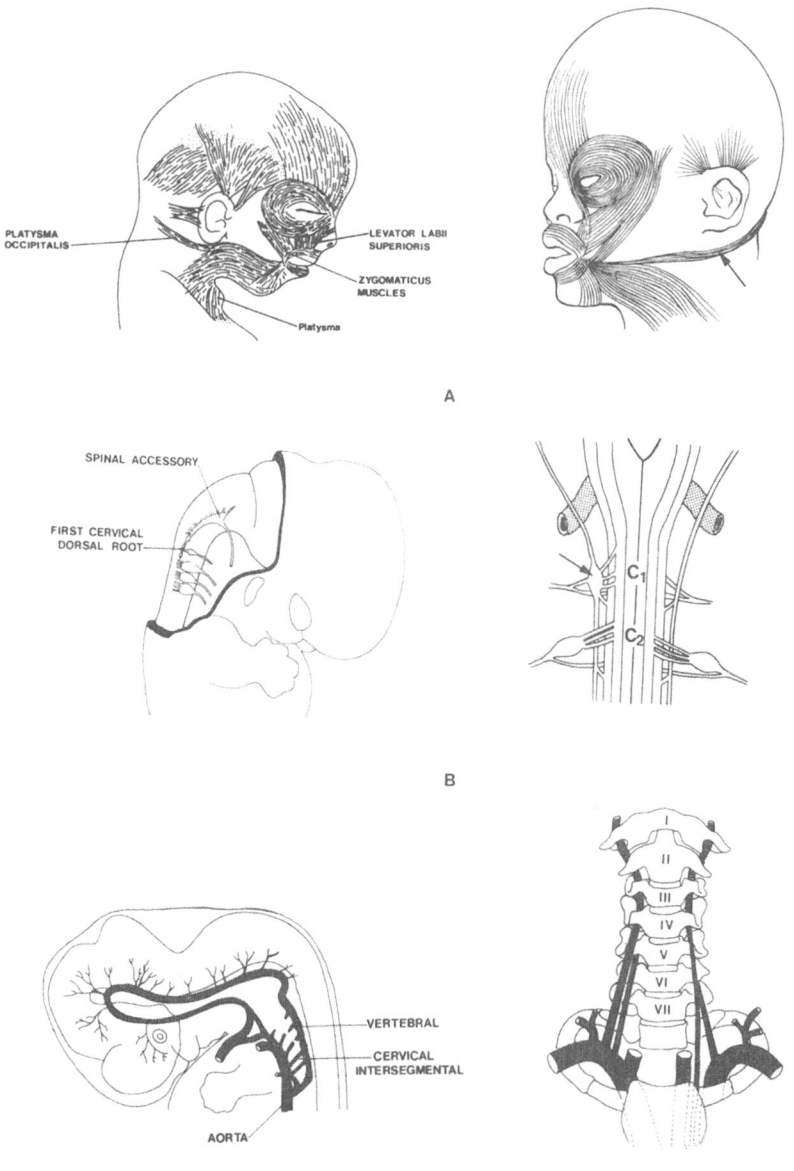

**Figure 10.6** Trisomy 21 'unique' features, shown on the right, which result from failure of embryonic structures, shown on the left, to regress. A, Occipital platysma muscle (arrow) which represents retention of embryonic muscle shown on the left. B, Ganglion (arrow) associated with $C_1$ and spinal accessory nerve which represents retention of small embryonic ganglion shown on the left. C, Multiple roots of vertebral artery which result from retention of cervical intersegmental arteries shown on the left

171

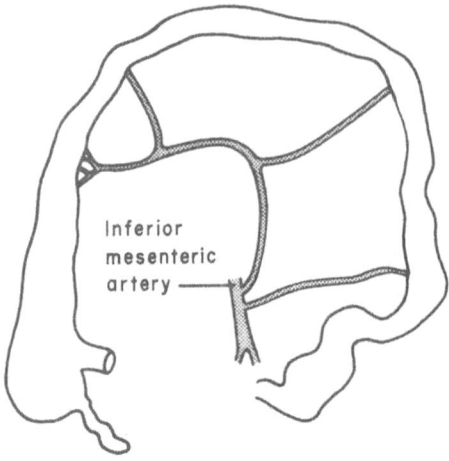

**Figure 10.7** Trisomy 21. Anomalous distribution of inferior mesenteric artery found in all 5 cases of trisomy 21

bodies had additional roots which contributed to the vertebral artery. These additional roots may represent the retention of one or more of the five pairs of usually transient cervical intersegmental arteries which early contribute to the development of the definitive vertebral arteries[24].

Any suggestion concerning a mechanism which might lead to the persistence of these structures can only be tentative. One possibility is a disturbance in the synchrony of developmental events involved in their regression. In the trisomy 19 mouse, for example, we have observed 1- and 2-day lags in the maturation of the external phenotype and haematopoietic systems from 13 to 17 gestational days respectively[25-27]. In the case of the occipital platysma muscle, an earlier or later appearance of its primordium with respect to the development of surrounding structures could result in its persistence and continued growth.

## COMPARISON OF TRISOMIES 13, 18 AND 21

Shapiro[28] reviewed and compared the dental anomalies in the Down syndrome to those in normals. Missing teeth, tooth size, abnormalities of tooth shape and eruption patterns were among the characteristics considered. The types of deviations in the Down syndrome population resembled those seen in the normal population but the frequencies were higher and the severity was greater in the former. He concluded that 'those units normally less stable are those that are most affected in Down's syndrome'.

In a study of palatal dimensions and hand dermatoglyphics Shapiro[29] showed that palate length and atd angle were the most variable from among several measurements in the Down syndrome and also in normals with, again, a higher incidence in the Down syndrome. These data support an hypothesis that the extra chromosomal material results in 'amplified developmental instability'.

If 'amplified developmental instability' results from extra chromosomal material, trisomies 13, 18 and 21 should share some of the anatomical variations we have observed. Waddington[30] suggested that the developing organism is buffered against variation. According to Shapiro[29], this buffering or 'canalization' is decreased by the presence of extra chromosomal material so that those features which are less buffered, i.e. those most commonly variant in the normal population, are those most likely to be altered in the Down syndrome.

We have presented the unique 'new' anatomical features observed in three human trisomy syndromes. Tables 10.3 and 10.4 indicate those deviations in the muscular system which two or all three of the syndromes share. We do not claim that the muscular system is the most affected system, but the data we have collected for this system are reasonably precise and lend themselves well to comparison among syndromes and among normals when limited to basic features such as absences, extra muscles, and variations in form and attachments. Other organ systems, such as the kidney,[34] the brain[35] and the heart[36] have been compared in a similar fashion from detailed autopsy reports.

Table 10.4 Muscle variations common to trisomies 13, 18, and 21

| Variation | Incidence/number of sides in | | | |
| | 13 | 18 | 21 | Normals* |
| --- | --- | --- | --- | --- |
| Anterior compartment of arm | 10/16 | 12/18 | 4/10 | 9% |
| Infraclavicular region | 6/16 | 13/18 | 0/10 | infrequent |
| Suprahyoid muscles | 13/16 | 18/18 | 1/10 | frequent |
| Extensor indicis | 14/16 | 8/18 | 0/10 | 5% |
| Extensors carpi radialis longus and brevis | 10/16 | 10/18 | 0/10 | frequent |

* Sources for incidences were Macalister[31], LeDouble[32] and Mortensen and Pettersen[33]

## Muscle absences

Muscle absences were most frequent in trisomy 13, almost as frequent in trisomy 18 and relatively infrequent in trisomy 21. Absences of peroneus tertius and plantaris are not included in Table 10.3 because they were features unique to trisomy 13. Inclusion of these makes trisomy 13 somewhat more characterized by muscle absences than trisomy 18. The percentages shown for normals are not necessarily reliable because of the unknown nature of the dissection room material from which they were derived. The psoas minor and the tendon of flexor digitorum brevis to the fifth toe are absent so frequently that they are not useful criteria for defining constellations of characteristics.

The frequencies of absence of palmaris longus, flexor tendons to the fifth digit in the hand and foot, psoas minor and palmaris brevis are consistent with the 'amplified developmental instability' hypothesis. Each of these affected muscles or tendons is characterized by late development. The best example is the palmaris longus muscle which was shown by Lewis[37] to be the last muscle to separate from the primordial superficial flexor mass in the forearm. It seems likely that this developmental history reflects its frequent absence in the normal population and that, if extra chromosomal material upsets the usual activities of primordial cells in some way, this muscle will be among the first to be affected. Following separation of the primordium into the other superficial flexors within the forearm, there may be nothing left to form a critical mass for the palmaris longus.

The stylohyoid may be a special case. Its high rate of absence in trisomy 18 may relate to vascular disturbances in the otomandibular region as previously discussed. It is rarely absent in normals but, when absence occurs, it is presumably due to its failure to separate from the posterior belly of digastricus. This mechanism is consistent with its low rate of absence in trisomy 13 as an example of developmental instability.

It is difficult to explain the presence of peroneus tertius and plantaris muscles in all of the cases of trisomy 18 and trisomy 21. These are among the most frequently absent muscles in normals with reported incidences of about 8.5 and 7% respectively[32]. The developmental timing, as determined by Bardeen[38], suggests that failure to separate from common mesodermal masses accounts for their frequent absence. In this sense, one would expect that the trisomy 18 and 21 cases would show some absences of these muscles as was the case with palmaris longus. The case of distal 13q trisomy further complicates the analysis because no muscle absences occurred. One must assume that the effect of aneuploidy, as it relates to muscle absences, is not totally non-specific.

The peroneus tertius muscle has been studied recently[39]. In this study, it was absent in 12 of 169 lower limbs for an incidence of 7.1%. Its attachments, particularly distally, were found to vary considerably from the usual textbook descriptions. The implication was made that it has significance in the evolution of bipedal locomotion as a factor in eversion of the foot so that the weight can be transferred toward the medial arch. Its presence is restricted to man and the gorilla. In other primates the foot is rarely in the everted position. It was postulated that it developed from the extensor digitorum longus but its anatomical relationships are distinct and it should not be considered a part of that muscle. A possible interpretation is that this muscle represents a relatively 'new' addition which, although of some importance to bipedal locomotion, is not sufficiently 'canalized' to prevent its absence in trisomy 13.

### Extra muscles

Table 10.3 indicates the extra muscles which were common to two or to all three of the trisomy syndromes. Trisomies 13 and 18 are both characterized by the presence of supernumerary muscles. Trisomy 21 is characterized in this respect only by the consistent presence of an occipital platysma muscle (Figure 10.6). This muscle is also consistently present in trisomy 18 but it is less developed. Poorly developed occipital platysma muscles were also found in about half of the trisomy 13 cases. The above indicates that in all three trisomies this muscle failed to regress completely suggesting a general developmental disturbance due to extra chromosomal material.

Figure 10.8 illustrates a radiocarpus muscle which is found occasionally in trisomies 13 and 18 and a peroneocalcaneus medialis muscle found in all three trisomies but with a significant incidence in trisomy 18. Peroneus quartus is another extra leg muscle that is found in both trisomies 13 and 18. Peroneus quartus and fibulocalcaneus medialis muscles were also found bilaterally in the distal 13q trisomy subject. The above three muscles are examples of increased incidence of extra muscles which have been reported in the euploid population with a very low incidence. We suggest that they represent non-specific variations due to a general disturbance in developmental stability.

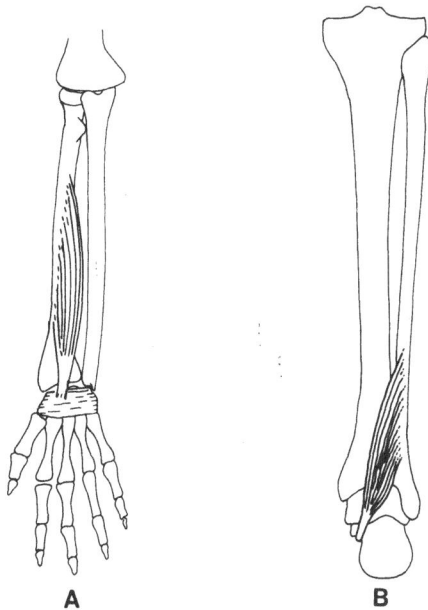

A                                                                  B

**Figure 10.8**   A, Radiocarpus muscle as seen in cases of trisomy 13 and 18. B, Peroneocalcaneus medialis muscle seen frequently in trisomy 18 and occasionally in trisomies 13 and 21

## Variations in muscles

Trisomies 13 and 18 are both characterized by muscle variations (Table 10.4). Some examples of these variations are shown in Figure 10.9. Trisomy 21 is rather stable in this respect except for the muscles of facial expression, as discussed above, and the anterior compartment of the arm wherein the biceps tends to deviate from its usual structure.

We believe that variations observed in trisomy 18 in the suprahyoid region, the anterior compartment of the arm and, to a lesser extent, the infraclavicular region relate in part to developmental field defects as previously discussed. In addition, there were numerous muscle variations on the radial side of the forearm, not indicated in Table 10.4, which are also attributable to a

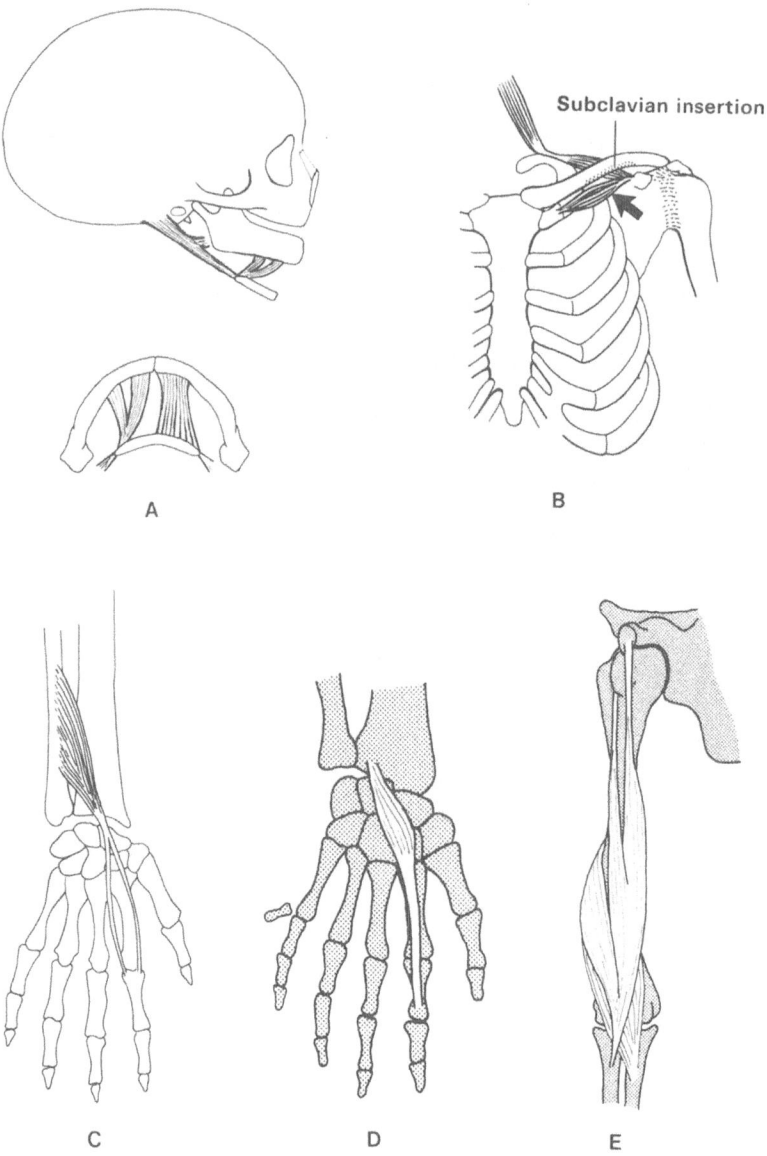

**Figure 10.9** Examples of muscle variations common to trisomies 13 and 18. A, Variations in suprahyoid muscles including absent stylohyoideus, slip from posterior belly of digastricus to middle pharyngeal constrictor and doubling of anterior digastric belly. B, Subclavius posticus muscle (arrow) illustrating one example of an infraclavicular region muscle variation common to trisomies 13 and 18. C, Typical extensor indicis variation in trisomy 18. D, Typical extensor indicis variation in trisomy 13. E, Multiple heads of biceps brachii in a case of trisomy 13. A broad range of variations was seen in the flexor compartment of the arm in trisomies 13 and 18 and, to a lesser extent, in trisomy 21

developmental field defect. Examples of these are shown in Figures 10.4B and 10.4C.

Extensor indicis varied frequently in both trisomies 13 and 18. In six of the trisomy 18 limbs its tendon was doubled but the origin from the ulna was normal. In trisomy 13 the tendency was for the origin to be displaced distally, often to the distal end of the radius or to the carpal bones, although extra tendons sometimes occurred also. In one case of trisomy 13 this muscle was bilaterally absent. Because this muscle varies in the normal population, developmental instability resulting from extra chromosomal material is the presumed cause in trisomies 13 and 18, but it is obvious from inspection of the dissected material that this instability is expressed differently in the two syndromes (Figure 10.9C and 9D). In both syndromes extensors carpi radialis longus and brevis tended to be fused at their origins with extra tendons at their insertions. In variations involving these muscles we cannot discriminate between the two syndromes.

## SUMMARY AND CONCLUSIONS

The results of our anatomical investigations of three human trisomy syndromes, with emphasis on the muscular system, suggest that many of the variations observed are non-specific and can be considered to be a result of a general disturbance in developmental stability due to extra chromosomal material as proposed for the Down syndrome by Shapiro[29]. Trisomy for the larger number 13 and 18 chromosomes results in more of these variations than does trisomy for the smaller number 21 chromosome.

Within this framework of 'amplified developmental instability' there exists a specificity for each syndrome (Table 10.1) and among the variations which two or all three of the syndromes share (Tables 10.3 and 10.4) there are tendencies for one or another of these syndromes to be more characterized by a particular variation than are the others. The unique genetic content of each particular chromosome presumably plays a role in this specificity.

### Acknowledgements

This work was supported by DHEW/PHS Grants GM20130 (Wisconsin Clinical Genetics Research Center) and GM00723-114 (Anatomy Training Grant) from the National Institute of General Medical Sciences.

### References

1 Barash, B. A., Freedman, L. and Opitz, J. M. (1970). Anatomic studies in the 18-trisomy syndrome. *Birth Defects: Orig. Art. Ser.*, **6**(4), 3. (New York: National Foundation)
2 Bersu, E. T. and Ramirez-Castro, J. L. (1977). Anatomical analysis of the developmental effects of aneuploidy in man – the 18-trisomy syndrome: I. Anomalies of the head and neck. *Am. J. Med. Genet.*, **1**, 173
3 Colacino, S. C. and Pettersen, J. C. (1978). Analysis of the gross anatomical variations found in four cases of trisomy 13. *Am. J. Med. Genet.*, **2**, 31
4 Ramirez-Castro, J. L. and Bersu, E. T. (1978). Anatomical analysis of the developmental effects of aneuploidy in man – the trisomy 18 syndrome: II. Anomalies of the upper and lower limbs. *Am. J. Med. Genet.*, **2**, 285

5 Pettersen, J. C., Koltis, G. G. and White, M. J. (1979). An examination of the spectrum of anatomic defects and variations found in eight cases of trisomy 13. *Am. J. Med. Genet.*, **3**, 183

6 Opitz, J. M., Herrman, J., Pettersen, J. C., Bersu, E. T. and Colacino, S. C. (1979). Terminological, diagnostic, nosological, and anatomical aspects of developmental defects in man. In Harris, H. and Hirschhorn, D. (eds.) *Advances in Human Genetics*. Vol. 9, pp. 71–164. (New York: Plenum Press)

7 Pettersen, J. C. (1979). Anatomical studies of a boy trisomic for the distal portion of 13q. *Am. J. Med. Genet.*, **4**, 383

8 Bersu, E. T., (1980). Anatomical analysis of the developmental effects of aneuploidy in man – the Down syndrome. *Am. J. Med. Genet.*, **5**, 399

9 Patau, K., Smith, D. W., Therman, E., Inhorn, S. L. and Wagner, S. P. (1960). Multiple congenital anomaly caused by an extra autosome. *Lancet*, **1**, 790

10 Patau, K., Therman, E., Smith, D. W. and Inhorn, S. L. (1961). Two new cases of $D_1$ trisomy in man. *Hereditas*, **47**, 239

11 Warkany, J., Passarge, E. and Smith, L. B. (1966). Congenital malformations in autosomal trisomy syndromes. *Am. J. Dis. Child.*, **112**, 502

12 Taylor, A. I., (1968). Autosomal trisomy syndromes: a detailed study of 27 cases of Edwards' syndrome and 27 cases of Patau's syndrome. *J. Med. Genet.*, **5**, 227

13 Escobar, J. I. and Yunis, J. J. (1974). Trisomy for the proximal segment of the long arm of chromosome 13. *Am. J. Dis. Child.*, **128**, 221

14 Miglet, A. W., Schuller, D., Ruppert, E. and Lim, D. J. (1975). Trisomy 18. A temporal bone report. *Arch. Otolaryngol.*, **101**, 433

15 Sando, I., Bergstrom, L., Wood. R. P. and Hemenway, W. G. (1970). Temporal bone findings in trisomy 18 syndrome. *Arch. Otolaryngol.*, **91**, 552

16 Pfeiffer, R. A. and Santelman, R. (1977). Limb anomalies in chromosomal aberrations. In Bergsma, D. and Lenz, W. (eds.) Morphogenesis and Malformations of the Limb. *Birth Defects: Orig. Art. Ser.*, **13**(1), 319 (New York: National Foundation)

17 Hsu, L. Y., Strauss, L., Dubin, E. and Hirschhorn, K. (1973). Prenatal diagnosis of trisomy 18. Pathologic findings in a 20-week conceptus. *Am. J. Dis. Child.*, **125**, 290

18 Machin, G. A. and Crolla, J. A. (1974). Chromosome constitution of 500 infants dying during the perinatal period. *Humangenetik*, **23**, 183

19 Rehder, H. (1976). Prenatal pathology of the Down's syndrome and Edwards' syndrome. In Boué, A. (ed.) *Prenatal Diagnosis*, pp. 117–130. (Paris: INSERM)

20 Ford, E. H. R. (1973). *Human Chromosomes*, p. 223. (New York: Academic Press)

21 Anson, B. J. (1963). *An Atlas of Human Anatomy*, p. 407. (Philadelphia: Saunders)

22 Gasser, R. F. (1967). The development of the facial muscles in man. *Am. J. Anat.*, **120**, 357

23 Streeter, G. L. (1905). The development of the cranial and spinal nerves in the occipital region of the human embryo. *Am. J. Anat.*, **4**, 83

24 Patten, B. M. (1968). *Human Embryology*. 3rd Edn. (New York: McGraw-Hill)

25 Bersu, E. T. (1979). Disruptions in the normal progression of phenotypic development caused by trisomy 19 in the mouse. *Teratology*, **19**, 19A

26 Bersu, E. T. (1980). Ontogeny of the retarded phenotype in the trisomy 19 mouse. *Teratology*, **21**, 28A

27 White, B. J., Tjio, J. H., Van de Water, I. C. and Crandall, C. (1974). Trisomy 19 in the laboratory mouse. II. Intrauterine growth and histochemical studies of trisomies and their normal littermates. *Cytogenet. Cell. Genet.*, **13**, 232

28 Shapiro, B. L. (1970). Prenatal dental anomalies in mongolism: comments on the basis and implications of variability. *Ann. NY Acad. Sci.*, **171**, 562

29 Shapiro, B. L. (1975). Amplified developmental instability in Down's syndrome. *Ann. Hum. Genet.*, **38**, 429

30 Waddington, C. H. (1942). The canalization of development and the inheritance of acquired characters. *Nature (London)*, **150**, 563

31 Macalister, A. (1875). Additional observations on muscular anomalies in human anatomy (third series) with a catalogue of the principal muscular variations hitherto published. *Trans. R. Irish Acad.*, **25**, 1

32 LeDouble, A. F. (1897). *Traité des variations du système musculaire de l'homme.* (Paris: C. Rheinwald)

33 Mortensen, O. A. and Pettersen, J. C. (1966). The musculature. In Anson, B. J. (ed.). *Morris' Human Anatomy*. 12th Edn., pp. 421–611. (New York: McGraw-Hill)

34 Egli, F. and Stalder, G. (1973). Malformations of kidney and urinary tract in common chromosomal aberrations. I. Clinical studies. *Humangenetik*, **18**, 1

35 Neuhäuser, G. and Usener, M. (1966). Hirnmissbildung und autosomal trisomy. *Z. Kinderheilk.*, **95**, 244

36 Polani, P. E. (1968). Chromosomal abnormalities and congenital heart disease. *Guy's Hosp. Rep.*, **117**, 323

37 Lewis, W. H. (1901). The development of the arm in man. *Am. J. Anat.*, **1**, 145

38 Bardeen, C. R. (1907). Development and variation of the nerves and the musculature of the inferior extremity and of the neighboring regions of the trunk in man. *Am. J. Anat.*, **6**, 259

39 Krammer, E. B., Lischka, M. F. and Gruber, H. (1979). Gross anatomy and evolutionary significance of the human peroneus. III. *Anat. Embryol.*, **155**, 291

# 11
# Current concepts in congenital adrenal hyperplasia

D. T. MININBERG, LENORE S. LEVINE AND MARIA I. NEW

Ambiguous genitalia in the newborn is a medical emergency; its treatment requires judgement, wisdom and, most of all, an understanding of the biology of sexual differentiation. We present information on the recent developments and current knowledge regarding congenital adrenal hyperplasia (CAH), the most common cause of ambiguous genitalia in the newborn.

Congenital adrenal hyperplasia is an inborn error of metabolism transmitted by an autosomal recessive gene. Females and males are affected with equal frequency. The frequency of this disorder has been estimated to be as low as 1 in 67 000 births[1] and as high as 1 in 5401 births[2]. The lower estimate is probably attributable to inaccurate diagnosis rather than to actual incidence.

## PATHOPHYSIOLOGY

During embryogenesis, the mesodermal, or cortical, element of the future adrenal proliferates at the 6th week, whereas the neuroectodermal, or medullary, element proliferates at the 7th week. Zonal differentiation and proliferation are later events; the glomerulosa and fasciculata are present at birth and the reticularis differentiates by the age of 3 years.

The normal adrenal gland synthesizes three classes of steroid hormones: glucocorticoids, mineralocorticoids and sex steroids. At least seven enzymes have been identified as necessary for the interrelated synthesis of these compounds: 21-hydroxylase, $3\beta$-hydroxysteroid dehydrogenase, $11\alpha$-hydroxylase, $17\alpha$-hydroxylase, cholesterol desmolase, and the 17–20 lyase. The relationships and interdependence of these enzymes are shown in Figure 11.1.

Common to all forms of congenital adrenal hyperplasia is a relative deficiency of cortisol that results in increased adrenocorticotropic hormone (ACTH) secretion secondary to decreased negative feedback to the hypothalamic–pituitary axis. Adrenal hyperplasia and excess production of the precursor hormones and their byproducts ensue. Enzyme defects produce clinically recognizable syndromes that depend upon the specific hormones that are deficient and those that are produced in excess.

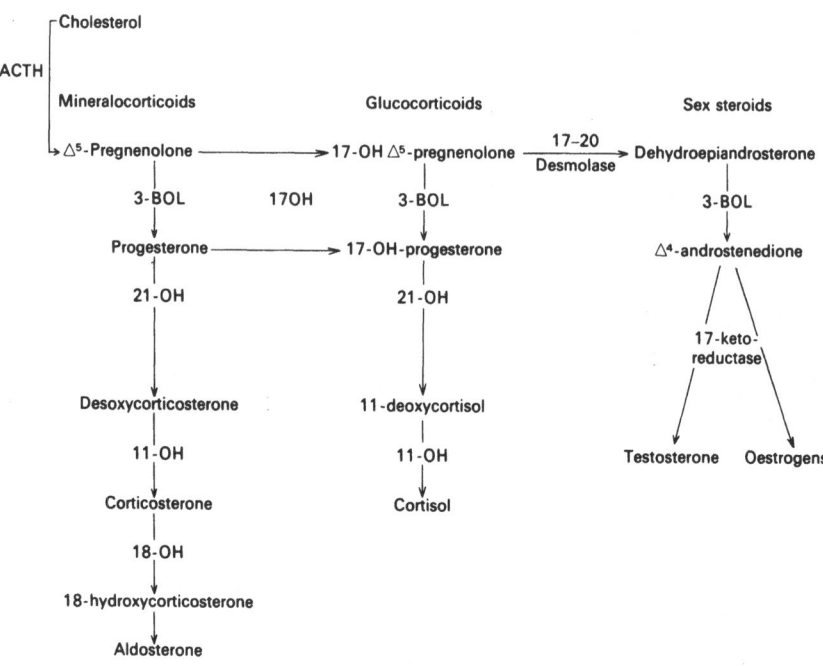

**Figure 11.1** A simplified scheme of adrenal steroid synthesis

*21-hydroxylase deficiency* is the most common clinical form of congenital virilizing adrenal hyperplasia and accounts for approximately 95 % of patients with CAH. The mild form of this disorder results in deficient cortisol production (Figure 11.1) and increased release of corticotropin via the negative feedback system. Consequently, 17OH-progesterone and progesterone accumulate. Inasmuch as sex hormone synthesis is not dependent on 21-hydroxylation, an excess of testosterone is produced, which results in virilization. Because the enzyme deficiency is mild, the block in synthesis of salt-retaining hormones from progesterone is not observed clinically except in times of severe stress. These patients present with virilization. Inasmuch as adrenocortical function begins in the 3rd month of gestation, the affected fetus is exposed to an overproduction of fetal adrenal androgens during the critical time of sexual differentiation. In females, the external genitalia are masculinized and present with a large clitoris and fusion of the labioscrotal folds. Rarely is the masculinization so profound that the urethra is penile[3-16]. The internal genitalia, uterus and fallopian tubes are normal because the abnormality of adrenal androgen production does not involve Müllerian inhibiting factor produced by the Sertoli cells of the fetal testis.

Males affected by the 21-hydroxylase deficiency do not manifest genital abnormalities and thus appear normal at birth. In the untreated male or female child the continuing overproduction of androgens manifests in an initially tall stature with subsequent premature closure of the epiphyses and eventual short stature.

The *severe form of 21-hydroxylase deficiency* results in the salt-losing type of adrenal hyperplasia accompanied by virilization. In these patients, the cortisol deficiency is greater than in patients with the simple virilization form of CAH and aldosterone is markedly deficient[3-5]. The aldosterone deficiency is present even when the patient is sodium depleted. This finding is in contrast to findings in the simple form of 21-hydroxylase deficiency in which aldosterone secretory deficiency has not been well documented. Elevated plasma renin activity, presumably a response to salt-wasting and hypoaldosteronism, is present in the salt wasters and to a lesser extent in patients with simple virilization[3-5].

*Hypertensive adrenal hyperplasia* is associated with defects in $11\beta$-hydroxylation. As can be seen in Figure 11.1, this defect results in decreased cortisol production and increased corticotropin production; the secretion of desoxycorticosterone, a potent salt-retaining hormone, increases. This increase results in salt and water retention, suppressed plasma renin activity, and decreased aldosterone secretion. The sex steroid pathway, with its end product of testosterone, is also stimulated by the high corticotropin levels and leads to virilization. Congenital adrenal hyperplasia attributable to $11\beta$-hydroxylase deficiency is pleomorphic; cases with and without hypertension have been described. The hypertension is thought to result from the excess of desoxycorticosterone and salt retention. This concept is supported by reports of hypertension secondary to exogenous desoxycorticosterone[17-19]. There have also been instances in which the manifest virilization associated with this deficiency has been delayed until puberty[20,21].

*3β-hydroxysteroid dehydrogenase deficiency (3-OL)* results in a block early in the synthetic pathways that affects the production of all three classes of steroids. The synthesis of cortisol, aldosterone and testosterone is markedly reduced[22]. Because the enzyme is also deficient in the testes[23], the deficiency results in incomplete virilization in the male. In the female, there is mild virilization which has been attributed to the overproduction of dehydroepiandrosterone, a weak androgen which is synthesized before blockade[24]. Dehydroepiandrosterone has been shown to virilize the female rat[24]. Inasmuch as this enzyme is found in the testes, it is interesting to speculate about a possible role of 3β-hydroxysteroid dehydrogenase deficiency in uncomplicated hypospadias.

The development of pubertal breast tissue suggests a fetal testosterone deficiency. According to Federman[25] a deficiency in fetal testosterone results in a failure to inhibit the breast anlage and inappropriate breast development at puberty. This concept is further supported by Neumann and Elger[26,27] and Neumann and Goldman[28] who showed that male rats treated *in utero* with a specific 3β-hydroxysteroid dehydrogenase inhibitor developed mammary glands at puberty. This development could be prevented by the administration of testosterone[26-28].

*17-hydroxylase deficiency* results in a deficiency in the synthesis of cortisol and sex steroids which results in sexual ambiguity in the male and lack of development of secondary sexual characteristics in the female, although a uterus is present. The external genitalia of the affected males are ambiguous and differentiation of the androgen dependent external genitalia is markedly incomplete. Inasmuch as the 17-hydroxylase deficiency does not interfere with production of the Müllerian inhibiting hormone, neither a uterus nor fallopian tubes are found in affected males. At puberty these males develop breasts, indicative of prenatal testosterone deficiency.

An oversecretion of corticotropin results in oversecretion of desoxycorticosterone, and salt and water retention with renin suppression. Because of the renin suppression, aldosterone secretion is decreased[29]. The elevated desoxycorticosterone levels and subsequent salt and water retention may result in the development of hypertension. This sequence has been substantiated by the work of Newton and Laragh[30] and Biglieri[31] who showed a decline in aldosterone secretion with corticotropin administration. Bledsoe et al.[32] have shown that a low sodium diet, probably mediated by an increase in plasma renin activity, increased the secretion of desoxycorticosterone and aldosterone.

Rovner and associates[29] have postulated that at least a part of the decrease in aldosterone secretion after the administration of large doses of corticotropin is mediated by a decrease in plasma renin activity. They also showed that if corticotropin is suppressed and desoxycorticosterone only mildly elevated, a low sodium diet causes a measurable increase in plasma renin activity with a concomitant rise in aldosterone excretion[29]. Other investigators, Goldsmith et al.[33], Mills et al.[34] and New[35] reported a similar phenomenon. However, deLange et al. (cited by Rovner et al.[29]) described five patients resistant to normalization of aldosterone biosynthesis.

*Lipoid adrenal hyperplasia* (cholesterol desmolase) results from an early block in steroid synthesis because of a cholesterol desmolase deficiency. This block prevents the conversion of cholesterol to pregnenolone (Figure 11.1) and thus results in deficiencies of all three classes of steroid hormones[36]. This deficiency is probably incompatible with life, unless adequate replacement therapy is given[37–44].

The male child with congenital adrenal hyperplasia poses a special set of problems, in that the genitalia are characteristically not ambiguous. This is certainly true in the 21-hydroxylase form of the syndrome, which accounts for 95% of cases. The genitalia will appear as normal male, with a normal degree of virilization. If anything, the infant might appear to be overly well virilized. The diagnosis in these instances will be dependent upon the biochemical determinations (see below), particularly the 17-hydroxyprogesterone.

The male infant with the less common forms of CAH, $3\beta$-hydroxysteroid dehydrogenase deficiency and 17-hydroxylase deficiency, will have ambiguous genitalia and insufficient virilization. This will precipitate further investigations and make the final diagnosis easier to ascertain. In all the above situations, a positive family history will also alert the physicians to a possible diagnosis.

*18-hydroxylase and 18-dehydrogenase enzyme defects* can be related to the discussion of congenital adrenal hyperplasia. In the 18-hydroxylase deficiency there is a syndrome of salt wasting, dehydration, poor weight gain and intermittent fevers. There is a defect in the conversion of corticosterone to 18-OH corticosterone, with eventual deficiency in aldosterone production. Growth and development are normal and genital development is normal as well. The increased secretion of corticosterone can be suppressed by the administration of desoxycorticosterone, which suggests a regulatory function for the renin–angiotensin system rather than the corticotropin axis. In fact, plasma renin activity was found to be high in this syndrome by Jean et al.[45]. It has been suggested that the clinical improvement with age is the result of a maturation of renal sodium conserving factors in the face of the continued adrenal enzyme defect[46].

*18-dehydrogenase deficiency* is characterized by a syndrome of failure to thrive, dehydration, hyponatraemia and hyperkalaemia[47–49]. These patients have short stature, but, unlike the 18-hydroxylase deficiency patients, they do not have frequent fevers. Cortisol and sex hormone secretions are normal. As in the cases of 18-hydroxylase deficiency, the renin–angiotensin system seems to exert a regulatory function. David et al.[48] and Rapaport et al.[49] found plasma renin activity to be markedly increased. The clinical picture is readily reversible by the administration of desoxycorticosterone and salt. It should be emphasized that in these two conditions there is no demonstrable hyperplasia of the adrenal gland despite adrenal enzyme defects.

## LABORATORY STUDIES

The specific enzyme defects causing the clinically diagnosed congenital adrenal hyperplasia can be distinguished by hormonal testing. The traditional laboratory determinations used were urinary 17-ketosteroid, 17-hydroxy-corticosteroid, and pregnanetriol concentrations. The urinary 17-ketosteroid determination measures the metabolites of the androgens including dehydroepiandrosterone, $\Delta^4$-androstenedione and testosterone. The lack of specificity of this determination has led to the development of more precise methods.

The urinary 17-hydroxycorticosteroids are metabolic products of cortisol and 11-desoxycortisol which can be measured by the Porter–Silber reaction[50]. This reaction measures the metabolic products of both cortisol and 11-desoxycortisol. Under most conditions 11-desoxycortisol is produced in such small amounts that the Porter–Silber reaction is a measure of cortisol production. This is not true in 11-hydroxylase deficiency, the hypertensive form of congenital adrenal hyperplasia. In this disorder there is an increased production of 11-desoxycortisol which is measured by the Porter–Silber reaction – a useful index of cortisol production.

Urinary pregnanetriol, the metabolic end product of 17-hy-droxyprogesterone has traditionally been measured by a colorimetric method[51,52]. Urinary testosterone may also be measured but this determination is not necessary to the diagnosis. The $3\beta$-hydroxysteroid dehydrogenase deficiency can readily be diagnosed by determining increased urinary levels of $\Delta^5$-steroids such as pregnenolone and dehydroepiandrosterone. These end products will be found in combination with pregnenetriol, a metabolite of 17-hydroxypregnenolone, and pregnenediol, a metabolite of pregnenolone.

Steroid secretion rates can be measured using laboratory techniques, such as the double isotope dilution derivative method for the simultaneous determination of the secretion rates of cortisol, 11-desoxycortisol, corticosterone, 11-desoxycorticosterone, and aldosterone[53]. Although these methods are somewhat tedious and restricted to specialized laboratories, the precise information obtained is invaluable in pinpointing the defect and directing therapy.

In the newborn, the diagnosis of 21-hydroxylase deficiency can be made by measuring serum 17-hydroxyprogesterone on the 3rd day of life. An exciting new development in the laboratory diagnosis of congenital adrenal hyperplasia has been the application of a microfilter paper method for 17-hydroxyprogesterone radioimmunoassay as a rapid screening test for congenital adrenal hyperplasia[54]. This method requires only 20 $\mu$l of blood which can be dried and stored on the filter paper for up to 21 days and allows easy transport of samples to the appropriate laboratory for testing. Cord blood levels of 17-hydroxyprogesterone are usually sufficiently elevated to allow diagnosis of the 21-hydroxylase deficiency at birth[55].

It has been shown that the normal 17-hydroxyprogesterone levels in the normal newborn are 1.2 ng/ml, between day 2 and 7[56]. It is important to allow 36 hours of time in the independent environment before making these determinations[56]. The levels of 17-hydroxyprogesterone in the CAH child are significantly higher[56]. The 17-hydroxyprogesterone level for a diverse group of newborns with CAH and a matched group are greater than 395 ng/ml and less than 1.1 ng/ml[54,57].

Readily available testing should help define the true incidence of congenital adrenal hyperplasia, which may be more common than phenylketonuria for which screening tests are mandated.

Other useful tests for the evaluation of an infant with ambiguous genitalia are the buccal smear for sex chromatin, karyotyping and a genitogram. This last procedure involves placing an occlusive tip catheter at the external urethral meatus and injecting contrast material in a retrograde fashion; this will show the urethra and a distinct vagina with a cervical 'dimple' in the vault in females with ambiguous genitalia attributable to *in utero* virilization.

Attempts have been made to diagnose congenital adrenal hyperplasia prenatally. If this could be done with suitable accuracy, prenatal therapy might be possible. To date, the accuracy of prenatal diagnosis is not great enough to allow this method to be recommended unequivocally[58,59].

Jeffcoate et al.[60] have reported elevated 17-ketosteroid and pregnanetriol levels in amniotic fluid in cases in which the child was born with congenital adrenal hyperplasia. These findings were not substantiated by Merkatz et al.[61]. Cathro et al.[62] attempt to establish the prenatal diagnosis by means of oestriol determinations. This work could not be confirmed by Nichols and Gibson[63], but they did find increased pregnanetriol levels in pregnancies in which the fetus was proven to have congenital adrenal hyperplasia. It has been suggested that determinations of the 16-hydroxylated metabolites of progesterone might be useful in the prenatal diagnosis of congenital adrenal hyperplasia[46]. Recently, the prenatal diagnosis of CAH has been made by measuring 17-hydroxyprogesterone in amniotic fluid and by HLA typing of amniotic cells[58,64].

This development has been confirmed and additionally substantiated by the investigations of Mouses et al.[65]. These latter authors also confirmed the usefulness of HLA typing for prenatal diagnosis of congenital adrenal hyperplasia.

## GENETICS

Congenital adrenal hyperplasia is established as an inborn error of metabolism transmitted by an autosomal recessive gene[66–68]. Males are affected as frequently as females[69]. The severity of the specific enzyme defect tends to be similar in a given family[23]. The salt-losing variety of 21-hydroxylase deficiency occurs in approximately 30–80 % of these patients[70,71]. A very high incidence of this salt-wasting variant has been found among the Yupik Eskimos in Alaska[72]. Historically, there is a considerable controversy concerning the genetic basis of the 21-hydroxylase deficiency. Bongiovanni and Eberlein[73] postulated that the same enzyme catalyses the 21-hydroxylation of 17-hydroxyprogesterone and progesterone. In contrast, Bryan and co-workers[74] propose a two-enzyme theory to account for the salt-wasting and non-salt-wasting forms of congenital adrenal hyperplasia. Levine and associates[75] have shown that the gene for congenital adrenal hyperplasia of the 21-hydroxylase deficiency type is located very close to the HLA-B locus. They also showed, by genetic recombination, that this gene can be separated from the HLA-A locus and from the locus for glyoxalase I-polymorphism[75,76]. This was done in a group of 34 unrelated families from New York and Zurich with a total of 48 patients, 48 siblings, and their parents. These studies established that no HLA-A, B or C antigens were found to be selectively increased in these subjects, and thus showed a random association between alleles of the HLA loci and the 21-hydroxylase locus[75,76]. These findings support the concept that 21-hydroxylase deficiency occurs as the result of multiple, unrelated mutations that are relatively common[75–78]. The genetic relationship is different from those in complement $C_2$ deficiency and in idiopathic haemachromatosis in which genetic linkage to HLA with genetic linkage disequilibrium is found[79,80]. It is of interest that other genitourinary anomalies, specifically vesicoureteral reflux and congenital ureteropelvic junction obstruction, have a relationship to the HLA locus[81] although the relationship is not so clearly established as in the case of the 21-hydroxylase deficiency. There was no difference in HLA antigens among patients with and without salt-wasting crises in the group of 34 unrelated patients studied.

The genetic linkage between HLA and congenital adrenal hyperplasia attributable to 21-hydroxylase deficiency provides a means to identify siblings of patients with congenital adrenal hyperplasia attributable to 21-hydroxylase deficiency who are heterozygous carriers and siblings who are homozygous normal on this locus[75]. Such family studies are identifying previously undiagnosed siblings. Price et al.[82] have reported the detection of a previously undiagnosed patient with 21-hydroxylase deficiency by means of HLA genotyping.

The identification of siblings who are heterozygous carriers or homozygous normal for the gene for 21-hydroxylase deficiency has been established biochemically by Lorenzen et al.[83]. They have shown that the rise in 17-hydroxyprogesterone in response to ACTH administration in obligate heterozygous parents and in prepubertal and early pubertal siblings hetero-zygous for congenital adrenal hyperplasia by HLA typing was higher than in the control population and in the homozygous unaffected siblings. This finding was not true in postmenarchal females in whom overlap in the hormonal response prevented clear differentiation of the heterozygous female from the control population[84]. In contrast, Grosse-Welde et al.[85] were able to differentiate heterozygous females more accurately using dexamethasone before cortrosyn administration.

'Acquired' adrenal hyperplasia with 21-hydroxylase deficiency presents with virilization and menstrual disturbances in later childhood or early adulthood. New et al.[86] recently showed that this disorder is genetically different from congenital adrenal hyperplasia. The genetic locus of the acquired form has no demonstrable relationship to the HLA locus[86].

In a similar manner, it has been established that the gene responsible for 11-hydroxylase deficiency is not linked to the HLA locus[87].

A recent study of 124 families in Italy has determined eight pedigrees, with 16 pubertal or postpubertal family members of either sex who had bio-chemical evidence of 21-hydroxylase deficiency, but who were without clinical symptoms of excess virilism, amenorrhoea or infertility. These family members were designated as having cryptic 21-hydroxylase deficiency. Within each generation, the family members with cryptic 21-hydroxylase deficiency were HLA identical. These findings support the concept of heterogeneity of 21-hydroxylase deficiency as a result of allelic variability at the locus for steroid 21-hydroxylase. The hormonal profile of these clinically asympto-matic family members of classical patients with CAH is similar to the clinically symptomatic patients with late onset adrenal hyperplasia, but differs from that of patients with classical congenital adrenal hyperplasia in the ratio of dehydroepiandrostenedione/$\Delta^4$ androstenedione[87].

## THERAPY

The natural history of untreated virilizing congenital adrenal hyperplasia is progressive virilization, advanced bone age, and tall stature with ultimately short stature because of premature closure of the epiphyses. In females, the in utero virilization results in the characteristically enlarged clitoris with fusion of the labioscrotal folds and a urogenital sinus (Figure 11.2).

The aims of therapy are to provide replacement of the deficient hormones. In the case of the most common 21-hydroxylase form of congenital adrenal hyperplasia, this means replacing cortisol, both to remedy the deficiency in excretion and to suppress corticotropin overproduction. Proper replacement will prevent excessive stimulation of the androgen pathway, avert the process of virilization, retard the advancing bone age and allow a resumption of normal growth. Historically, the adequacy of therapy has been measured by clinical observation and the determination of urinary 17-ketosteroid and pregnanetriol levels[88,89]. The discovery of new radioimmunoassays to determine adrenal steroids has led to attempts to improve and simplify the monitoring of these patients. The determination of 17-hydroxyprogesterone levels requires standardization of the collection times and a broadening of the

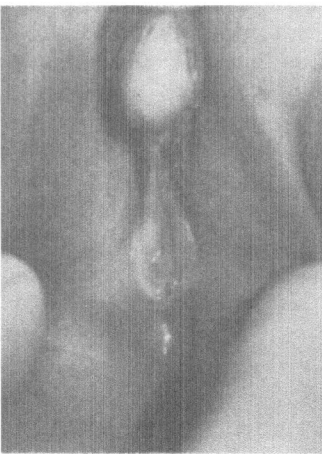

**Figure 11.2** A 2-year-old girl with CAH presenting with the typically enlarged clitoris, fusion of the labioscrotal folds and a urogenital sinus

acceptable range of values[90,91]. Testosterone and 17$\beta$-oestradiol were also studied and felt to be unreliable indicators[90]. Korth-Schutz and associates[92] carried out longitudinal studies of testosterone, $\Delta$-androstenedione, dehydroepiandrosterone, and urinary 17-ketosteroids in a group of 19 girls and 17 boys with congenital adrenal hyperplasia. They used celite chromatography for separation of the serum steroids[90]. They found that, compared to the normal adrenal gland, the hyperplastic adrenal gland in congenital adrenal hyperplasia secretes more $\Delta^4$-androstenedione than other androgens measured. Hence the $\Delta^4$-androstenedione determination is the most sensitive serum indicator of the adequacy of adrenal suppression in congenital adrenal hyperplasia. These findings are in accord with those of Horton and Frasier[93] and Rivarola *et al.*[94].

In congenital adrenal hyperplasia, serum testosterone is derived predominantly from peripheral conversion of $\Delta^4$-androstenedione. Thus, testosterone is a good index of therapeutic control in the prepubertal child and in pubertal females in whom the major source of testosterone is $\Delta^4$-androstenedione from the adrenal[95]. However, in the pubertal male, gonadal secretion of testosterone is too great to permit this steroid to be a useful guide of therapeutic control[89]. Korth-Schutz and associates[92] found testosterone in the low normal range in a group of poorly controlled pubertal males with congenital adrenal hyperplasia. These findings are in agreement with those of Rivarola *et al.*[94] and Prader *et al.*[96].

Dehydroepiandrosterone is not an accurate index of therapeutic control; it was low in all treated patients whether in good or poor control[91]. McKenna *et al.*[97], in contrast, found elevated levels of these metabolites of dehydroepiandrosterone in treated patients with congenital adrenal hyperplasia who were out of control. These differences may have been methodological. They also found that plasma levels of pregnenolone and 17-hydroxypregnenolone were lower than normal in treated patients with congenital adrenal hyperplasia[97].

The above considerations suggest that $\Delta^4$-androstenedione in addition to the traditional urinary 17-ketosteroid excretion is a valuable index of therapeutic control in the treated patient with congenital adrenal hyperplasia[98].

Cortisol replacement is useful therapy in $11\beta$-hydroxylase deficiency as well, because it suppresses corticotropin production and thus prevents excessive accumulation of androgens and desoxycorticosterone. This suppression prevents progressive virilization and hypertension[29].

Patients with the salt-wasting forms of congenital adrenal hyperplasia attributable to severe 21-hydroxylase deficiency, 3$\beta$-OL-dehydrogenase deficiency or lipoid adrenal hyperplasia require replacement with both cortisol and either desoxycorticosterone or 9$\alpha$-fluorohydrocortisone and dietary salt. The efficacy of this therapy in the severe 21-hydroxylase deficiency can be followed by the same parameters as in the mild deficiencies. In patients with the 3$\beta$-OL-dehydrogenase deficiency and lipoid adrenal hyperplasia both cortisol and either desoxycorticosterone or 9$\alpha$-fluorohydrocortisone are required. Measurement of plasma renin activity provides a good index of the sodium balance[99]. The 17-hydroxylase deficiency also requires cortisol to suppress the corticotropin and desoxycorticosterone. In forms of congenital adrenal hyperplasia in which gonadal hormones are decreased (cholesterol desmolase, 3$\beta$-hydroxysteroid dehydrogenase and 17-hydroxylase) replacement therapy with sex steroids is also necessary at puberty.

## SURGERY

In the 21- and 11-hydroxylase deficiencies, surgical therapy is restricted to female children. The aim of surgical repair should be to remove the redundant erectile tissue, preserve the sexually sensitive glans clitoris and provide an exteriorized vagina that will function adequately for menstruation and intromission. The preservation of the glans clitoris is believed to be important to allow full physical enjoyment of normal sexual activity. From the psychosexual point of view, patients who have had clitorectomy show ambivalence toward sexual activity and sexual inhibition[100,101].

Vaginoplasty is almost always required in these girls. Before any rational surgical plan can be evolved, the precise location of the vaginal orifice in relation to the urethra and common urogenital sinus must be established. This anatomic localization can be accomplished by the use of voiding cystourethrogram or retrograde urethrogram. It must be emphasized that a negative radiographic examination does not exclude the presence of a vagina. Endoscopy is an additional aid to the precise localization of the vagina. This procedure is easily and safely accomplished in these young girls if adequate precautions are taken to supplement their replacement therapy. If the vagina is quite close to the external meatus, which is to say the urogenital sinus is short, then a simple 'cutback' procedure will suffice nicely[102-104]. The most commonly used vaginal reconstructive procedure is a vaginoplasty. In this repair a flap of perineal skin is brought down and in to form part of the floor of the vaginal canal. This manoeuvre permits repair of the vagina when the urogenital sinus is longer than the cutback procedure would allow. The flap technique is easy enough to perform and yields good cosmetic and functional results. The procedure can be undertaken in early childhood, at the time of clitoroplasty. If this is done, there is rarely a need for later revision[102,105].

In the rare instances in which the vagina is deeply recessed, i.e. a long urogenital sinus, a pull-through vaginoplasty must be performed. This more extensive dissection requires mobilization of the deeply recessed vagina with care being taken in the area of the bladder neck and internal sphincter mechanism. Skin flaps are developed in the perineum as well as superiorly and at each lateral aspect in order to provide adequate length to the vaginal canal. This procedure is arduous and tedious, but can also be undertaken between the ages of 1 and 2 years[102].

Historically, clitorectomy has been the treatment offered for the enlarged clitoris of the adrenogenital syndrome[106]. As has been discussed above, excision gives an unsatisfactory functional result. It has been stated that the cosmetic result of clitorectomy is acceptable and desirable, but this position seems at variance with the facts, considering normal female genital anatomy. It is conceivable to encounter cases in which the clitoris is so large as to make clitorectomy a desirable solution. To date, the senior author has not encountered any such children.

One proposed alternative is clitoral recession, which is advocated for girls with a small clitoris. This operation buries the erectile tissue of the clitoris at the mons pubis and leaves the glans clitoris exposed to function as a sexual organ[102,107]. Our only experience with the procedure has been the subsequent need to revise two clitoral recessions that were excruciatingly painful with erection during normal sexual excitation. The cosmetic results are such that there is a visible prominence at the mons pubis. In a similar manner, clitoral plication has been proposed[108] in an apparent attempt to avoid the pubic bulging that is cosmetically unattractive.

The method of dealing with the enlarged clitoris of the adrenogenital syndrome that has yielded the best functional and cosmetic results has been corporal resection[109-112]. In this procedure, the erectile tissue of the clitoris is removed, but the glans is preserved. Along with the glans, a strip of ventral skin with a blood supply is preserved when possible. The dorsal neurovascular bundle is retained as well. In cases in which the glans is large, the corners of the glans can be trimmed off, leaving a functionally intact and cosmetically satisfactory situation. The skin over the glans will often partially or completely slough in 7–10 days; however the glans will re-epithelialize and be quite satisfactory. Two of our patients operated by this method were 17 and 20 years of age and had been sexually active before surgery. They reported similar pleasant sensations postoperatively, without the obviously embarrassing clitoral erection.

## PROGNOSIS

In the common forms of congenital adrenal hyperplasia, the 21-hydroxylase and 11$\beta$-hydroxylase deficiencies, the prognosis with early diagnosis and therapy is good. If suitable replacement therapy is given, growth and development are normal. These patients have the potential to be fertile. The cosmetic and functional results of properly planned surgery are good. There has not been enough experience with the less common forms of adrenal hyperplasia to allow prognostic statements.

## References

1 Childs, B., Grumbach, M. M. and Van Wyck, J. J. (1956). Virilizing adrenal hyperplasia: A genetic and hormonal study. *J. Clin. Invest.*, **35**, 213
2 Prader, A. and Gurtner, H. P. (1955). Das Syndrome des Pesudohermaphroditismus masculinus bei kongenitaler Hyperplasie ohne Androgenbeiproduktion (adrenaler Pseudo-hermaphroditismus masculinus). *Helv. Paediatr. Acta*, **10**, 397
3 Godard, C., Riondel, A. M., Veyrat, R., Megevand, A. and Muller, A. F. (1971). Plasma renin activity and aldosterone secretion in congenital adrenal hyperplasia. *Pediatrics*, **41**, 883
4 Imai, M. Y., Igarashi, Y. and Sokaba, H. (1968). Plasma renin activity in congenital virilizing adrenal hyperplasia. *Pediatrics*, **41**, 897
5 Simpoulas, A. P., Marshall, J. R., Delea, C. S. and Bartter, F. C. (1971). Studies on the deficiency of 21 hydroxylation in patients with congenital adrenal hyperplasia. *J. Clin. Endocrinol. Metab.*, **32**, 438
6 Ceinger, L. E., Zapata, G. C., Ely, R. S. and Kelley, V. C. (1958). Female pseudohermaphro-ditism with penile urethra: report of unusual case of congenital adrenal hyperplasia. *Am. J. Dis. Child.*, **95**, 410
7 Bentinck, R. C., Lisser, H. and Reilly, W. A. (1956). Female pseudohermaphroditism with penile urethra, masquerading as precocious puberty and cryptorchidism: Case report. *J. Clin. Endocrinol. Metab.*, **16**, 412
8 Jeune, M. and Bertrand, J. (1959). Pseudo hermaphrodisme féminin avec virilisation total (Urètre pénien) par hyperplasie surrénale: à propos d'un cas. *Sem. Hôp. Paris*, **35**, 2131
9 Jirasek, J. E. (1971). *Development of the Genital System and Male Pseudohermaphroditism.*, (Baltimore and London: Johns Hopkins UP)
10 Matheson, W. J. and Ward, E. M. (1954). Hormone sex reversal in females. *Arch. Dis. Child.*, **29**, 22
11 Maxted, W., Baker, R., McCrystal, H. and Fitzgerald, E. (1965). Complete masculinization of the external genitalia in congenital adrenal hyperplasia. Presentation of two cases. *J. Urol.*, **94**, 266
12 Peris, L. A. (1960). Congenital adrenal hyperplasia producing female hermaphroditism with phallic urethra. *Obstet. Gynecol.*, **16**, 156
13 Perloff, W. H., Conger, K. B. and Levy, L. M. (1953). Female pseudohermaphroditism: description of two unusual cases. *J. Clin. Endocrinol. Metab.*, **13**, 783
14 Prader, A. (1955). Vorkommen männliche aussere Genitalentwicklung und Salzverlust-Syndrom bei Mädchen mit kongenitalem adrenogenitem Syndrom. *Helv. Paediatr. Acta*, **10**, 398
15 Reilly, W. A., Hinman, F. Jr., Pickering, E. and Crane, J. T. (1958). Phallic urethra in female pseudohermaphroditism. *Am. J. Dis. Child.*, **95**, 9
16 Hamilton, W. and Brush, M. G. (1964). Four clinical variants of congenital adrenal hyperplasia. *Arch. Dis. Child.*, **39**, 66
17 Bergstrand, C. G., Burke, G. and Plantin, L. O. (1959). Corticosteroid excretion pattern in infants and children with adrenogenital syndrome. *Acta Endocrinol.*, **30**, 500
18 Chaptal, J., Jean, R., Christol, P. and Bonnet, H. (1959). Augmentation des 17-hydroxycorticoïdes plasmatiques et urinaires dans un cas d'hyperplasie congénitale des surrénales sans hypertension artérielle: identification des stéroïdes anormaux relevant d'un déficit de la 11-hydroxylase. *Ann. Endocrinol.*, **20**, 323
19 Gandy, M. H., Keutmann, E. H. and Izzo, A. J. (1960). Characterization of urinary steroids in adrenal hyperplasia: isolation of metabolites of cortisol, compound S and desoxycorticos-terone from a normotensive patient with adrenogenital syndrome. *J. Clin. Invest.*, **39**, 364
20 Dyrenfurth, I., Sybulski, S., Notchev, V., Beck, J. C. and Venning, E. H. (1958). Urinary corticosteroid excretion patterns in patients with adrenocortical dysfunction. *J. Clin. Endocrinol. Metab.*, **18**, 391

21 Gabrilove, J. L., Sharma, D. C. and Dorfman, R. I. (1965). Adrenocortical 11-hydroxylase deficiency and virilism first manifest in the adult woman. *N. Engl. J. Med.*, **272**, 1189

22 Bongiovanni, A. M. (1962). The adrenogenital syndrome with deficiency of 3-β-hydroxysteroid dehydrogenase. *J. Clin. Invest.*, **41**, 2089

23 New, M. I. (1968). Congenital adrenal hyperplasia. *Pediatr. Clin. North Am.*, **15**, 395

24 Goldman, A. S. (1970). Virilization of the external genitalia of the female rat fetus by dehydroepiandrosterone. *Endocrinology*, **87**, 432

25 Federman, D. D. (1968). *Abnormal Sexual Development*, p. 119. (Philadelphia: Saunders)

26 Neumann, F. and Elger, W. (1966). The effect of the anti-androgen 1, methylene-6-chloro-Δ⁴,⁶pregnadiene-17-ol-3, 2 dioneacetate (cyproterone acetate) on the development of the mammary glands of male fetal rats. *J. Endocrinol.*, **36**, 347

27 Neuman, F. and Elger, W. (1966). Permanent changes in gonadal function and sexual behavior as a result of early feminization of male rats by treatment with an anti-androgenic steroid. *Endocrinologie*, **50**, 209

28 Neumann, F. and Goldman, H. S. (1970). Prevention of mammary gland defects in experimental congenital adrenal hyperplasia due to inhibition of 3-hydroxysteroid dehydrogenase in rat. *Endocrinology*, **86**, 1169

29 Rovner, D. R., Conn, J. W., Cohen, E. L., Berlinger, F. G., Kem, D. C. and Gordon, D. L. (1979). 17-Hydroxylase deficiency. A combination of hydroxylation defect and reversible blockade in aldosterone synthesis. *Acta Endocrinol.*, **90**, 490

30 Newton, M. A. and Laragh, J. H. (1968). Effects of glucocorticoid administration on aldosterone excretion and plasma renin in normal subjects, essential hypertension and in primary aldosteronism. *J. Clin. Endocrinol. Metab.*, **28**, 1014

31 Biglieri, E. G., Schambelen, M. and Seaton, P. E. Jr. (1969). Effect of adrenocorticotropin on desoxycorticosterone, corticosterone and aldosterone excretion. *J. Clin. Endocrinol. Metab.*, **29**, 1090

32 Bledsoe, T., Island, D. P. and Liddle, G. W. (1966). Studies on the mechanism through which sodium depletion increases aldosterone biosynthesis in man. *J. Clin. Invest.*, **45**, 524

33 Goldsmith, O., Solomon, D. H. and Horton, R. (1967). Hypogonadism and mineralocorticoid excess: the 17-hydroxylase deficiency syndrome. *N. Engl. J. Med.*, **277**, 673

34 Mills, I. H., Wilson, R. J., Tait, A. D. and Cooper, H. R. (1967). Steroid metabolic studies in a patient with 17-hydroxylase deficiency. *J. Endocrinol.*, **38**, 19

35 New, M. I. (1970). Male pseudohermaphroditism due to 17-hydroxylase deficiency. *J. Clin. Invest.*, **49**, 1930

36 Degenhart, H. J. (1971). A study of the cholesterol splitting enzyme system in normal adrenals and in adrenal lipoid hyperplasia. *Acta Paediatr. Scand.*, **60**, 611

37 Brutachy, P. (1921). Hochgradige Lipoid Hyperplasie beider Nebennieren mit Herdformigen Kalkalbagerungen bei einem Fall von Hypospadias penisscrotalis und doppelsitigem Kryptorchismus mit unechter akzess-orischer Nebenniere am rechten Hoden. (Pseudo hermaphroditismus masculinus externus). *Frankfürt J. Pathol.*, **24**, 203

38 Dhom, G. (1958). Zur Morphologie und Genese der kongenitalen Nebennierenrinden-Hyperplasie beim männlichen Scheiznzruitter. *Zentralbl. Allg. Pathol.*, **97**, 23

39 O'Doherty, N. N. (1964). Lipoid adrenal hyperplasia. *Guy's Hosp. Rep.*, **113**, 368

40 Prader, A. and Siebennmann, R. E. (1957). Nebennieren-insuffizienz bei kongenitaler Lipoid Hyperplasia der Nebennieren. *Helv. Paediatr. Acta*, **12**, 569

41 Sandison, A. T. (1955). A form of lipoidosis of the adrenal cortex in an infant. *Arch. Dis. Child.*, **30**, 538

42 Tilp, A. (1913). Hochgradige Verfettung der Nebennieren Sauglings. *Verh. Dtsch. Ges. Pathol.*, **16**, 305

43 Zahn, J. (1948). Über Intersexualität und Nebennieren-rindenhyperplasie. *Schweiz. Med. Wochenschr.*, **78**, 480

44 Camacho, A. M., Kowarski, A., Migeon, C. J. and Braugh, A. J. (1968). Congenital adrenal hyperplasia due to a deficiency of one of the enzymes involved in the biosynthesis of pregnenolone. *J. Clin. Endocrinol. Metab.*, **28**, 153

45 Jean, R., Legrand, J. C., Meylan, R. and Astrue, J. (1969). Hypoaldostéronisme primaire par anomalie probable de la 18-hydroxylation. *Arch. Fr. Pediatr.*, **26**, 769

46 New, M. I. and Levine, L. S. (1973). Congenital adrenal hyperplasia. In Harris, H. and Hirschhorn, K. (eds.) *Advances in Human Genetics*. (New York, London: Plenum Press)

47 Vecsei, P., Nolten, W., Purjesz, S. T. and Wolff, H. P. (1967). Studies on the secretion and metabolism of desoxycorticosterone (DOC). *Acta Endocrinol.*, **119**, 141 (Abstr.)
48 David, R., Golan, S. and Drucker, W. (1968). Familial aldosterone deficiency: enzyme defect, diagnosis and clinical course. *Pediatrics*, **41**, 403
49 Rapoport, R., Dray, F., Legrand, J. C. and Royer, P. (1968). Hyperaldostéronisme congénital familial par défaut de la 18-OH déhydrogenase. *Pediatr. Res.*, **2**, 456
50 Silber, R. H. and Porter, C. C. (1954). Determination of 17, 21-dehydroxy 20-ketosteroids in urine and plasma. *J. Biol. Chem.*, **210**, 923
51 Bongiovanni, A. M. (1953). Detection of pregnanediol and pregnanetriol in urine of patients with adrenal hyperplasia, suppression with cortisone (preliminary report). *Bull. Johns Hopkins Hosp.*, **92**, 244
52 Bongiovanni, A. M. and Clayton, G. W. (1954). Simplified method for determination of pregnanediol and pregnanetriol in urine. *Bull. Johns Hopkins Hosp.*, **94**, 180
53 New, M. I., Seaman, M. P. and Peterson, R. E. (1969). A method for the simultaneous determination of the secretion rates of cortisol, 11-desoxycortisol, corticosterone, 11-desoxycorticosterone and aldosterone. *J. Endocrinol. Metab.*, **29**, 514
54 Pang, S., Hotchkiss, J., Drash, A. L., Levine, L. S. and New, M. I. (1977). Microfilter paper method for 17-hydroxyprogesterone radioimmunoassay: Its application for rapid screening for congenital adrenal hyperplasia. *J. Clin. Endocrinol. Metab.*, **45**, 1003
55 Jenner, M. R., Grumbach, M. M. and Kaplan, S. L. (1970). Plasma 17-OH-progesterone in maternal and umbilical cord plasma in children and in congenital adrenal hyperplasia (CAH): application to neonatal diagnosis of CAH. *Program and Abstracts, Society for Pediatric Research, Annual Meeting*, Atlantic City
56 Hughes, I. A., Riad-Fahmy, D. and Griffiths, K. (1979). Plasma 17-OH progesterone concentrations in newborn infants. *Arch. Dis. Child.*, **54**, 347
57 Peterson, K. E. and Christensen, T. (1979). 17-hydroxyprogesterone in normal children and congenital adrenal hyperplasia. *Acta Paediatr. Scand.*, **68**, 205
58 New, M. I. (1977). Present status of prenatal diagnosis of congenital adrenal hyperplasia. In Lee, P. A., Plotnick, L. P., Kowarski, A. A. and Migeon, C. J. (eds.) *Congenital Adrenal Hyperplasia.* (Baltimore: University Park Press)
59 New, M. I. (1972). Adrenogenital syndrome. In Dortman, A. (ed.) *Antenatal Diagnosis.* (Chicago: University of Chicago Press)
60 Jeffcoate, T. N. A., Fliegners, J. R. N., Russell, S. H., Davis, J. C. and Wade, A. P. (1965). Diagnosis of the adrenogenital syndrome. *Lancet*, **2**, 553
61 Merkatz, I. R., New, M. I., Peterson, R. E. and Seaman, M. P. (1969). Prenatal diagnosis of adrenogenital syndrome by amniocentesis. *J. Pediatr.*, **75**, 977
62 Cathro, D. M., Bertrand, J. and Coyle, M. G. (1969). Antenatal diagnosis of adrenocortical hyperplasia. *Lancet*, **1**, 732
63 Nichols, J. and Gibson, G. G. (1969). Antenatal diagnosis of the androgenital syndrome. *Lancet*, **2**, 1068
64 Duchon, M., Owens, R. P., Merkatz, I. R., Nitowsky, H. W., Sachs, G. and Dupont, B. (1979). Prenatal diagnosis of congenital adrenal hyperplasia (21-hydroxylase deficiency) by HLA typing. *Lancet*, **1**, 1107
65 Mouses, E. S., Holcombe, J. H., Tulchansky, D., Rich, R. R. and Riccardi, V. M. (1979). Prenatal diagnosis of congenital adrenal hyperplasia. *Am. J. Hum. Gen.*, **4**, 201
66 McKusick, V. A. (1975). *Mendelian Inheritance in Man.* 4th edn. (Baltimore: Johns Hopkins UP)
67 Knudson, A. G. Jr. (1957). Mixed adrenal disease of infancy. *J. Pediatr.*, **39**, 408
68 Bentinck, R. C., Hinman, F. Sr., Lisser, H. and Traut, H. F. (1952). The familial congenital adrenogenital syndrome: report of two cases and review of the literature. *Postgrad. Med.*, **11**, 301
69 Baulieu, E. E., Peillow, F. and Migeon, C. J. (1967). Adrenogenital syndrome. In Eisenstein, A. B. (ed.) *The Adrenal Cortex.* (Boston: Little, Brown)
70 Cohen, J. M. (1969). Salt-losing congenital adrenal hyperplasia. *Pediatrics*, **44**, 621
71 Rimion, D. and Schimke, R. N. (1971). *Genetic Disorders of Endocrine Glands.* (St. Louis: Mosby)
72 Hirschfeld, A. J. and Fleshman, J. K. (1969). An unusually high incidence of salt-losing congenital adrenal hyperplasia in the Alaskan Eskimo. *J. Pediatr.*, **75**, 492

73 Bongiovanni, A. M. and Eberlein, W. R. (1958). Defective steroidal biogenesis in congenital adrenal hyperplasia. *Pediatrics*, **21**, 661

74 Bryan, G. K., Kliman, D. and Bartter, F. C. (1965). Aldosterone production in salt losing congenital adrenal hyperplasia. *J. Clin. Invest.*, **44**, 957

75 Levine, L. S., Zachman, M., New, M. I., Prader, A., Pollack, M. S., O'Neill, G. J., Yang, S. Y., Oberfield, S. E. and Dupont, S. (1978). Genetic mapping of the 21-hydroxylase deficiency gene within the HLA linkage group. *N. Engl. J. Med.*, **229**, 911

76 New, M. I., Dupont, B., Pang, S., Pollack, M. S. and Levine, L. S. (1981). An update of congenital adrenal hyperplasia. *Prog. Horm. Res.*, **37**, 105

77 Yang, S., Levine, L. S., Zachman, M., New, M. I., Prader, A., Oberfield, S. E., O'Neill, G. J., Pollack, M. S. and Dupont, B. (1978). Mapping of the 21-hydroxylase deficiency gene within the HLA linkage group. *Transplant Proc.*, **10**, 753

78 Dupont, B., Oberfield, S. E., Smithwick, E. M., Lee, T. D. and Levine, L. S. (1977). Close genetic linkage between HLA and congenital adrenal hyperplasia (21-hydroxylase deficiency). *Lancet*, **2**, 1309

79 Weitkamp, L. R., Bryson, M. and Bacon, T. E. (1978). HLA and congenital hyperplasia linkage confirmed. *Lancet*, **1**, 931

80 Fu, S. M., Kunkel, H. G., Brushman, H. P., Allen, F. J. and Fotino, M. (1975). Mixed lymphocyte culture determinants and CZ deficiency: LD-7a associated with CZ deficiency in four families. *J. Exp. Med.*, **142**, 495

81 Bamford, A., Eddleston, A. L. W. F., Kennedy, L. A., Batchelor, J. R. and Williams, R. (1977). Histocompatibility antigens as markers of abnormal iron metabolism in patients with idiopathic haemochromatosis and their relatives. *Lancet*, **1**, 327

82 Sengar, D. P. S., Rashid, A. and Wolfish, N. M. (1979). Familial urinary tract anomalies: Association with the major histocompatibility complex in man. *J. Urol.*, **121**, 194

83 Price, D. A., Klonda, P. T. and Harris, R. (1978). HLA congenital adrenal hyperplasia linkage confirmed. *Lancet*, **1**, 930

84 Lorenzen, F., Pang, S., New, M. I., Dupont, B., Chow, D. and Levine, L. S. (1979). Hormonal phenotype and HLA genotype in families of patients with congenital adrenal hyperplasia (21-hydroxylase deficiency) *Pediatr. Res.* **13**, 1356

85 Brautbar, C., Levine, C., Sack, J., Bendule, A., Moses, S. and Dupont, B. (1979). No linkage between HLA and congenital adrenal hyperplasia due to 11-hydroxylase deficiency. *N. Engl. J. Med.*, **300**, 205

86 Grosse-Welde, H., Weill, J., Albert, E., Schloz, S., Bedlingmaier, F., Sippel, W. G. and Know, D. (1979). Genetic linkage studies between congenital adrenal hyperplasia and the HLA blood group system. *Immunogenetics*, **8**, 41

87 New, M. I., Lorenzen, F., Pang, S., Gunzler, P., Dupont, B. and Levine, L. S., (1979). 'Acquired' adrenal hyperplasia with 21-hydroxylase deficiency is not the same genetic disorder as congenital adrenal hyperplasia. *J. Clin. Endocrinol. Metab.*, **48**, 358

88 Levine, L. S., Dupont, B., Lorenzen, F. *et al.* (1980). Cryptic 21-hydroxylase deficiency in families of patients with classical adrenal hyperplasia. *J. Clin. Endocrinol. Metab.*, **51**, 1316

89 Bongiovanni, A. M. and Root, A. W. (1963). The adrenogenital syndome. *N. Engl. J. Med.*, **268**, 1283

90 Lippe, B. M., La Franchi, S. H., Lavin, N., Parlow, A., Cyotups, J. and Kaplan, S. A. (1974). Serum 17-hydroxyprogesterone, progesterone, estradiol, and testosterone in the diagnosis and management of congenital adrenal hyperplasia. *J. Pediatr.*, **85**, 782

91 Hughes, I. A. and Winters, J. S. D. (1976). The application of a serum 17-OH progesterone radioimmunoassay to the diagnosis and management of congenital adrenal hyperplasia. *J. Pediatr.*, **88**, 766

92 Korth-Schutz, S., Levine, L. S. and New, M. I. (1976). Serum androgens in normal prepubertal and pubertal children with precocious adrenarche. *J. Clin. Endocrinol. Metab.*, **42**, 117

93 Korth-Schutz, S., Viris, R., Saenger, P., Chow, D. M., Levine, L. S. and New, M. I. (1978). Serum androgen as a continuing index of adequacy of treatment of congenital adrenal hyperplasia. *J. Clin. Endocrinol. Metab.*, **46**, 452

94 Horton, R. and Frasier, S. D. (1967). Androstenedione and its conversion to plasma testosterone in congenital adrenal hyperplasia. *J. Clin. Invest.*, **46**, 1003
95 Rivarola, M. A., Saez, J. M. and Migeon, C. J. (1967). Studies of androgens in patients with congenital adrenal hyperplasia. *J. Clin. Endocrinol. Metab.*, **27**, 624
96 Solomon, I. L. and Scheoen, E. V. (1975). Blood testosterone values in patients with congenital virilizing adrenal hyperplasia. *J. Clin. Endocrinol. Metab.*, **40**, 355
97 Prader, A., Zachman, M. and Illig, R. (1977). Normal spermatogenesis in adult males with congenital adrenal hyperplasia after discontinuation of therapy. In Lee, P. A., Plotnick, L. P., Kowarski, A. A. and Migeon, C. J. (eds.) *Congenital Adrenal Hyperplasia.* (Baltimore: University Park Press)
98 McKenna, T. J., Jennings, A. S., Liddle, G. W. and Burr, I. M. (1976). Pregnenolone, 17-OH pregnenolone, and testosterone in plasma of patients with congenital adrenal hyperplasia. *J. Clin. Endocrinol. Metab.*, **42**, 918
99 Cavallo, A., Corn, C., Bryan, G. T. and Meyer, W. J. III (1979). The use of plasma androstenedione in monitoring therapy of patients with congenital adrenal hyperplasia. *J. Pediatr.*, **95**, 33
100 Rosler, A., Levine, L. S., Schneider, B., Novogroder, M. and New, M. I. (1977). The interrelationship of sodium balance, plasma renin activity and ACTH in congenital adrenal hyperplasia. *J. Clin. Endocrinol. Metab.*, **45**, 500
101 Money, J. and Schwartz, M. (1972). Dating, somatic friendships, and sexuality in 17 early treated adrenogenital females, aged 15–25. In Lee, P. A., Plotnick, J. P., Kowarski, P. V. and Migeon, C. J. (eds.) *Congenital Adrenal Hyperplasia.* (Baltimore: University Park Press)
102 Money, J. and Ehrhardt, A. A. (1972). *Man and Women, Boy and Girl Differentiation and dimorphism of Gender Identity from Conception to Maturity.* (Baltimore: Johns Hopkins UP)
103 Hendren, W. H. and Crawford, J. D. (1969). Adrenogenital syndrome: The anatomy of the anomaly and its repair. Some new concepts. *J. Pediatr. Surg.*, **4**, 49
104 Jones, H. W. Jr. (1960). Revision of the urogenital sinus. *Fertil. Steril.*, **11**, 157
105 Jones, H. W. Jr. and Verekauf, B. S. (1960). Surgical treatment in congenital adrenal hyperplasia. Age at operation and other prognostic factors. *Obstet. Gynecol.*, **36**, 1
106 Fortunoff, S., Lattimer, J. K. and Edson, M. (1964). Vaginoplasty-technique for female pseudohermaphrodites. *Surg. Gynecol. Obstet.*, **118**, 545
107 Gross, R. E., Randolph, J. and Crigler, J. F. Jr. (1966). Clitorectomy for sexual abnormalities: Indication and techniques. *Surgery*, **59**, 300
108 Lattimer, J. K. (1961). Relocation and recession of the enlarged clitoris with preservation of the glans: an alternative to amputation. *J. Urol.*, **86**, 113
109 Stefan, H. (1967). Surgical reconstruction of the external genitalia in female pseudohermaphrodites. *Br. J. Urol.*, **39**, 347
110 Spence, H. W. and Allen, T. D. (1973). Genital reconstruction in the female with the adrenogenital syndrome. *Br. J. Urol.*, **45**, 126
111 Barinka, L., Stauratjero, M. and Toman, M. (1968). Plastic adjustment of female genitals in adrenogenital syndrome. *Acta Chir. Plast.*, **10**, 99
112 Kumar, H., Kiefer, J. H., Rosenthal, I. E. and Clark, S. S. (1974). Clitoroplasty experience during a 19-year period. *J. Urol.*, **111**, 81

# 12
# Epidemiology of birth defects in twins

P. M. LAYDE AND L. D. EDMONDS

## INTRODUCTION

In theory, epidemiological studies of birth defects in twins are powerful tools in the elucidation of the causes of congenital malformations. Despite the clinical and public health importance of birth defects[1], virtually nothing is known of their aetiology[2]. A few environmental teratogens such as thalidomide and rubella virus have been identified. In addition, a handful of genetic causes of congenital malformations, mostly single-gene lesions, are well established. Together, these account for only a small proportion of birth defects. Not only are the specific determinants of most birth defects unknown, the relative importance of genetic and environmental causes is unclear. The appeal of twin studies is that they potentially can help separate the genetic and environmental influences on the aetiology of birth defects.

The comparison of the concordance rate for various malformations between monozygotic (MZ) and dizygotic (DZ) twin pairs should clarify the genetic contribution. All twin pairs have a relatively similar intrauterine environment. MZ twins are also identical genetically, while DZ twins are no more similar genetically than other siblings. While this approach ignores the uneven distribution of egg cytoplasm between MZ twins and localized differences in the intrauterine environment, these limitations do not totally negate the worth of concordance studies in twins. If genes play a important causal role for a given malformation, MZ twins should have a higher concordance rate than DZ twins for that malformation.

Another reason for studying birth defects in twins is that they have significantly higher rates than singletons for many malformations. Despite the infrequency of twins, they have an important impact on the overall rates of certain malformations.

This brief review focusses on the epidemiological approaches taken to studying birth defects in twins, on the difficulties with these types of studies and on some of the results of these studies.

# EPIDEMIOLOGICAL STUDIES OF BIRTH DEFECTS IN TWINS

## The problems of numbers

Because of the rarity of both twinning and individual congenital malformations, many births must be monitored to estimate the incidence of specific malformations in twins reliably. For example, Hay and Wehrung estimated that 40 000 births must be studied to detect one twin with a cleft lip and/or palate[3]. Just estimating the incidence of cleft lip and/or palate for twins in general obviously requires a very large series; looking for differences in the malformation rates for MZ and DZ twins necessitates a much larger series. The type of concordance study discussed in the introduction needs even larger numbers; there must be a sizeable number of both MZ and DZ twin pairs in which at least one twin has the malformation.

To make the comparison of malformation rates for singletons and MZ and DZ twins meaningful, equal attention must be paid to diagnosing defects in each of these groups. The anecdotal reporting of case series, the alternative to major epidemiological studies, potentially introduces major biases. Twins concordant for various defects may be overrepresented in such reports. Alternatively, MZ twins discordant for a given trait may be selectively reported because of the supposed curiosity they represent.

## The major studies

Few studies have been done with unbiased ascertainment of birth defects for a series of births large enough to satisfy the purposes of twin studies. None of the studies are perfect, but they provide the best data currently available on the epidemiology of birth defects in twins.

The National Cleft Lip and Palate Intelligence Service (NIS)[3] was a study conducted from 1961 to 1966 in the United States and was based on birth defects reported on birth certificates. From participating states, the NIS received all birth certificates recording congenital malformations and a 1 % systematic sample of birth certificates of normal infants. This study is the largest study of congenital malformations in twins yet conducted; the birth certificates studied represent over 10 million births, including almost 200 000 twins. The major drawback of the NIS is the well-known underreporting of congenital malformations on birth certificates[4].

The Collaborative Perinatal Study (CPS)[5,6] was much smaller than the NIS, but its intensive follow-up procedures, including special physical examinations of all infants, ensured excellent ascertainment of birth defects. The CPS was a prospective study of the offspring of 56 000 women identified early in pregnancy. While only 1230 twins were studied, the complete ascertainment of birth defects and the determination of zygosity by gross and microscopic placental examination coupled with the comparison of nine blood-type systems makes the study unique.

The Metropolitan Atlanta Congenital Defects Program (MACDP) is an ongoing birth defects surveillance system conducted by the Center for Disease Control[7]. Birth defects diagnosed in the approximately 25 000 babies born per year in metropolitan Atlanta are identified by several methods including hospital records, paediatric referrals and vital records. While most major congenital malformations are believed to be ascertained, no special physical examinations are performed, and some minor malformations are undoubtedly missed. The report on congenital malformations in twins from the MACDP was based on over 200 000 births during the period 1969–1976 and included 4490 twins.

## Zygosity estimation

As mentioned above, zygosity of twins was directly determined only in the CPS. The NIS and MACDP estimated the number of MZ and DZ twins by the Weinberg formula[8]. All twin pairs were classified as same-sexed or opposite-sexed. All opposite-sexed twin pairs are, of course, DZ. Since DZ twin pairs should be equally likely to be of the same or opposite sex, an equal number of same-sexed and opposite-sexed twin pairs were assumed to be DZ. The other same-sexed twin pairs were considered MZ. Similarly, for the presentation of results that follows, the incidence of congenital malformations for DZ twins is assumed to be the same whether they are same-sexed or opposite-sexed. Accordingly, any differences in the incidence of a malformation between the same- and opposite-sexed twin pairs is assumed to be due to a different rate for those same-sexed twins who are monozygotic. James[9] recently suggested that the assumption underlying the Weinberg method – that DZ twins are equally likely to be same-sexed or opposite-sexed – may not be entirely valid. If his belief that slightly more than 50 % of DZ twins are of the same sex is true, the calculations in this paper based on Weinberg's formula underestimate the true differences in malformation rates and concordance rates between MZ and DZ twins.

## THE INCIDENCE OF BIRTH DEFECTS IN TWINS

Despite the differences in the definition and ascertainment of birth defects, the three studies show a similar pattern in the relative proportion of singletons and DZ and MZ twins with congenital malformations (Table 12.1). MZ twins are at an appreciably higher risk than singletons of being malformed, while the risk for DZ twins is virtually identical to that for singletons.

Data on specific categories of defects support the notion that DZ twins are

Table 12.1  Incidence* of infants with congenital malformations† – singletons and MZ and DZ twins

| Study | Incidence | | |
|-------|-----------|---------|---------|
|       | Singletons | DZ twins | MZ twins |
| CPS   | 71.0(1.0) | 77.8(1.1) | 107.2(1.5) |
| NIS   | 5.76(1.0) | 5.48(1.0) | 7.75(1.3) |
| MACDP | 33.47(1.0) | 31.64(0.9) | 61.84(1.8) |

* Per 1000 infants. Relative risk compared to singletons in parentheses
† Major malformations only

similar to singletons, while MZ twins are in some way different. The NIS data (Table 12.2) show that DZ twins have rates very close to those for singletons for most malformations. MZ twins have higher rates than singletons for cleft lip, anencephaly, hydrocephaly, congenital heart disease, clubfoot, and reduction deformities. Down syndrome less frequently affected MZ twins than singletons. Because MACDP was a smaller study, its zygosity-specific rates for individual defects are rather unstable (Table 12.3). Overall, however, the pattern is similar; DZ twins have rates similar to those for singletons for most defects, while the rates for MZ twins are more discrepant from those for singletons (generally higher). Unfortunately, the CPS data are too sparse to permit calculation of zygosity-specific rates for individual malformations.

The higher rates of birth defects in MZ twins, coupled with data from animal experiments showing that certain teratogens produce MZ twinning as

**Table 12.2** Selected congenital malformations in twins by zygosity. Incidence and relative risk compared to singletons. NIS birth certificate data*

| Defect | Singletons | DZ twins | MZ twins |
|---|---|---|---|
| Cleft lip | 29.5(1.0) | 26.1(0.9) | 44.7(1.5) |
| Cleft lip and palate | 47.5(1.0) | 39.9(0.8) | 43.2(0.9) |
| Cleft palate | 34.0(1.0) | 35.3(1.0) | 30.3(0.9) |
| Anencephaly | 23.2(1.0) | 21.5(0.9) | 66.4(2.9) |
| Spina bifida | 62.5(1.0) | 46.0(0.7) | 76.5(1.2) |
| Hydrocephaly | 30.2(1.0) | 30.7(1.0) | 57.7(1.9) |
| Congenital heart disease | 58.9(1.0) | 49.1(0.8) | 111.9(1.9) |
| Clubfoot | 127.3(1.0) | 150.3(1.2) | 176.1(1.4) |
| Polydactyly | 89.3(1.0) | 87.4(1.0) | 99.6(1.1) |
| Reduction deformities | 30.2(1.0) | 26.1(0.9) | 41.8(1.4) |
| Down syndrome | 43.5(1.0) | 35.3(0.8) | 14.3(0.3) |

\* Rates per 100 000 births. Relative risks compared to singletons in parentheses. Zygosity-specific rates estimated (see text for details)

**Table 12.3** Selected congenital malformations in twins by zygosity. Incidence and relative risk compared to singletons. MACDP data*

| Defect | Singletons | DZ twins | MZ twins |
|---|---|---|---|
| Cleft Lip + / − Cleft palate | 10.3(1.0) | 25.0(2.4) | 12.1(1.2) |
| Cleft palate | 6.1(1.0) | 0.0(0.0) | 15.7(2.6) |
| Anencephaly | 7.3(1.0) | 26.8(3.7) | 24.4(3.3) |
| Spina bifida | 11.7(1.0) | 9.0(0.8) | 0.0(0.0) |
| Encephalocoele | 4.0(1.0) | 0.0(0.0) | 15.7(3.9) |
| Hydrocephaly | 5.8(1.0) | 40.2(6.9) | 0.0(0.0) |
| Tetralogy of Fallot | 1.9(1.0) | 0.0(0.0) | 15.7(8.3) |
| Ventricular septal defect | 15.0(1.0) | 26.8(1.8) | 16.5(1.1) |
| Patent ductus arteriosus | 19.6(1.0) | 26.8(1.4) | 298.9(15.2) |
| Clubfoot | 42.4(1.0) | 30.4(0.7) | 10.3(0.2) |
| Polydactyly | 45.7(1.0) | 10.7(0.2) | 114.8(2.5) |
| Down syndrome | 10.1(1.0) | 4.4(0.4) | 0.0(0.0) |

\* Rates per 10 000 births. Relative risks compared to singletons in parentheses. Zygosity-specific rates estimated (see text for details)

well as other malformation[5,6], suggest that MZ twinning may itself be considered a type of congenital malformation[6].

## CONJOINED TWINS

Partly because of the high rate of congenital malformations in MZ twins, we studied the rate of birth defects in conjoined twins. Conjoined twins are believed to be related to MZ twins, except that the fertilized ovum separates a few days later in gestation – too late for the embryos to become completely autonomous[10].

Conjoined twins appear to have a very high incidence of a variety of congenital malformations. Many of the defects, such as omphalocoele for omphalopagus twins, are obviously related to the mechanical disruption of conjoining. Other defects such as imperforate anus may be due to the disruption of normal migration of the embryonic tissues as a direct result of conjoining. Recent work suggests that a large proportion of conjoined twins also have malformations not obviously related to the site of conjoining.

In a study of 81 sets of conjoined twins identified in the United States through the nationwide birth defects monitoring programme (BDMP)[11], the high rates that were found for several defects appeared to be related to a generalized failure of embryonic migration and fusion in the conjoined twins. Among the 81 sets of twins, the most common defects not obviously associated with fusion were neural tube defects (6), orofacial clefts (6), imperforate anus (6), diaphragmatic hernia (6), congenital heart disease (6), cryptorchidism (3) and ambiguous genitalia (3).

## CONCORDANCE RATES

As already discussed, small numbers are an even bigger problem in studying differential concordance rates between MZ and DZ twins. In the CPS, in 34 % of the MZ twin pairs in which at least one twin had any malformation, both twins had some malformation. The corresponding figure for DZ twin pairs was 7 %. Similar differences in the MZ and DZ concordance rates were found in CPS for their other broad categories of major, minor, major and minor, and single or multiple malformations.

We pooled the available data on concordance rates for MZ and DZ twins from the NIS, CPS, and MACDP (Table 12.4). In all defect categories concordance rates are higher for MZ than in DZ twins. This is consistent with the belief that all the defects have at least some genetic component in their aetiologies. If genetic factors were all that were necessary to cause malformations, however, we might intuitively expect that virtually all MZ twins pairs would be concordant. Smith showed, however, that if the risk for the disease in the general population is low (as it is for all individual birth defects) and if disease causation is multifactorial, then only diseases with very high heritabilities will produce even moderately high concordance rates in MZ twins[12]. Estimates of heritability can be made from MZ concordance rates, from DZ concordance rates (or sibling concordance rates in general), and,

**Table 12.4  Estimated concordance rates for selected malformations by zygosity – pooled NIS, CPS and MACDP data**

| Defect | % Concordance* | |
|---|---|---|
| | MZ | DZ |
| Cleft lip + / – cleft palate | 15.8(57) | 2.3(87) |
| Cleft palate | 33.3(18) | 4.5(44) |
| CNS malformations† | 5.4(112) | 1.8(110) |
| Congenital heart disease† | 7.2(69) | 3.3(60) |
| Clubfoot | 22.5(111) | 2.9(208) |
| Polydactyly | 48.2(56) | 5.3(114) |
| Reduction deformities† | 7.4(27) | 0.0(32) |
| Down syndrome | 80.0(5) | 0.0(47) |
| Hypospadias‡ | 16.7(12) | 0.0(4) |
| Patent ductus arteriosus‡ | 35.0(20) | 0.0(1) |
| Haemangioma | 6.3(16) | 0.0(11) |

\* Total number of pairs with at least one infant affected in parentheses
† Based on NIS data only
‡ Based on MACDP and CPS data only

most importantly, from the difference in MZ and DZ concordance rates. For example, using the formula provided by Smith, we made the following estimates of heritability of clubfoot from the pooled data:

| Based on | Concordance rate | Heritability |
|---|---|---|
| MZ | 22.5% | 0.88 |
| DZ | 2.9% | 0.87 |
| The MZ/DZ difference | – | 0.88 |

In this case the three estimates are in close agreement. Where they disagree, the heritability estimate from the MZ–DZ difference is preferable since the other estimates ignore the possibility that twins share common environmental factors. The close agreement of the three estimates of heritability for clubfoot suggests that environmental factors play a small role in the aetiology of this malformation.

Concordance data for MZ and DZ twins should be interpreted cautiously, since inferences from them are based on the assumption that all twins, whether MZ or DZ, share equally similar intrauterine environments. In fact, while all DZ twins have separate amnions and chorions, most pairs of MZ twins share at least a common chorion[13]. Consequently, most MZ twin pairs may share a more similar intrauterine environment than do DZ twin pairs.

### Acknowledgements

The authors thank J. David Erickson, DDS, PhD, Harriet Aufdemorte and Angela Carollo for their contributions to this manuscript.

# References

1 Flynt, J. W. (1973). Extent of the birth defects problem. *Pediatr. Ann.*, October, pp. 10–16

2 Warkany, J. (1971). *Congenital Malformations: Notes and Comments*, pp. 55–129. (Chicago: Year Book Medical)

3 Hay, S. and Wehrung, D.A. (1970). Congenital malformations in twins. *Am. J. Hum. Genet.*, **22**, 662

4 Mackespang, M., Hay, S. and Lundi, A. S. (1972). Completeness and accuracy of reporting of malformations on birth certificates. *Health Services Rep.*, **87**, 43

5 Myrianthopoulos, N. C. (1976). Congenital malformations in twins: *Acta Genet. Med. Gemellol.*, **25**, 331

6 Myrianthopoulos, N. C. (1975). Congenital malformations in twins: epidemiologic survey. *Birth Defects. Orig. Art. Ser.*, **11**(8), 1

7 Layde, P. M., Erickson, J. D., Falek, A. and McCarthy, B. J. (1980). Congenital malformations in twins. *Am. J. Hum. Genet.*, **32**, 69

8 Bulmer, M. G. (1970). *The Biology of Twinning in Man*, pp. 68–74. (Oxford: Clarendon Press)

9 James, W. H. (1971). Excess of like sexed pairs of dizygotic twins. *Nature (London)*, **232**, 277

10 Benirschke, K. and Kim, C. K. (1973). Multiple pregnancy. *N. Engl. J. Med.*, **288**, 1276, 1329

11 Edmonds, L. D. and Layde, P. M. (1980) Conjoined twins in the United States, 1970–1977. *Teratology.* (In press)

12 Smith, C. (1970). Heritability of liability and concordance in monozygous twins. *Ann. Hum. Genet.*, **34**, 85

13 Edwards, J. H. (1968). The value of twins in genetic studies. *Proc. R. Soc. Med.*, **61**, 227

# 13
# Prenatal diagnosis of genetic diseases by amniocentesis

NANCY E. SIMPSON

## INTRODUCTION

Prenatal diagnosis is now possible for a considerable number of genetic diseases and/or birth defects using a variety of techniques.

(1) The traditional technique is by *amniocentesis* in the middle of the second trimester (about 16 weeks) and from the fluid or cultured cells it is possible to:

   (a) Diagnose chromosomal defects from banded karyotypes after culturing the fetal cells from the fluid;

   (b) Diagnose open neural tube defects (NTDs) by measurement of the $\alpha_1$-fetoprotein (AFP) in conjunction with ultrasonography. Although still in the research stage, several other tests – rapidly adhering (RA) cells, concanavalin A binding and cholinesterase measurements in the amniotic fluid – also help to diagnose NTDs,

   (c) Diagnose biochemical defects when the product (or absence of product) of the disease gene (or that of a closely linked gene) can be assayed either in amniotic fluid or cells or by recombinant DNA molecules when the appropriate probes are available;

   (d) Determine sex for X-linked diseases when the biochemical defect is unknown or impossible to assay in amniotic cells or fluid.

(2) *Fetal blood sampling* is used when the gene defect cannot be detected in amniotic fluid or cells, but can be in blood. Blood can be collected from the umbilical cord or chorionic plate during visualization through a fetoscope or blind withdrawal of blood from the placenta after localization by ultrasonography.

(3) *Fetal imaging* can be accomplished by radiography, ultrasonography, amniography and fetoscopy; these methods are used in conjunction with amniocentesis in some instances, and alone in others.

By far the greatest proportion of prenatal diagnosis is done by amniocentesis and it is now accepted as an effective and safe procedure with minimal risks to mother and fetus, albeit some risk[1-7]. The demand for such service by pregnant woman is increasing exponentially in North America and some parts of Western Europe, and is straining resources in many instances, beyond reasonable limits. The hazards of amniography and fetoscopy are considerably greater than of amniocentesis and the risk and burden of the potential defect or genetic disease has to be weighed against the hazards of the procedures with even more care than for amniocentesis.

The primary purpose of prenatal diagnosis is to make parents aware that a pregnant woman is carrying a fetus destined to have a serious genetic disease or defect that will result in suffering for a child and/or a serious burden to the family. The woman then can elect to have an abortion and thereby the immediate family will be relieved of the predictable burden. Whether the practice of amniocentesis will reduce the burden of defects at the population level or to society is not so clear and is difficult to evaluate; but is a goal for which geneticists strive.

## DIAGNOSIS BY AMNIOCENTESIS

### Chromosomal

The greatest number of women receiving prenatal diagnoses by amniocentesis are those at risk for having a chromosomal abnormality[1-7]. Selection of those at risk is made by maternal age or family history of a chromosomal defect, the former being the greater proportion of the two. The major concern is that of having a child with Down syndrome (trisomy 21), a chromosomal defect resulting in a child with severe mental retardation who has an average life span of at least 40 years. Fetuses with most of the other autosomal chromosomal defects have an even more serious and less viable condition and the fetuses do not appear to reach term anything like as often as do those with trisomy 21. If infants with the other chromosome abnormalities live, they do so for only a few months and on rare occasions, a few years. Those fetuses with sex chromosomal abnormalities, on the other hand, are quite frequent but when they come to term they usually have a less severe condition than Down syndrome that is not always associated with mental retardation. The risk for having a child with Down syndrome begins to rise when the mother is 35 years of age (Figure 13.1) and increases as the age of the mother goes up. The curves of the frequency of Down syndrome by the age of the mother for annual intervals both at the time of amniocentesis[7] and per livebirth[8] are given in Figure 13.1. The frequencies are greater at the time of amniocentesis than at birth. The difference between the frequencies has been explained by a variety of reasons, the most likely being that of fetal loss between about 16 weeks[8] and term, although this is not the only reason[9].

### *Effectiveness of prenatal diagnosis in elimination of Down syndrome*

It is difficult to estimate how effective a programme of prenatal diagnosis of Down syndrome can be in reducing the population frequency. Risks for the

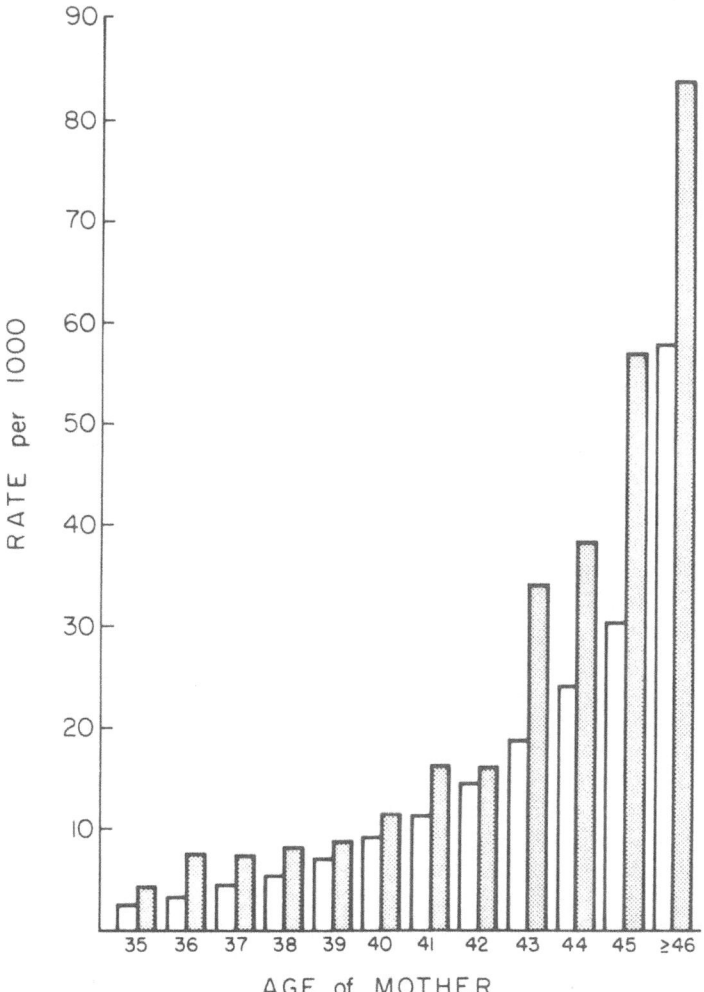

**Figure 13.1**  A comparison of the frequencies of trisomy 21 or Down syndrome at the time of genetic amniocentesis and among livebirths. ▨ = rate of trisomy 21 per 1000 genetic amniocenteses (9500). Data are from one Canadian, one European and two collaborative studies and summarized in Hamerton and Simpson[7]. □ = estimated rates of Down syndrome per 1000 livebirths in New York State from Hook and Chambers[8]

syndrome at different ages of the mother may be changing in at least some populations[10]. The proportion of births to women greater than 35 years fluctuates over time and the proportion of women at risk having amniocentesis and selectively aborting fetuses with Down syndrome is variable between populations and over time. The risk for having a second child with any chromosomal abnormality is 1.7 %[11] whereas about 1 % is the recurrence risk usually quoted for a full trisomy 21 Down syndrome when the parents have a normal karyotype. Parents who are balanced carriers of translocations involving chromosome 21 (particularly mothers) are at a greater risk of having

207

a child with Down syndrome than those parents with a normal karyotype. The risk for carrier mothers is about 15 % and for carrier fathers about 5 %[12]. The translocation carriers are easily identified by a karyotype from a lymphocyte culture. The contribution of Down syndrome cases due to a translocation carrier parent or a family in which there has been recurrence of the syndrome to the prevalence in a population, however, is very small ($\sim 2.5 \%$)[13].

When considering the effectiveness of amniocentesis in reducing the prevalence of Down syndrome, the age of the mother is the major consideration. It has been estimated that if all of the women 35 years of age and over and those who had had a previous child with Down syndrome had prenatal diagnosis by amniocentesis and were prepared to abort their trisomy 21 fetuses, less than one third of Down syndrome would be eliminated in future years in the UK[13]. Such reduction of the frequency would still be economical[14]. It would require 146 amniocenteses to detect 1 Down syndrome[13]. If the age of the mother is less than 35 years the cost of screening would be greater than the economic benefit in the UK[14] but not in the USA[15]. Practically speaking, no screening programme encompasses 100 % of pregnant women > 35 years old: at present only about 15 %–20 % in some parts of Canada[7] and a few years ago less than 20 % in the UK > 40 years of age[16,17]. Although the percentage of older pregnant women wishing to have prenatal diagnosis is increasing[18], there will always be women who do not wish to have the procedure or who would not be willing to abort the abnormal fetuses.

In order to make more of an impact on reducing the prevalence of Down syndrome, a non-invasive method of screening for high risk mothers needs to be found. Boué[19] has suggested that studies in France indicate that patterns of movement seen by the real-time scanner in Down syndrome and normal fetuses differ. If Boué's data were confirmed and were sufficiently distinctive, a real-time scan might be a screening test for the selection of high risk mothers. Sampling of nucleated fetal cells from the circulation of pregnant women[20,21] is another possibility for a non-invasive screening test but is still in the developmental stage. Such a technique could hopefully replace screening by amniocentesis for at least the chromosomal abnormalities.

### Efficiency of diagnosing chromosomal abnormalities

The efficiency of diagnosing chromosomal abnormalities by amniocentesis is high. The main reason for failure to diagnose a chromosomal abnormality is that the cells do not grow. Most studies report[1-3,5,6] that about 4 % fail to grow on the first amniocentesis, although this varies according to experience in the laboratory and whether or not samples have been mailed to the laboratory. The failure rate is about 13 % for mailed samples[5]. Successful diagnosis is considerably better after a repeated amniocentesis when the first culture fails. Other reasons for failure to diagnose are: no fluid was obtained at amniocentesis or something went wrong with the test and, in the case of chromosomal abnormalities, the spreads were not of good enough quality to analyse. When the cells grow and a good spread is made, the chromosomes can be seen clearly and a supernumerary chromosome 21, particularly, can be identified with ease.

There are, however, a few things that can go wrong.

(1) Occasionally, (about 3 per 1000 amniocenteses) the cells from the mother rather than the fetus are grown. In the case of a male fetus, maternal contamination can be ruled out because of the presence of Y chromatin in uncultivated amniotic cells or a Y chromosome in the karyotype (the latter should always be done) and possibly by the ratio of testosterone to FSH[7]. When the fetus is female, maternal contamination is sometimes ruled out by comparison of parental chromosome markers and those of presumed fetal cells, or by HLA typing of the parental and presumed amniotic cells[7]. All of this is, of course, time consuming and expensive, and in the case of HLA typing requires very specialized expertise. The latter test would be used only in cases of mosaicism that are difficult to interpret.

(2) Interpretation of mosaicism is of most concern and is another source of error, although fortunately it occurs rarely. Because it occurs rarely, there are scant data to serve as a guide in determining whether the mosaicism is due to a few maternal cells (partial contamination), a cultural artifact, or true mosaicism in the fetus. Identification of maternal cells can be made as described in (1) above. If cells from only a single primary clone of fetal cells have a different karyotype from all the rest, it is not likely to represent true mosaicism[7]. If a true mosaic is found it is difficult to predict the phenotype unless there has been previous documentation, e.g. a mosaic for trisomy 21. A labelling mistake at any step in the processing of the sample can also contribute to error and rigid protocols are in existence to avoid this kind of error. Finally, a chromosomal abnormality may be too small and be missed or not visible with the present techniques. Prediction of the phenotype may be difficult when there is a small chromosomal variation such as an apparent balanced reciprocal translocation that is, in fact, aneusomy of recombination, but such events are exceptionally rare.

## Neural tube defects (NTDs)

The diagnosis of neural tube defects (NTDs) has been primarily based on a 'raised' level of $\alpha_1$-fetoprotein (AFP) in amniotic fluid and is effective for open, but not for closed, lesions. The proportion of women having amniocentesis when at risk for a NTD compared to other reasons is dependent on the population frequency which varies from about 1 to 8 per 1000[7] in different parts of the world.

### Efficiency of α-fetoprotein (AFP) determinations in amniotic fluid for diagnosis of open NTDs

The probability of prenatally diagnosing an open NTD from AFP in amniotic fluid is dependent on:

(1) the prevalence of NTDs in the population,

(2) whether the serum AFP was high after correction by ultrasonography for dates and exclusion of multiple births,

(3) whether there was a previous infant with a NTD,

(4) the cut-off level of AFP (usually expressed in multiples of the median (MOMs) because of the great variability between laboratories),

(5) whether the defect is anencephaly or spina bifida,

(6) the gestational age at the time of testing.

Table 13.1 summarizes data from the UK collaborative study[22] and shows that the percentage of false positive tests (i.e. the percentage of normals with a positive test) increases with gestational age and decreases as the cut-off level is raised. On the other hand, the percentage of false negatives with open spina bifida does the opposite – decreases with gestational age and increases as the cut-off level increases. Ideally, the combination of false negatives and positives should be minimized and when this is done the best combination requires a higher cut-off level as gestational age increases (see boxes in Table 13.1). There are fewer false negatives for anencephaly than for spina bifida and

Table 13.1  Percentage of false positive and negative $\alpha_1$-fetoprotein (AFP) determinations in amniotic fluid for open spina bifida in fetuses from singleton pregnancies with corrected gestational age by ultrasonography*

| Gestation (weeks completed) | N | | Cut-off levels in multiples of the median (MOM) | | | | |
|---|---|---|---|---|---|---|---|
| | | | 2.0 | 2.5 | 3.0 | 3.5 | 4 |
| 13–15 | 3279 | FP† | 1.9 | 0.7 § | 0.4 | 0.3 | 0.3 |
| | 23 | FN‡ | 4.0 | 4.0 | 9.0 | 13.0 | 17.0 |
| 16–18 | 7858 | FP | 3.7 | 1.2 | 0.7 | 0.5 | 0.5 |
| | 74 | FN | 1.0 | 1.0 | 1.0 | 8.0 | 14.0 |
| 19–21 | 1561 | FP | 7.4 | 3.3 | 1.5 | 1.0 | 0.6 |
| | 21 | FN | 5.0 | 5.0 | 5.0 | 5.0 | 5.0 |
| 22–24 | 407 | FP | 9.8 | 5.2 | 3.2 | 2.0 | 1.5 |
| | 5 | FN | 0 | 0 | 0 | 0 | 0 |

\* Modified from the UK collaborative study, 1979[22]

† FP = false positives: the percentage of normals with a positive test

‡ FN = false negatives: the percentage of open spina bifida fetuses with a negative test

§ ☐ = Selected MOM that is best for gestational age, i.e. the fewest false positives and negatives

there is a greater chance that anencephaly will be detected during the ultrasonography when verifying gestational age and a singleton pregnancy. In practice the best time to measure AFP in amniotic fluid is at 17 weeks[7], a compromise giving few false positive or negative tests when 3 MOMs is the cut-off level (at 16–18 weeks[7], Table 13.1). This gives time for chromosome analysis in centres in which it is feasible and allows time to abort an abnormal fetus before 20 weeks.

### Other tests for open NTDs
Although the measurement of AFP in amniotic fluid is a very good predictive test of NTDs, it can be seen from the foregoing discussion that there are problems. As well as being documented in the UK collaborative study[21] they

have been outlined in Milunksy and Alpert[23,24]. The problems and pitfalls[23,24] have led to attempts to develop adjunct and/or better methods of diagnosing NTDs using amniotic fluid.

Three methods look promising, at least, as adjunct methods. Rapidly adhering cells (RA) may be helpful[25] but require laboratories with cytological expertise since the test is based on the number and morphology of amniotic fluid cells that adhere to glass after short term culture. The cells are characteristically found in cases of neural tube defect and are not found in some of the other defects that produce high AFP in amniotic fluid, such as exomphalos and a traumatized placenta. Others[26,27] have looked at the differential binding of AFP isoproteins to lectins such as concanavalin A. The most interesting development, however, has been studies of cholinesterase in amniotic fluid[28-31]. Although most reports refer to the cholinesterase as acetylcholinesterase, our experience has been that both cholinesterase (serum or pseudocholinesterase) and acetylcholinesterase are raised in amniotic fluid from pregnancies carrying an open NTD (Simpson, unpublished data). If cholinesterase determinations turn out to be as promising as they look and as diagnostic as AFP measurements, the methodology will be more convenient for routine laboratories than AFP measurements. Buamah et al.[32] have suggested that even bloodstained amniotic fluid samples from pregnant women carrying normal fetuses can be distinguished from open NTDs, exomphalos, and intrauterine deaths by an adapted electrophoretic method of Smith et al.[28] but our experience has not been so clearcut (Simpson, unpublished data).

## Prescreening for women at risk of carrying a fetus with an open NTD by AFP levels in serum

Identification of the woman at risk for an open NTD is considerably more effective than for the chromosomal abnormalities because it is possible to prescreen by measuring AFP in their serum. A raised AFP in maternal serum followed by ultrasonography can be used as prescreening tests. About two thirds of those women with high serum AFP are carrying multiple fetuses or the gestational age is incorrect and, when corrected, the AFP is not indicative of a NTD[33]. Both the multiple births and incorrect gestational age usually can be detected by ultrasonography. Anencephaly and sometimes spina bifida will also be detected at this time by ultrasonography[22]. Figure 13.2 compares the false negatives for anencephaly and open spina bifida fetuses in singleton pregnancies that are corrected for gestational age at different cut-off levels of serum AFP from the UK collaborative study[33], showing that it is necessary to use a low cut-off level to reduce the false negatives. The false negatives due to anencephaly are virtually eliminated by the ultrasonography but the proportion of open spina bifida fetuses missed even at 2 MOMs is 9%, with a false positive rate of 7%. If 7% for the false positive rate is considered too high at 2.5 MOMs the false positive rate is lowered to 3%, but the false negative rate for spina bifida goes up to 21%. However, one third of the false positive tests are in the normal range when the test is repeated after the multiple births and underestimated gestational ages are eliminated. A repeat serum test is therefore very desirable before going to amniocentesis. Besides multiple births

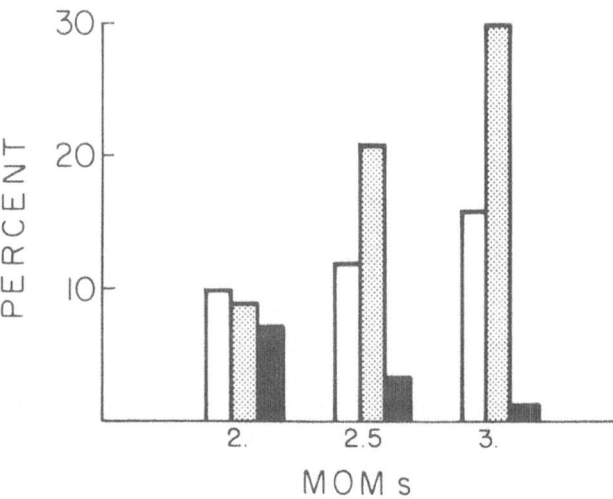

**Figure 13.2**  Percentage of false negative serum $\alpha_1$-fetoprotein (AFP) tests at 16–18 weeks for anencephaly and open spina bifida and percentage of false positive tests for three different cut-off levels in multiples of the median (MOMs). Data are for singleton pregnancies with corrected gestational ages using ultrasonography. □ = false negative for anencephaly. ▨ = false negative for open spina bifida. ■ = false positive. (Data are from UK collaborative study in Hamerton and Simpson[7])

and incorrect gestational ages, there are a number of known reasons for false positive serum tests – exomphalos, congenital nephrosis, intrauterine death or an impending spontaneous fetal loss. The latter is of particular concern since the amniocentesis may precipitate the impending abortion. Reasons for the false negative tests are not clear; they can be reduced by lowering the cut-off level at the expense of the false positive rate. False negative tests for anencephaly and some spina bifida can be eliminated by ultrasonography but as yet there is no other serum test that will improve the false negative rate. In practice, after identifiable reasons for false positive serum tests are eliminated, a second serum test is still positive, and the woman wishes it, amniocentesis is performed and the AFP is measured in the amniotic fluid.

Table 13.2 summarizes the factors influencing the probability of a woman carrying a singleton fetus with an open spina bifida, from data in the UK collaborative study[7]. The probability is dependent on:

(1) The reason for the amniocentesis – the highest probability being after prescreening by serum AFP levels and ultrasonography, the next highest being in women who had had a previous infant or fetus with a NTD and the lowest probability when the amniotic AFP was measured incidentally and the reason for the amniocentesis was chromosomal, biochemical or for sexing,

(2) the population prevalence of open spina bifida,

(3) the cut-off level of AFP in the amniotic fluid.

212

Table 13.2 Amniotic fluid (AFP): 16–18 weeks gestation probability of having a singleton fetus with open spina bifida according to prevalence of open spina bifida, reason for amniocentesis and cut-off level (MOM)

| Reason for amniocentesis | Prevalence of open spina bifida per 1000 | Probability of NTD by amniotic fluid AFP cut-off level (MOM) | |
|---|---|---|---|
| | | 3.0 | 4.0 |
| Single serum AFP ≥ 2.5 | 1 | 90% | 94% |
| X median and ultrasound | 2 | 95% | 97% |
| to correct 'dates' | 3 | 96% | 98% |
| Previous infant with a | 1 | 67% | 80% |
| neural tube defect | 2 | 83% | 90% |
| | 3 | 88% | 93% |
| Other | 1 | 20% | 33% |
| | 2 | 33% | 50% |
| | 3 | 40% | 60% |

Modified from the report of the international conference[7]; the data are from the UK collaborative study[22]

## Biochemical defects

### Diagnosis of inborn errors of metabolism by enzyme levels

Over fifty of the classical metabolic disorders with known enzyme defects have been successfully diagnosed prenatally. They are listed in Table 13.3 and the working group at the recent international conference estimated that 340–450 affected fetuses have been diagnosed out of 1600–2000 fetuses[7]. Classically, the inborn errors of metabolism that have been diagnosed are those in which the biochemical defect is known and can be measured in amniotic cells (or fluid). Most families are identified first by having had an affected child who has been extensively investigated to assure the diagnosis and particular enzyme that needs to be measured in the amniotic fluid or cells from a subsequent pregnancy. Since all the biochemical disorders are individually rare and the tests are extremely specialized, it is not practicable to screen populations for all these disorders by amniocentesis. On the other hand, once a family is ascertained and the disease identified, the risk is high (25% since usually we are dealing with two heterozygous parents and a recessive disorder or an X-linked recessive disorder) and therefore prenatal diagnosis can be of great value.

Measurement of enzymes from cells involves a number of difficulties.

(1) There are some partial enzyme deficiencies that may mimic heterozygous levels and therefore make diagnosis uncertain,

(2) Microtechniques are needed to handle reduced numbers of cells so that either growing time can be shortened or amniocentesis can be earlier, thereby enabling the biochemist to produce a result in sufficient time for a therapeutic abortion if the test indicates an abnormal fetus.

(3) More sensitive methods involving fluorogenic, chromogenic or radio-labelled substrates need to be developed to be used in the micro-methods.

**Table 13.3   Prenatal diagnosis of metabolic diseases**

| Disease | Diagnostic test |
|---|---|
| *Lipid metabolism* | |
| Tay–Sachs | hexosaminidase A |
| Sandhoff | hexosaminidase A and B |
| Gaucher | $\beta$-glucosidase |
| $G_{ML}$-gangliosidosis | $\beta$-galactosidase |
| Niemann–Pick | sphingomyelinase |
| Fabry | $\alpha$-galactosidase |
| Metachromatic leucodystrophy | arylsulphatase A |
| Krabbe | galactosylceramide $\beta$-galactosidase |
| Wolman | acid esterase |
| Farber | ceramidase |
| Familial hypercholesterolaemia | LDL cell-surface receptor |
| | |
| *Mucopolysaccaridoses and related disorders* | |
| Hurler (MPS IH) | $\alpha$-iduronidase |
| Scheie (MPS IS) | $\alpha$-iduronidase |
| Hunter (MPS II) | sulphoiduronate sulphatase |
| Sanfilippo A (MPS IIIA) | heparan sulphate sulphamidase |
| Sanfilippo B (MPS IIIB) | $\alpha$-$N$-acetylglucosaminidase |
| Maroteaux–Lamy (MPS VI) | arylsulphatase B |
| Mucolipidosis II (I-cell disease) | lysosomal hydrolases |
| Mucolipidosis III | lysosomal hydrolases |
| Mucolipidosis IV | EM ultrastructure |
| Fucosidosis | $\alpha$-fucosidase |
| Mannosidosis | $\alpha$-mannosidase |
| Sialidosis (variant) | $\alpha$-neuraminidase and $\beta$-galactosidase |
| | |
| *Amino acid metabolism* | |
| Argininosuccinic aciduria | argininosuccinase |
| Cystinosis | 3 & 5S-cystine uptake |
| Citrullinaemia | argininosuccinate synthetase |
| Homocystinuria | cystathionine synthetase |
| Maple syrup urine disease | $\alpha$-keto acid decarboxylase |
| Methylmalonic acidaemia | |
| $\quad B_{12}$-responsive | deoxyadenosyl-$B_{12}$ synthesis |
| $\quad B_{12}$-non-responsive | methylmalonyl-CoA mutase |
| Propionic acidaemia | propionyl-CoA decarboxylase |
| | |
| *Carbohydrate metabolism* | |
| Glycogen storage Type II | $\alpha$-glucosidase |
| Glycogen storage Type IV | amylo-$(1,4 \rightarrow 1,6)$ transglucosidase |
| Galactosaemia | gal-l-P uridylyl transferase |
| | |
| *Blood* | |
| Haemophilia A | coagulation/Factor VIII antigen |
| Sickle cell | $\beta$-chain synthesis/Hpa I restriction analysis |
| Homozygous $\beta$-thalassaemia | $\beta$-chain synthesis |
| Homozygous $\alpha$-thalassaemia | cDNA hybridization |
| $\beta°\delta°$-thalassaemia | cDNA hybridization |
| Chronic granulomatous disease | NBT/superoxide formation |
| | |
| *Miscellaneous* | |
| Combined immunodeficiency | adenosine deaminase |
| Lesch–Nyhan syndrome | HGPRT |
| Menke's disease | copper uptake |

(*continued*)

Table 13.3 (*continued*)

| Disease | Diagnostic test |
|---------|-----------------|
| Xeroderma pigmentosum | DNA repair |
| Hypophosphatasia | alkaline phosphatase |
| Acute intermittent porphyria | uroporphyrinogen I synthetase |
| Congenital adrenal hyperplasia | |
| (21-hydroxylase) | HLA-B linkage |
| Lysosomal acid phosphatase deficiency | acid phosphatase |
| Congenital nephrosis | α-fetoprotein |

From *Prenatal Diagnosis*, special issue 1980[7], with permission of John Wiley & Sons Ltd

Because of the small numbers of cases diagnosed for any one disease, the effectiveness of prenatal diagnosis of the classical biochemical defects is not nearly as well documented as for the diagnosis of chromosomal abnormalities and neural tube defects.

### Diagnosis of genetic disease by the 'New' genetics

Until recently, the diagnosis of a number of disorders that required whole blood, such as the haemoglobinopathies[7,34-36] or blood coagulation defects[37,38], were possible technically only when a fetal blood sample was obtained. Fetal blood sampling by fetoscopy or placental aspiration, however, has more risk for the mother and fetus than amniocentesis[7,39]. Therefore, elegant alternative techniques have been developed making it possible to diagnose some of the haemoglobinopathies[40-44] from the amniotic cells. Using DNA extracted from the fetal amniotic cells and annealed to radioactively-labelled specific complementary DNA (cDNA) probes, Kan *et al.*[40] diagnosed an α-thalassaemia-1 (heterozygous α-thalassaemia with two α-globin genes) in a fetus at risk for the fatal condition, homozygous α-thalassaemia (no α-globin genes). This was done by assaying the number of α-globin genes in the DNA from the amniotic cells. Normal persons have four α-globin genes. The DNA extracted from the cells obtained by amniocentesis was annealed in solution to a $^{32}$P-labelled α-cDNA probe, and the amount of radioactivity recovered was compatible with two α-globin genes indicating α-thalassaemia-1. The genotype was confirmed from cord blood at birth. The method, however, has some technical problems and requires the growth of a considerable number of fetal fibroblasts[41].

The mapping of restriction endonuclease sites in DNA from persons having either mutant, deleted or partially deleted globin genes has also proved to be a successful technique for diagnosis by amniocentesis[41-44]. Different restriction endonucleases cleave DNA at specific nucleotide sequences cutting the DNA into segments having different sequences and lengths. The different sizes of fragments can be separated by agarose gel electrophoresis and overlaid with the radioactively-labelled cDNA probe which hybridizes only with the DNA to which the probe is complementary. After autoradiography the fragments are visualized. When the different lengths of DNA hybridize with the same probe they are known as restriction fragment length polymorphisms (RFLPs)[45]. Orkin *et al.*[41] have demonstrated that α- and β-thalassaemias can be diagnosed using the α- and β-cDNA probes respectively. In the case of a full

H                                        215

or partial gene deletion, the restriction fragments are smaller than the normal gene and can therefore be detected by the difference in size of the molecule hybridizing to the probe. Theoretically, with the appropriate probe, this method could be used to diagnose any disease caused by a gene deletion. It requires less DNA per test than hybridization in solution, making repeated tests for confirmation possible; there may be less non-specific hybridization and the purity of the probe may not be as important for hybridization in solution[41].

A modification of the mapping method has now been used to detect structural gene defects; sickle cell haemoglobin and $\beta$-thalassaemia have been diagnosed from DNA hybridization using a $\beta$-cDNA probe[43]. This was possible because there is a polymorphism for the length of the HpaI restriction endonuclease fragment containing the $\beta$-chain haemoglobin gene. The sickle cell mutant gene (Hb$^s$) in the $\beta$-chain occurs more often in one fragment, 70–80 % of the time in the larger fragment, i.e. the 13 kilobase (Kb) band[43,44], whereas the normal haemoglobin gene for the $\beta$-chain occurs more often in a 7 or 7.6 kilobase band. It was possible to diagnose sickle cell disease from the cells contained in 16 ml of amniotic fluid using this method[43]. On the other hand, it is possible to diagnose only 60 % of the pregnancies at risk because not all of the Hb$^s$ genes are in a 13 Kb fragment. In order to establish that the sickle cell gene is on the 13 Kb fragment in both parents, the fragments would have to be identified in a previous child (or fetus) with sickle cell disease and either both 13 Kb or both 7 (or 7.6) Kb fragments would need to contain the Hb$^s$ gene for the family to be informative. A similar approach has been used[44] to identify the Hb S-O$^{Arab}$ diseases using restriction fragments generated by another restriction endonuclease (EcoRl).

If a great many restriction length polymorphisms can be found using random cDNA probes, as suggested by Botstein et al.[45], the chances of establishing linkages between the RFLPs and single gene defects for which the biochemical defect is unknown will be greatly increased. Such linkages could be useful in prenatal diagnosis in the future, although there will always be families that will not be informative; the number of non-informative families will depend on the tightness of the linkage and the extent of the polymorphism. In the past, the linkage between secretor factor and myotonic dystrophy has been used in prenatal diagnosis by measuring secretor factor in the amniotic fluid[46]. Myotonic dystrophy is an example of a disease in which the biochemical defect is not understood but the linkage is useful. Recently the linkage between HLA-B and congenital adrenal hyperplasia due to 21-hydroxylase deficiency has also been used in prenatal diagnosis[47,48]. Although linkages will not be as good as knowing the biochemical defect because of the uninformative families, the linkages will make it possible to diagnose the linked diseases in some families that would not otherwise have the opportunity of prenatal diagnosis.

The most common burdensome disease for which the basic defect is uncertain is cystic fibrosis. Some promise in diagnosing the disease by assaying protease activity in amniotic fluid has been shown[49,50] but there are uncertainties[51]. Potentially, the linkage of a RFLP with cystic fibrosis such as that for sickle cell disease could be useful, at least, for some families.

*Prescreening for biochemical defects*

There are not many biochemical defects that can be prescreened at present. Two examples of non-invasive screening tests which can be used, however, are that for Tay–Sach's disease and that for sickle cell anaemia. Heterozygotes for both these diseases can be identified. Since Tay–Sach's disease is clustered in Ashkenazi Jews and sickle cell anaemia in blacks, the populations screened can be limited to those with a relatively high risk of having the gene in question.

## Sex determination

The determination of sex has been made for certain X-linked diseases with a high burden when they cannot be diagnosed *in utero*. The most effective way is from a karyotype with fluorescent banding and this should always be done. X-linked muscular dystrophies and haemophilias have been the prime candidates for this approach[1–6]. Although attempts have been made to diagnose Duchenne muscular dystrophy[52,53] they have required fetal blood sampling and are not as yet considered reliable[7,54]. The potential of studies on muscle biopsy has been examined by Emery and Burt[55], but this work is still in the research stage. A non-invasive B-scan by ultra-sound imaging may solve the prenatal diagnosis of muscular dystrophies[56]. Success in the diagnosis of classical haemophilia (factor VIII deficiency) has been reported[37,38] using an immunoradiometric assay on fetal plasma and amniotic fluid mixtures, obtained by fetoscopy. Again, this still involves the hazards of fetoscopy. If some of the burdens from X-linked diseases that do not have a known basic defect can be solved by the 'new' genetics, fetal sexing and selective abortion of males will become a thing of the past for prenatal diagnosis.

## Risks of the procedure of amniocentesis

Early amniocentesis for prenatal diagnosis of genetic diseases, with the implications for selective abortion of fetuses with defects of a predictable nature, was carefully monitored by the collaborative studies in the United States, Canada, the United Kingdom and western Europe. Other individual studies have also evaluated the risks to both mother and fetus and have been reviewed by a group with expertise in the field[7]. The US and UK collaborative studies were the only ones with matched controls and their conclusions, at first sight, were not in agreement. Essentially the differences are reflected in the titles of their reports; the title of the US study is 'Mid-trimester amniocentesis for prenatal diagnosis: *Safety* and accuracy', whereas that of the UK study is 'An assessment of the *hazards* of amniocentesis' (present author's italics).

*Risks to the fetus*

The major 'hazard' reported in the British study was that of an increase in fetal loss before 28 weeks following the procedure, whereas the American study reported no such increase that was statistically significant. An effort to explain this discrepancy was made by a group of experts at a recent international conference[7]. The group concluded that the main reason for the difference was that there were 41 % of women at risk for NTDs in the UK study and it is known that they are more likely to have a fetal loss. The American study was done before NTDs were being diagnosed by AFP levels and therefore women at risk for fetal loss because of an NTD would not have been included in their study. When women at risk for NTDs in the British study were removed the excess fetal loss was not statistically significant. Other studies without controls have not reported a great many fetal losses but it is impossible to estimate the 'excess', if any, in the absence of controls. Undoubtedly there are some fetal losses due to the procedure, particularly caused by amnionitis. The percentage of excess fetal losses, however, is probably not greater than 1 %, as estimated by the international conference group[7].

Other 'hazards' reported by the UK collaborative study were an increased incidence over controls (1 %) in unexplained respiratory difficulties in the neonate, particularly between 34 and 37 weeks gestation, and an increased incidence of talipes and possibly of congenital dislocation of the hips. These data are somewhat difficult to interpret and more are needed to establish whether these defects are a result of the procedure. The conference group[7] has suggested that this problem be solved by animal studies, as it is becoming increasingly difficult to obtain suitable controls for such studies.

### Risks to the mother
Risks to the mother are usually of minor concern, consisting mainly of abdominal cramps and discomfort. Amnionitis occurs in about 1 per 1000[7] and occasionally leads to fetal loss. Only 1 maternal death in an estimated 100 000 amniocenteses was known to the international conference group.

## CONCLUSIONS

The first successful attempts to determine a fetal karyotype from cultured amniotic cells were made in the mid-1960s[57]. Prenatal diagnosis of chromosomal defects soon followed and since the late 1960s the number of diseases or defects that can be or have been prenatally diagnosed by amniocentesis has increased exponentially. Because decisions involving human lives are made from the results of the procedure, careful multidisciplinary assessment of the safety of the procedure itself has been made as groups began to use it, and as it became possible to diagnose a given genetic disease the efficacy of the test for a particular disease has been evaluated. Diagnosis of chromosomal defects, neural tube defects and many biochemical defects is well established and extremely reliable. The use of recombinant DNA techniques has extended our ability for diagnosing diseases by amniocentesis that previously required blood sampling of the fetus and hence fetoscopy or placental aspiration with considerable risk to the mother and fetus. Although sex determination is still made for some X-linked recessive diseases and consequently some normal males are aborted, the DNA recombinant techniques may give us the tools to reduce the number of such losses even when the basic defect is unknown.

Investigators and those who deliver amniocentesis services from many disciplines have met at various conferences, issued guidelines, and reported on their consensus in various problem areas. Many of these deliberations are reviewed in detail in the document produced by an international group that met in autumn 1979. The recommendations from this group are published[7] and cover the areas of methodology of the procedure, service planning, problems in diagnosis, suggestions for future research and ethical, legal and societal considerations.

## Acknowledgements

The author is grateful to Dr Jeannette Holden, Queen's University, for her careful reading of the manuscript and to Mrs Lauretta Parks for typing the manuscript.

## References

1 Lowe, C. U., Alexander, D., Bryla, D. and Seigel, D. (1978). The NICHD amniocentesis registry, the safety and accuracy of mid-trimester amniocentesis. *NIH Publication* 78–190. (Washington, D. C.; Department of Health, Education & Welfare)

2 Medical Research Council (1977). Diagnosis of genetic disease by amniocentesis during the second trimester of pregnancy: A Canadian study. (Ottawa: Minister of Supply & Services, Canada)

3 An assessment of the hazards of amniocentesis; report to the medical research council by their working party on amniocentesis (1978). *Br. J. Obstet. Gynaecol.*, **85** (Suppl.), 2, 1

4 Murken, J. D. (1979). In Murken, J. D. and Stengel-Rutowski, S. (eds.) *Deutschen Forschungsgemeinschaft, 13. Informationsblatt.* (Munich)

5 Golbus, M. S., Loughman, W. D., Epstein, C. J., Halbasch, G., Stephens, J. D. and Hall, B. D. (1979). Prenatal genetic diagnosis in 3000 amniocenteses. *N. Engl. J. Med.*, **300**, 157

6 Hsu, L. Y. F., Kaffe, S., Yahr, F., Serotkin, A., Giordano, F., Godmilow, L., Kim, H. J., David, K., Kereny, T. and Hirschhorn, K. (1978). Prenatal cytogenetic diagnosis: first 1000 successful cases. *Am. J. Med. Genet.*, **2**, 365

7 Hamerton, J. L. and Simpson, N. E. (eds.). (1980). Prenatal diagnosis – past, present and future. *Prenatal Diagnosis*, special issue, Dec. (Chichester: Wiley)

8 Hook, E. B. and Chambers, C. M. (1977). Estimated rates of Down syndrome in live births by one year maternal age intervals for mothers aged 20–49 in a New York State study – implications of the risk figures for genetic counselling and cost-benefit analysis of the prenatal diagnosis programs. *Birth Defects: Orig. Art. Ser.*, **13** (3A), 141. (New York: National Foundation)

9 Ferguson-Smith, M. A. (1976). Prospective data on risk of Down syndrome in relation to maternal age. *Lancet*, **2**, 252

10 Lowry, R. B., Renwick, D. H. C., Jones, D. C. and Trimble, B. K. (1976). Down syndrome in British Columbia, 1952–73: incidence and mean maternal age. *Teratology*, **14**, 29

11 Mikkelson, H. (1979). Previous child with Down syndrome and other chromosome aberration. In Murken, J. D., Stengel-Rutkowski, S. and Schwinger, E. (eds.) *Prenatal Diagnosis: Proceedings of the 3rd European Conference on Prenatal Diagnosis of Genetic Disorders*, April 12–14, 1978, Munich, p. 22. (Stuttgart: Enke)

12 Omenn, G. S. (1978). Prenatal diagnosis of genetic disorders. *Science*, **200**, 952

13 Forster, D. P. (1977). The antenatal detection of Down syndrome: some demographic aspects. *J. Ment. Def. Res.*, **21**, 263

14 Hagard, S. and Carter, F. A. (1976). Preventing the birth of infants with Down syndrome: a cost-benefit analysis. *Br. Med. J.*, **1**, 753

15 Stein, Z. A., Gusser, M., Kline, J. and Warburton, D. (1977). Amniocentesis and selective abortion for trisomy 21 in the light of the natural history of pregnancy and fetal survival. In Hook, E. B. and Porter, I. H. (eds.) *Population Genetic Studies in Humans*, pp. 257–274. (New York: Academic Press)

16 Forster, D. P. and Davison, C. M. (1977). Medical care aspects of the prenatal diagnosis of chromosomal abnormalities. *Soc. Sci. Med.*, **11**, 593

17 Polani, P. E., Alberman, E., Berry, A. C., Blunt, S. and Singer, J. D. (1976). Chromosomal abnormalities and maternal age. *Lancet*, **2**, 516

18 Dallaire, L., Rudd, N., Doran, T., Greenberg, C. R., Hamerton, J. L., Baird, P. A. and Thibault, O. (1980). Le diagnostic prénatal. In *Cahiers de Bioéthique*. Vol. 2, pp. 33–84. (Quebec: Les Presses de l'Université Laval)

19 Boué, J., Morer, I., Laisney, V. and Boué, A. (1979). Diagnostic prénatal: resultats de 1532 ponctions amniotiques et étude prospective de 1023 cas. *Nouv. Presse Med.*, **8**, 2949

20 Herzenberg, L. A., Bianchi, D. W., Schroder, J., Cann, H. M. and Iverson, G. M. (1979). Fetal cells in the blood of pregnant women: detection and enrichment by fluorescence-activated cell sorting. *Proc. Natl. Acad. Sci. USA*, **76**, 1453

21 Schroder, J., Turunea, O., Lundqvist, C. and de la Chapelle, A. (1979). Maternal blood as a source of fetal cells for prenatal diagnosis. In Murken, J. D., Stengel-Rutkowski, S. and Schwinger, E. (eds.) *Prenatal Diagnosis: Proceedings of the 3rd European Conference on Prenatal Diagnosis of Genetic Disorders*, April 12–14, 1978, Munich, p. 243. (Stuttgart: Enke)

22 Second report of UK collaborative study on alpha-fetoprotein in relation to neural tube defects (1979). *Lancet*, **2**, 651

23 Milunsky, A. and Alpert, E. (1976). Prenatal diagnosis of neural tube defects: 1. Problems and pitfalls: analysis of 2495 cases using the alpha-fetoprotein assay. *Obstet. Gynecol.*, **48**, 1

24 Milunsky, A. and Alpert, E. (1976). Prenatal diagnosis of neural tube defects: II. Analysis of false positive and false negative alpha-fetoprotein results. *Obstet. Gynecol.*, **48**, 6

25 Gosden, C. M., Eason, P., Fotherington, A., Ross, A., Bowden, C., Barron, L., Brown, S. and Brock, D. H. J. (1979). Amniotic fluid cell morphology and alpha-fetoprotein in the early antenatal diagnosis of congenital abnormalities. In Murken, J. D., Stengel-Rutowski, S. and Schwinger, E. (eds.) *Prenatal Diagnosis: Proceedings of the 3rd European Conference on Prenatal Diagnosis of Genetic Disorders*, April 12–14, 1978, Munich, p. 100. (Stuttgart: Enke)

26 Smith, C. M., Kelleher, P. C., Belanger, L. and Dallaire, L. (1979). Reactivity of amniotic fluid alpha-fetoprotein with concanavalin A in diagnosis of neural tube defects. *Br. Med. J.*, **1**, 920

27 Norgaard-Pedersen, B., Toftager-Larsen, K., Philip, J. and Hindersson, P. (1980). Concanavalin A reactivity pattern of human amniotic fluid AFP examined by crossed affino-immunoelectrophoresis. A definite test for neural tube defect? *Clin. Genet.*, **17**, 355

28 Smith, A. D., Wald, N. J., Cuckle, H. S., Stirrat, G. M., Bobrow, M. and Lagercrantz, H. (1979). Amniotic fluid acetylcholinesterase as a possible diagnostic test for neural tube defects in early pregnancy. *Lancet*, **2**, 685

29 Chubb, I. W., Pilowsky, P. M., Springell, H. J. and Pollard, A. C. (1979). Acetylcholinesterase in human amniotic fluid: an index of fetal neural development? *Lancet*, **2**, 688

30 Smith, A. D. (1979). Amniotic fluid acetylcholinesterase and neural tube defects: plea for standardization. *Lancet*, **2**, 307

31 Chubb, I. W., Pilowsky, P. M., Hodgson, A. J. and Pollard, A. C. (1979). Acetylcholinesterase in blood-contaminated amniotic fluid. *Lancet*, **2**, 1148

32 Buamah, P. K., Evans, L. and Ward, M. (1980). Amniotic fluid acetylcholinesterase isoenzyme patterns in the diagnosis of neural tube defects. *Clin. Chim. Acta*, **102**, 147

33 Report of UK collaborative study on alpha-fetoprotein in relation to neural tube defects: Maternal serum alpha-fetoprotein measurement in antenatal screening for anencephaly and spina bifida in early pregnancy (1979). *Lancet*, **1**, 1323

34 Alter, B. P., Modell, C. B., Fairweather, D., Hobbins, J. C., Mahoney, M. J., Frigoletto, F. D., Sherman, A. S. and Nathan, D. G. (1976). Prenatal diagnosis of hemoglobinopathies: a review of 15 cases. *N. Engl. J. Med.*, **295**, 1437

35 Kan, Y. W., Golbus, M. S., Trecartin, R. F., Filly, R. A., Valenti, C., Furbetta, M. and Coa, A., (1977). Prenatal diagnosis of β-thalassaemia and sickle cell anaemia. *Lancet*, **1**, 269

36 Kan, Y. W., Golbus, M. S., Klein, P. and Dozy, M. T. (1975) Successful application of prenatal diagnosis in a pregnancy at risk for homozygous β-thalassemia *N. Engl. J. Med.*, **292**, 1096

37 Firshein, S. I., Hoyer, L. W., Lazarchick, J., Forget, B. G., Hobbins, J. C., Clyne, L. P., Pitlick, F. A., Muir, W. A., Merkatz, I. R. and Mahoney, M. J. (1979). Prenatal diagnosis of classic haemophilia. *N. Engl. J. Med.*, **300**, 937

38 Mibashan, R. S., Rodeck, C. H., Thumpston, J. K., Edwards, R. J., Singer, J. D., White, J. M. and Campbell, S. (1979). Plasma assay of fetal factors VIIIC and IX for prenatal diagnosis of haemophilia. *Lancet*, **1**, 1309

39 Alter, B. P. (1979). Prenatal diagnosis of hemoglobinopathies and other hematologic diseases. *J. Pediatr.*, **95**, 4

40 Kan, Y. W., Golbus, M. S. and Dozy, A. M. (1976). Prenatal diagnosis of α-thalassemia. Clinical application of molecular hybridization. *N. Engl. J. Med.*, **295**, 1165

41 Orkin, S. H., Alter, B. P., Altay, C., Mahoney, M. J., Lazarus, H., Hobbins, J. C. and Nathan, D. G. (1978). Application of endonuclease mapping to the analysis and prenatal diagnosis of thalassemias caused by globin gene deletion. *N. Engl. J. Med.*, **299**, 166

42 Orkin, S. H., Old, J., Weatherall, D. J. and Nathan, D. G. (1979). Partial deletion of β-globin gene DNA in certain patients with β-thalassemia. *Proc. Natl. Acad. Sci. USA*, **76**, 2400

43 Kan, Y. W. and Dozy, A. M. (1978). Antenatal diagnosis of sickle cell anaemia by DNA analysis of amniotic-fluid cells. *Lancet*, **2**, 910

44 Phillips, J. A., Scott, A. F., Kazazian, H., Smith, K. D., Tetten, G. and Thomas, G. H. (1979). Prenatal diagnosis of hemoglobinopathies by restriction endonuclease analysis: pregnancies at risk for sickle cell anemia and S-0[Arab] disease. *Johns Hopkins Med. J.*, **145**, 57

45 Botstein, D., White, R. L., Skolnick, M. and Davis, R. W. (1980). Construction of a genetic linkage map in man using restriction fragment length polymorphisms. *Am. J. Hum. Genet.*, **32**, 314

46 Insley, J., Bird, G. W. G., Harper, P. S. and Pearce, G. W. (1976). Prenatal prediction of myotonic dystrophy. *Lancet*, **1**, 806

47 Wurster, K. G., Clemens, G. E., Schunter, F. and Heilbronner, H. (1979). Possible prenatal diagnosis of adrenogenital syndrome by HLA typing of cultured amniotic cells. *Dtsch. Med. Wochenschr.*, **104**, 526

48 Pollack, M. S., Maurer, D., Levine, L. S., New, M. I., Pang, S., Duchon, M. A., Owens, R. P., Merkatz, I. R., Nitowsky, H. M., Sachs, G. and Dupont, B. (1979). HLA typing of amniotic cells: the prenatal diagnosis of congenital adrenal hyperplasia (21-OH-deficiency type). *Transplant. Proc.*, **11**, 1726

49 Walsh, M. M. and Nadler, H. I. (1979). Methylumbelliferyl-guanidinobenzoate reactive proteases in amniotic fluid: possible marker for cystic fibrosis. *Lancet*, **1**, 622

50 Brock, D. J. H. and Hayward, C. (1979). Methylumbelliferyl-guanidinobenzoate reactive proteases and prenatal diagnosis of cystic fibrosis. *Lancet*, **1**, 1245

51 Dann, L. G. (1979). Arginine esterase in amniotic fluid: possible marker for cystic fibrosis. *Lancet*, **2**, 907

52 Emery, A. E. H., Burt, D., Dubowitz, V., Rocker, I., Donnai, D., Harris, R. and Donnai, P. (1979). Antenatal diagnosis of Duchenne muscular dystrophy. *Lancet*, **1**, 847

53 Mahoney, M. J., Haseltine, F. P., Hobbins, J. C., Banker, B. Q., Caskey, C. T. and Golbus, M. S. (1977). Prenatal diagnosis of Duchenne's muscular dystrophy. *N. Engl. J. Med.*, **297**, 968

54 Golbus, M. S., Stephens, J. D., Mahoney, M. J., Hobbins, J. C., Haseltine, F. P., Caskey, C. T. and Banker, B. Q. (1979). Failure of fetal creatinine phosphokinase as a diagnostic indicator of Duchenne muscular dystrophy. *N. Engl. J. Med.* **300**, 860

55 Emery, A. F. H. and Burt, D. (1980). Intracellular calcium and pathogenesis and antenatal diagnosis of Duchenne muscular dystrophy. *Br. Med. J.*, **280**, 355

56 Heckmatt, J. Z., Dubowitz, V. and Leeman, S. (1980). Preliminary communication detection of pathological change in dystrophic muscle with B-scan ultrasound imaging. *Lancet*, **1**, 1389

57 Steele, M. W. and Breg, W. R. (1966). Chromosome analysis of human amniotic-fluid cells. *Lancet*, **1**, 383

# 14
# Congenital anomalies: the role of ultrasound

## F. A. MANNING AND I. LANGE

## INTRODUCTION

In recent years advances have been made in intrapartum and antepartum detection of the fetus at risk for intrauterine death or damage. Those advances have resulted in a progressive and dramatic fall in perinatal mortality; for example, as observed in Manitoba, the perinatal mortality rate has fallen from 25 per 1000 in 1976 to a current figure of 14.3 per 1000 in 1979[1]. Major life-threatening congenital anomalies occur in approximately 1 % of all pregnancies[2,3]. The falling trend in perinatal mortality resulting from perinatal asphyxia has caused large relative increases in the contribution of congenital anomalies to perinatal mortality. In some studies, congenital anomalies are a cause of up to 40 % of all perinatal deaths[1]. The highly aggressive nature of modern obstetrics with its associated high induction and high caesarean section rates has undoubtedly been a major factor in the falling perinatal mortality. Obstetricians pay a price for this aggressive approach and nowhere is it more apparent than when maternal morbidity occurs as a result of aggressive treatment for the infant with unrecognized lethal congenital anomalies. Developing in parallel with this new obstetric dilemma are dramatically improved methods for identification and treatment of some major fetal congenital anomalies. In the present day before intervention the obstetrician must first ascertain whether the fetus is normal or not and secondly whether the anomaly, if present, is amenable to therapy. These major obstetric challenges have stimulated intensive research in methods for accurate and reliable antepartum screening of major congenital anomalies. In this chapter, the role of the various modalities of ultrasound in detecting structural and functional anomalies is discussed.

## ULTRASOUND METHODS FOR FETAL EVALUATION

Ultrasound was first introduced in obstetrics by Donald in 1958[4]. Since its introduction, this tool has undergone major development and has become an integral part of modern obstetrics. All diagnostic ultrasound systems operate in a frequency range of 2.25–10 MHz and deliver extremely low energy levels to tissues ($<$ 10 milliwatts per cm$^2$). At this time there is no evidence that diagnostic ultrasound has any deleterious effects on human tissues. Therefore, this method appears to be an ideal way to obtain information about fetal structural and functional integrity.

All diagnostic ultrasound instruments produce high frequency sound by piezoelectric excitation of synthetic crystals. Ultrasound systems are classified by the electronic method by which the returning echoes are analysed and displayed (Table 14.1). B-mode ultrasound is the most common form of

**Table 14.1  Ultrasound methods for detecting congenital anomalies *in utero***

| Method of ultrasound | Application |
|---|---|
| 1  *B-Mode ultrasound* | |
| (a) Static (compound) | Structural assessment of fetal organ systems |
| (b) Dynamic (real time) | Functional and structural assessment |
| 2  *Doppler* | |
| (a) Non-focussed | Detection of fetal heart rate and arrythymias. Detection of abnormal biophysical coupling |
| (b) Pulsed focussed | Evaluation of specific target structure function. E.g., mitral valve motion; detection of electromechanical coupling in fetal heart |
| 3  *Time/motion mode* | Evaluation of target motion e.g., recording fetal cardiac structure motion |

ultrasound imaging in obstetrics. In this system the returning echoes are detected and converted to an electrical signal that is proportionate in strength to the amplitude of the returning echo. This information is then converted to a spot on an oscilloscope, the intensity of which varies with the amplitude of the returning echo. By using a storage oscilloscope, a servomechanical linkage and a transducer it is possible to create an ultrasound image of a given target – for example, the fetus. This method of B-mode ultrasound display, called *static* or compound B-mode ultrasound scanning, is the method by which most fetal anomalies have been detected. Recently, dynamic B-mode ultrasound methods have become available. These methods create a *dynamic* ultrasound image by firing a series of transducers arranged in a linear pattern (linear array) or by mechanically or electronically rotating a single transducer through a prescribed arc (sector scanner). These systems, termed real time B-mode ultrasound, represent a huge advance in fetal assessment since they permit both structural and functional evaluation of fetal tissues. In extrauterine medicine, diagnoses are established by observing both structure and function of an organ system. The introduction of real time ultrasound now permits a similar form of examination to be conducted in the fetus *in utero*. Secondly, real time ultrasound systems are relatively inexpensive, mobile and reasonably simple to operate. These characteristics have begun to create an environment in which screening of all pregnancies for fetal congenital anomalies becomes a practical consideration. Most of the examples of ultrasound detection of fetal congenital anomalies described in this chapter were identified using a real time ultrasound method.

Two other forms of ultrasound play minor roles in detection of fetal anomalies. Doppler ultrasound detects the frequency of returning echoes and the modification of these frequencies by moving structures within the sound beam. The frequency change produced by moving structures are converted to a proportioned electrical signal which can be used to drive a strip chart recorder. Doppler ultrasound is primarily used in obstetrics to record fetal heart rate and rhythm. Time–motion mode (T/M mode) ultrasound utilizes light sensitive paper moved rapidly across a screen displaying returning ultrasound echoes in a B-mode context. This method provides information on the frequency and dynamic range of a given moving structure (e.g. cardiac valve motion).

## SYSTEMATIC CLASSIFICATION OF DETECTABLE STRUCTURAL AND FUNCTIONAL ANOMALIES

In this centre, we have applied a systematic approach to fetal ultrasound examination for detection of congenital anomalies. This method is derived from classic methods for extrauterine physical examination. Firstly, the fetal lie and attitude are determined. The fetus is then examined systematically with specific reference to major anatomical grouping (e.g. head and neck) and system function. This approach is important to the function of all ultrasound screening units since it ensures that in each patient examined for whatever indication, a search for major anomalies is undertaken. Examples of congenital anomalies that may be reliably detected by ultrasound are listed in this chapter by system rather than by frequency of occurrence.

### Central nervous system anomalies

*Hydrocephalus*
In our experience the diagnosis of hydrocephalus and its variants (hydrancephaly, porencephaly) is best diagnosed by assessment of relative head size to abdominal/thoracic diameter and by careful evaluation of intracranial structure. In some centres a biparietal diameter (BPD) of > 10.5 cm is reported as diagnostic of hydrocephalus. We have observed a fetus at term with a BPD of 10.8 cm who was normal at delivery and demonstrated no evidence of hydrocephalus at follow-up. The ratio of head circumference to abdominal circumference varies with gestational age but usually by not greater than 1 cm. In our experience a fetal head circumference that is 2 cm or more than abdominal circumference is highly suggestive of hydrocephalus. The definitive diagnosis of this anomaly requires careful evaluation of intracranial structure. Alteration of intracranial architecture with hydrocephalus varies with the site of occlusion and the degree and duration of the obstruction. With early hydrocephalus dilatation of the anterior horn of the lateral ventricles and the third ventricle appears. In the normal fetus and newborn the third ventricle is very small and frequently not visualized well (Figure 14.1). Parathalamic vessels are usually visualized easily

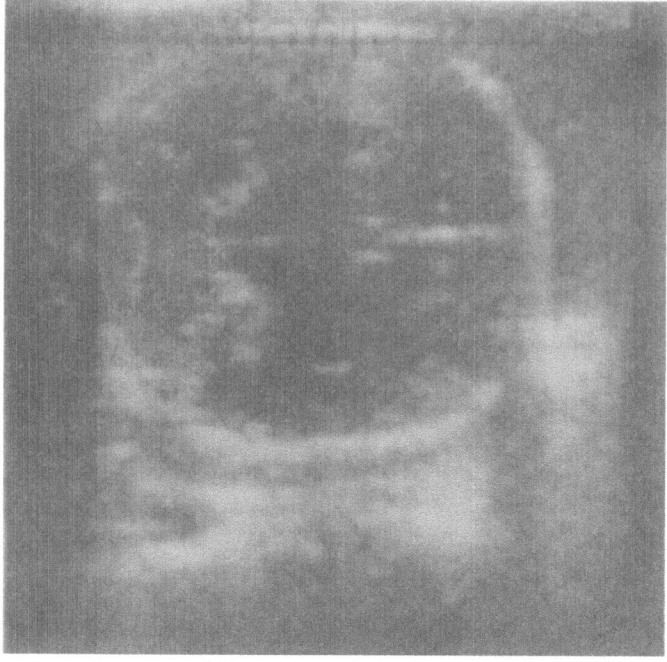

**Figure 14.1** Transverse scan of a normal fetal head at 36 weeks gestation demonstrating normal falx, a portion of the thalamus, and the third ventricle. Note the third ventricle in the normal fetus is slit-like and does not measure more than 5 mm

and course symmetrically along the base of the third ventricle. Displacement of these vessels laterally is an early sign of dilatation of the third ventricle. Care must be taken not to confuse the echolucent thalamus with a dilated third ventricle (Figure 14.2). We have prevented therapeutic abortion in a fetus in whom an erroneous diagnosis of early hydrocephalus was made. Review of the scan demonstrated a normal echogenic thalamus that was diagnosed as a dilated third ventricle. Dilatation of the anterior horn of the lateral ventricle is easily detected. With more severe hydrocephalus, intracranial architecture becomes markedly distorted. The entire intracranial contents become very sonolucent due to fluid distension of the lateral ventricle and midline shifts of the falx may be noted (Figure 14.3). With severe disease the middle cerebral artery becomes shifted laterally often to extreme degrees. With hydrocephalus affecting the fourth ventricle the posterior fossa is well visualized. The cerebellar tonsils are remarkably echogenic and visualization of these structures confirms fourth ventricle malformation (Figure 14.4). Confirmation of fourth ventricle dilatation suggests Arnold–Chiari or Dandy–Walker malformations which carry an extremely poor prognosis.

Prognostication in infants with hydrocephalus in our experience is difficult. Intrauterine estimation of cortical mantle thickness is possible using static B-mode ultrasound, although artifacts may arise because of incorrect gain

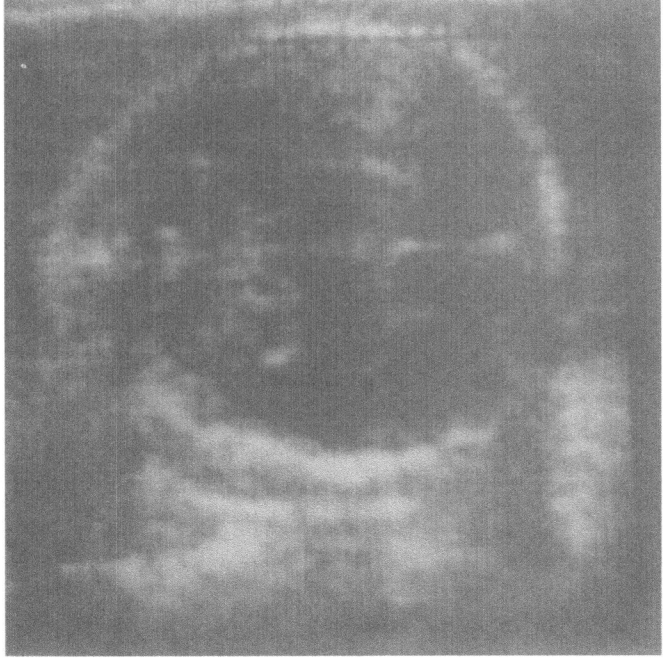

**Figure 14.2** Transverse scan of a normal fetal head at 37 weeks. Note the echolucent thalamus and the lateral wall of the lateral ventricle. The thalamus should not be confused with a dilated third ventricle

settings and reverberation echo. Intrauterine computerized axial tomography (CAT) scanning has been of little additional help in establishing fetal prognosis. In this centre, consultation with neurosurgery is an integral part of decisions regarding obstetrical management of hydrocephalus. It is important to determine the level and aetiology of c.s.f. flow destruction since surgical correction is possible in some instances (e.g. aqueduct stenosis), but hopeless in others (e.g. Dandy–Walker syndrome).

## Anencephaly

The diagnosis of this lethal congenital anomaly can be made with great certainty by ultrasound scanning techniques. We have detected this lesion from as early as 12 weeks gestation and as late as 41 weeks. The characteristic feature of this lesion is the absence of a fetal cranium and normal intracranial architecture. In the first and early second trimester the diagnosis may be difficult because of relatively small fetal size and marked mobility. This problem is easily overcome if the observer identifies the fetal thorax (fetal heart) and spine, then rotates the transducer to produce a longitudinal image of the fetal spine. The spine is then followed cranially to the region of the fetal head. When the scan is done in this fashion, the inability to observe the fetal head is evidence of anencephaly. As in all instances of screening for congenital

I

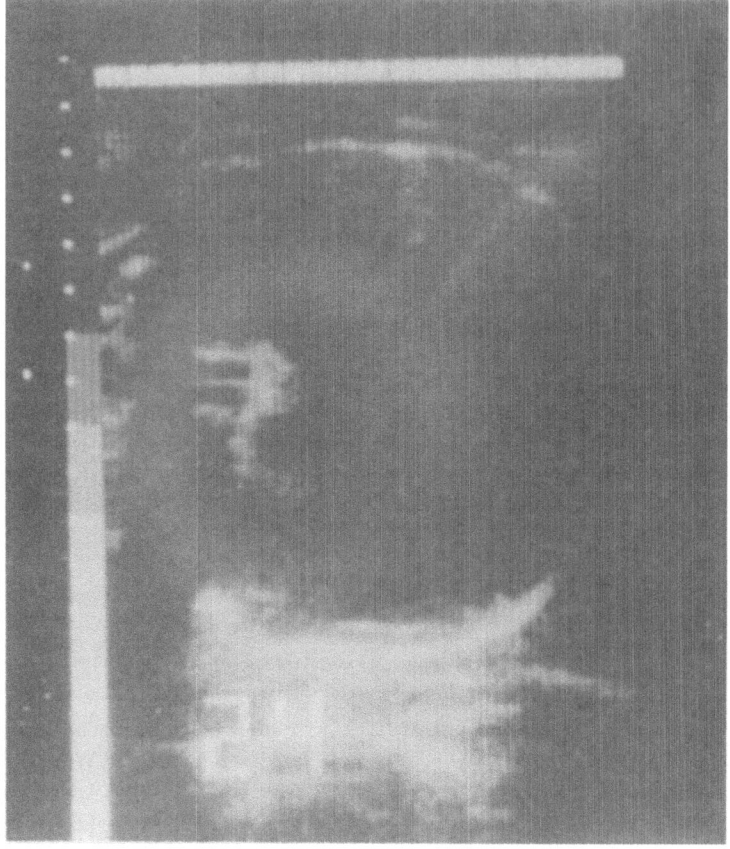

**Figure 14.3** Grossly abnormal fetal head in transverse plane at 35 weeks gestation. The fetal BPD measures 10.8 cm and no normal intracranial structures can be identified. The upper portion of the fourth ventricle is visualized and grossly dilated

anomalies, the skill and experience of the scanner are critical to the accuracy of the observation. Inexperienced observers are encouraged to obtain a second opinion prior to intervention. A classic example of anencephaly at 34 weeks gestation is shown in Figures 14.5A, 14.5B. Note the associated meningomyelocoele with the 34-week fetus. The latter findings occur in up to 15% of fetuses with anencephaly and can occasionally be confused with a fetal cranium.

### Spinal neural tube defects
Ultrasound detection of variants of spinal neural tube defects remains a very difficult area and is best done with combined real time and static ultrasound. Techniques for detecting these defects depend upon painstakingly careful longitudinal and transverse scanning of the entire fetal spine. This technique is laborious and time consuming and even in the hands of experts carries a

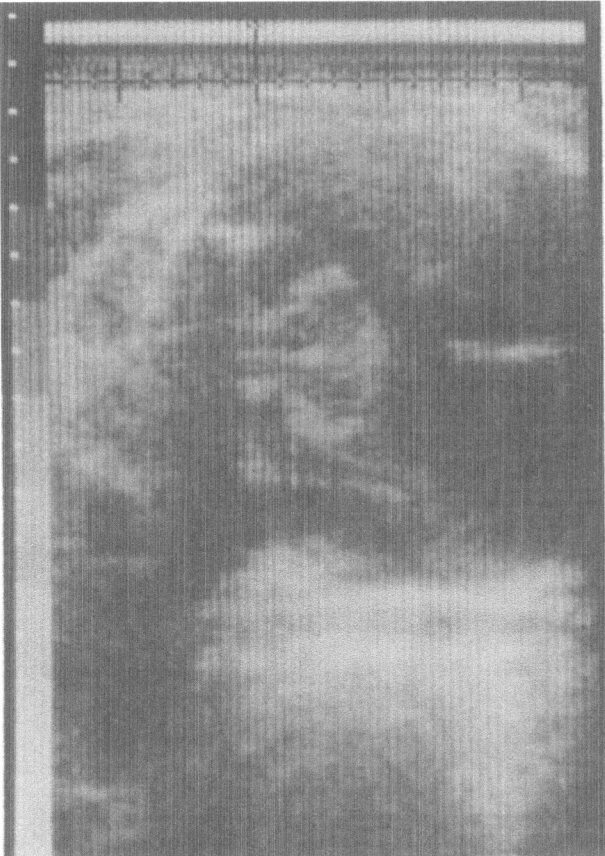

**Figure 14.4** Transverse scan of the fetal head at 34 weeks gestation. The lateral ventricles are grossly dilated and the dilatation of the temporal horn helps visualize the posterior fossa. The obese bulbous structures are the echogenic cerebellum. This fetus had Dandy–Walker syndrome

significant false negative rate ($\sim 20\%$)[5]. Specific lesions that may be detected include bony defects of the dorsal spine and herniation of the dura and spinal cord (Figure 14.6). In infants with meningomyelocoele or meningocoele the fluid filled membranes can be detected by gently rocking the fetus and observing membrane undulation. This phenomenon looks highly similar to movement of the amniotic membrane in diamniotic twins. Oligohydramnios is frequently seen in later gestation in these infants and precludes application of this technique. Careful evaluation of upper cervical and lower lumbar and sacral vertebral bodies is extremely difficult and contributes heavily to the false negative rate. In contrast the false positive rate of detection is extremely low and hence observation of a defect is highly reliable. Estimation of the effect of the lesion on subsequent extrauterine life is difficult. In early gestation ( < 20 weeks) even major degrees of spinal neural tube defects do not seem to

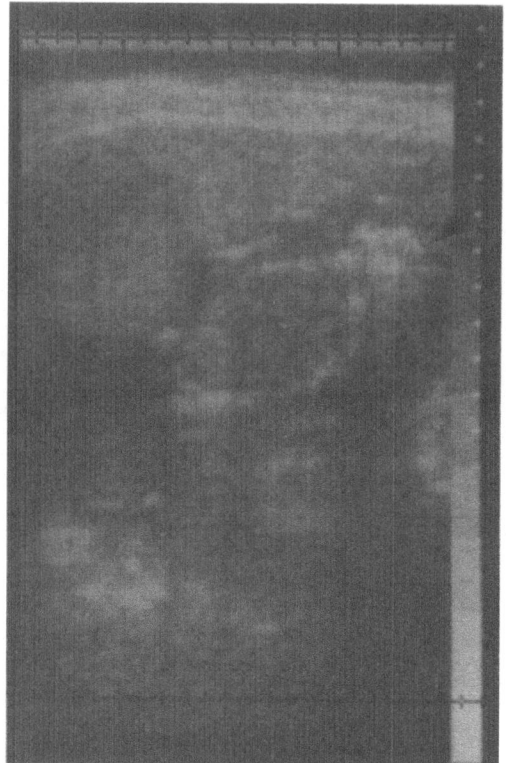

A

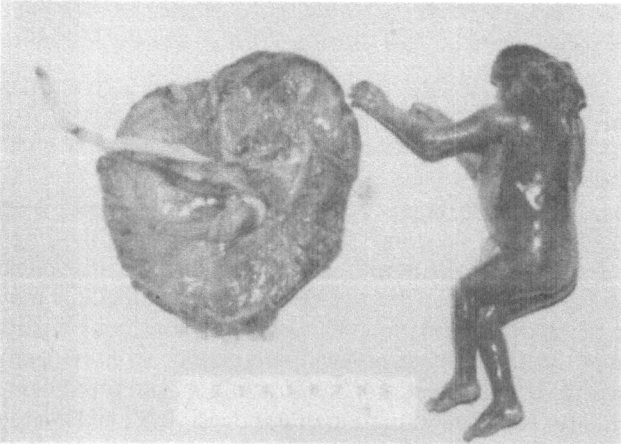

B

**Figure 14.5** A, Longitudinal scan of a fetus with anencephaly. The fetal calvarium is absent and only the base of the skull is noted. B, The same fetus at delivery. Note the associated cervical meningomyelocoele

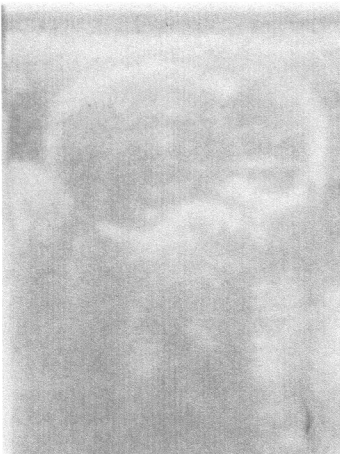

**Figure 14.6** Encephalocoele/cervical meningomyelocoele in a fetus at 27 weeks gestation. This defect was detected as a result of a scan done for fetal biophysical profile scoring and represents an *en passant* finding

affect fetal lower limb movement. We have observed one fetus with severe meningomyelocoele who demonstrated eight lower limb movements in 15 min at 16 weeks gestation. In later gestation ( > 24 weeks) decreased fetal lower limb movements and positioning of limbs in extension or semiextension may be a significant prognostic feature. Also in late pregnancy the fetal bladder emptying *in utero* may be useful in determining the degree of functional defect.

In our experience ultrasound scanning of population at risk for spinal neural tube defects before 20 weeks is of limited value and does not replace maternal plasma and amniotic α-fetoprotein (AFP) determination. We have at least two examples of fetuses in whom elevated AFP confirmed meningomyelocoele at 14 weeks; ultrasound scanning did not detect this lesion until 16 weeks in one fetus and not at all in the second. In contrast, the identification of a spinal neural tube by ultrasound precludes the need for amniocentesis for AFP determination.

## Cervical lesions

Major congenital anomalies in the cervical region are unusual. We have seen one fetus with a large multicystic lesion in the region of the anterior neck producing marked deflexion of the fetal head. This lesion was characterized by the multiple echogenic areas of varying size (Figure 14.7). The absence of flow characteristics within this lesion suggested it was not of venous or arterial origin. Differential diagnosis of such a lesion would include neonatal goitre, haemangioma and cystic hydroma. This fetus died at delivery and examination confirmed the presence of a huge cystic hygroma.

231

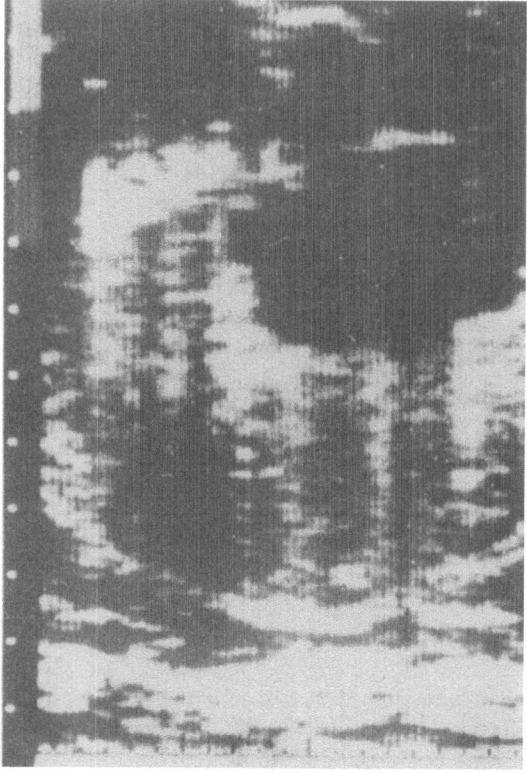

**Figure 14.7** Transverse scan of a mass in the region of the fetal neck. The mass was anterior and multicystic. The fetus died *in utero* and at autopsy had a cystic hygroma with mediastinal extension

## Cardiopulmonary and thoracic anomalies

### Detection of congenital heart disease

Ultrasound detection of major congenital defects of the fetal heart *in utero* is in its infancy but is likely to become a progressively more important aspect of fetal medicine. The key to recognizing abnormalities of the fetal heart lies in an understanding of the normal anatomy and then a systematic evaluation of each chamber of the heart. *In utero* the right ventricle predominates and is larger and thicker than the left ventricle. The relative right ventricular predominance results in clockwise rotation of the heart such that the right ventricle and atrium are anterior. In the normal fetus the maximal heart diameter in the transverse and anteroposterior (AP) plane does not exceed 55 % of the total thoracic diameter.

The right atrium is readily identified by following the inferior vena cava to its origin with the right atrium. In some instances the intra-atrial area may be visualized and the foramen ovale noted. The tricuspid valve may be easily seen once the position of the right atrium is confirmed. Next, the region of the endocardial cushion is visualized and the root of the intraventricular system noted and followed. In dynamic modes it is possible to detect large ventricular septal defects (VSDs) when they involve the muscular portion of the intraventricular septum. In the normal fetus the septum moves towards the left during systole. Paradoxal movement can suggest ductal atresia or stenosis. Both ventricles should be easily identified (Figures 14.8, 14.9). The lethal congenital anomaly, hypoplastic left ventricle syndrome, is identified in this manner. In the normal fetus the aortic and pulmonic roots are easily identified. In the abnormal heart, rotation and dilatation make identification of these areas much more difficult. The aortic arch and thoracic portion are easily visualized. Failure to identify these structures suggest the possibilities of aortic band syndromes and coarctation. Visualization of the ductus arteriosus *in utero* has not been possible at this time.

Cardiomegaly and hyperdynamic chamber motion suggest the presence of either overload resulting from closed loop circulation (e.g. VSD) or obstruction of outflow trait (Figure 14.10). We have observed an example of left atrial hypertrophy and dilatation occurring as a result of fetal left atrial myoma. The presence of any arrhythmia should suggest associated structural defects of the fetal heart particularly when atrioventricular dissociation is observed. This disorder usually presents as fetal bradycardia and can be confirmed by either abdominal ECG recording or by observing a dissociation of atrial and ventricular motion by real time scanning[7].

Fetal heart failure *in utero* can be detected by dynamic ultrasound scanning. Characteristic findings include cardiomegaly, hyperdynamic precordium, hepatic venous pulsations and hepatomegaly. In the absence of Rh isoim-

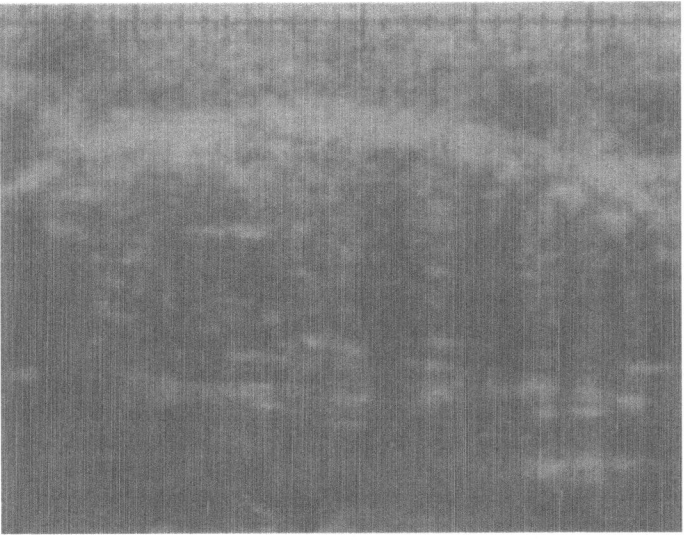

**Figure 14.8** Longitudinal scan of a normal fetus at 34 weeks (sagittal scan). The inferior vena cava and aorta are visualized. The right and left ventricle are visualized

233

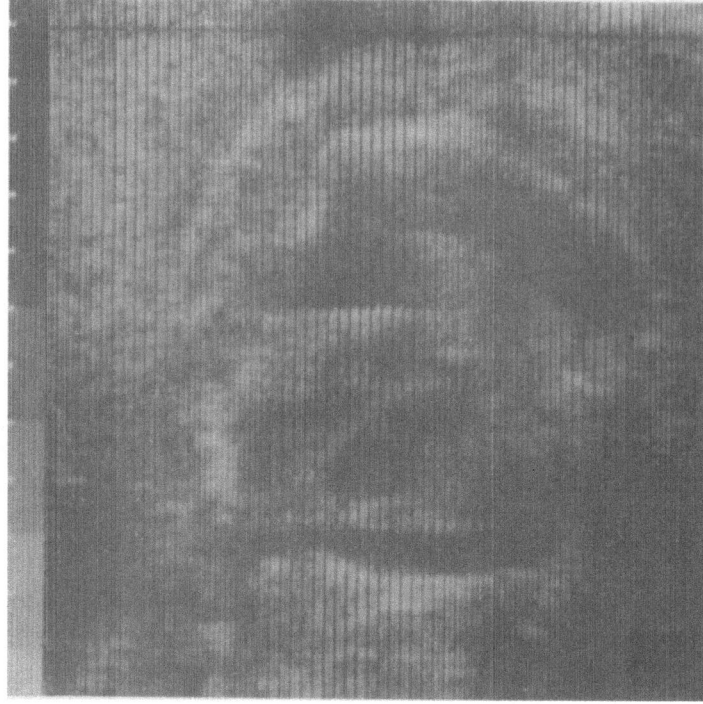

**Figure 14.9** Transverse scan of the fetal heart demonstrating all four cardiac chambers. The right ventricle lies anterior and the endocardial cushion is noted. The heart is moderately enlarged and bilateral pleural effusions are noted

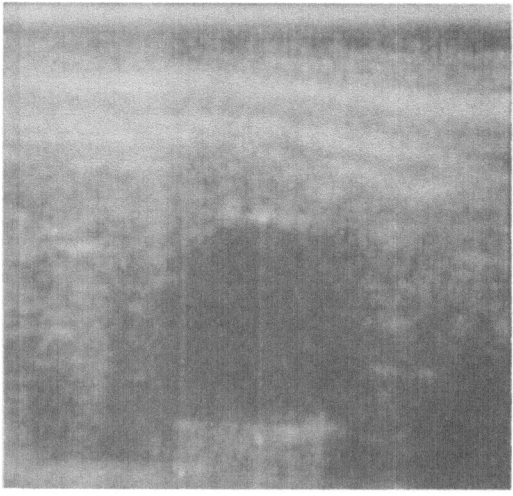

**Figure 14.10** Longitudinal scan of a fetus with gross cardiomegaly at 37 weeks gestation. The cardiac/thoracic ratio in this fetus was 0.77. The infant died within 2 h of delivery and at autopsy pulmonary atresia was noted

munization the observation of fetal heart failure should strongly suggest the presence of a major cardiac anomaly.

The role of direction Doppler ultrasound for determination of electromechanical coupling of the fetal heart, abdominal ECG to evaluate fetal arrhythmia and time–motion ultrasound to evaluate fetal cardiac motion remains to be determined.

## Thoracic cavity

In the normal fetus the thoracic cavity is relatively sonolucent, since the fetal lungs are distended with fluid whereas the heart is blood filled. The normal chest is cone-shaped. In most fetuses the lower limits of the chest cavity are easily identified by recognition of the diaphragm. The latter is seen best on the right side down the dome of the liver. The AP diameter of the lower thorax is similar to the maximal diameter of the upper abdomen. Dynamic ultrasound mode reveals episodic fetal breathing movements in 90 % of fetuses and these are characterized by their symmetrical motion[8]. Abnormalities of shape can be useful in detecting such disorders as asphyxiating thoracic dystrophy, and severe forms of intrauterine flexure disorder such as hemivertebrae and scoliosis. In the normal fetus the lung tissue is not visualized. In two fetuses we have identified bilateral pleural effusion *in utero*[9]. In these instances the equivalent of intrauterine 'atelectasis' is present and the collapsed lung and pulmonic root easily visualized (Figure 14.11A, 14.11B). One of these examples of isolated pleural effusion was secondary to intrauterine chylothorax and the other due to a variant of non-immune hydrops. Both fetuses had immediate bilateral thoracocentesis at delivery and survived. In the normal fetus the hemidiaphragms are at equal level in the transverse plane. Since the left hemidiaphragm is not easily visualized, its position can be accurately determined by extrapolation of the dome of the right hemidiaphragm. This technique allows for accurate assessment of the position of the stomach and gut in relationship to the dome of the left diaphragm, allowing diaphragmatic hernia to be identified. Other characteristics of ultrasound findings with diaphragmatic hernia include presence of multicystic areas in the left chest cavity, right sided mediastinal shift, asymmetric fetal breathing movements and a scaphoid abdomen. Intrauterine detection of this anomaly can modify immediate neonatal care and contribute to improved survival.

## Gastrointestinal tract anomalies

In the normal fetus the abdomen is slightly protuberant such that the maximal diameter (AP) at the level of the liver is within 1 cm of the BPD. The ductus venosus forms a reproducible landmark at which to obtain the maximal diameter and circumference of the fetal abdomen. Normograms are available for these measures[10]. The stomach is easily recognized in the left upper quadrant and in our experience is present in 99.6 % of all observations (Figure 14.12). Failure to observe the stomach fluid bubble on two separate examinations at least 24 h apart is very suggestive evidence of complete oesophageal atresia. We have observed this association on two separate

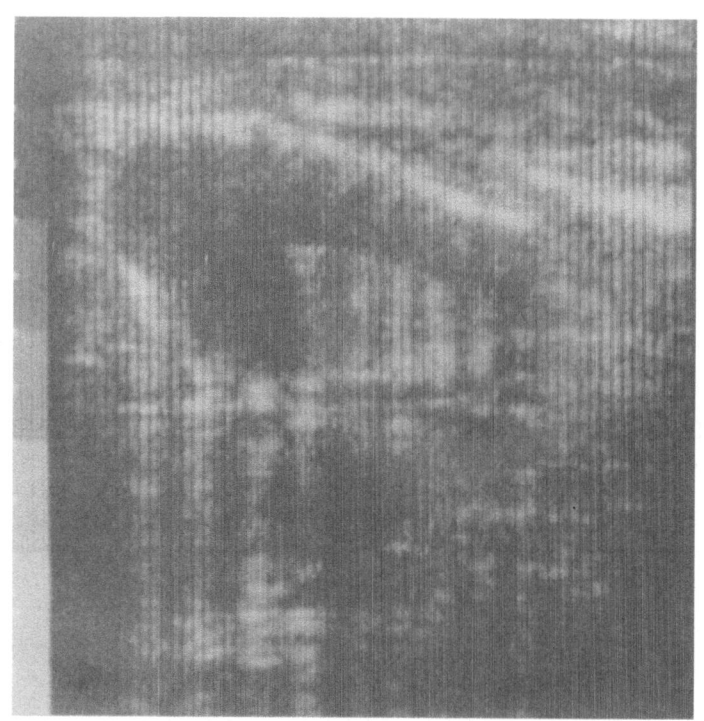

A

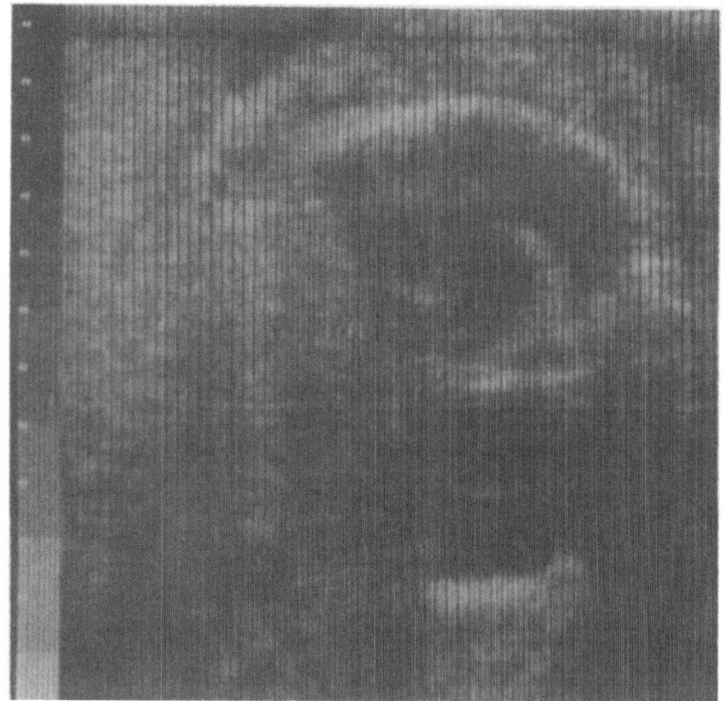

B

236

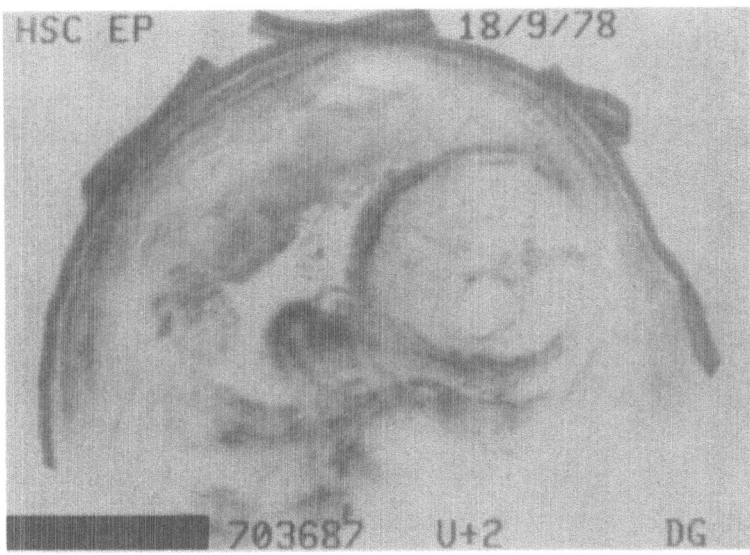

**Figure 14.12** Transverse scan of abdomen of a normal fetus at 34 weeks gestation demonstrating a normal fetal stomach and upper duodenum

occasions. In contrast, a persistently enlarged stomach or the presence of two large echogenic areas in the left upper quadrant suggest an obstructive disorder. This latter sign, referred to as a 'double bubble sign', is highly suggestive of duodenal obstruction secondary either to atresia or to an annular pancreas. In fetuses with this defect, careful cardiac scanning is indicated since disorders are linked in infants with trisomy 21. An isolated enlarged stomach is suggestive of pyloric stenosis and determination of fetal sex (male) can help confirm this diagnosis. All high obstructions of the gastrointestinal tract are usually associated with a degree of hydramnios. Lower gastrointestinal obstruction usually presents with dilated colon particularly in the transverse section. These findings often indicate meconium ileus (Figure 14.13). Hydramnios is not usually associated with lower GI tract obstruction. Interestingly, anal atresia is not associated with any recognizable dilatation of low gut. The latter finding is in keeping with the observation that lower sigmoid colon and rectum are markedly depressed *in utero*.

Gastroschisis and omphalocoele are easily identified and strict attention is paid to the position of gut in relationship to the base of the liver and the normal curve of the anterior abdominal wall. Mesenteric cysts may occur *in utero*, and in clinical practice are rarely differentiated from urinary tract anomalies. Ascites is easily identified (Figure 14.14). Congenital anomalies are

**Figure 14.11A** Longitudinal scan of a 38-week fetus referred for amniocentesis prior to repeat caesarean section. Bilateral pleural effusions are evident and 'atelectasis' of the right lung is noted. Note the depression of the right diaphragm. At delivery 150 cc of fluid was removed from the right pleural space and 90 cc from the left and the newborn survived
**Figure 14.11B** Transverse scan of the same fetus again demonstrating fluid in both pleural cavities. Note the moon-shaped echo in the right lung

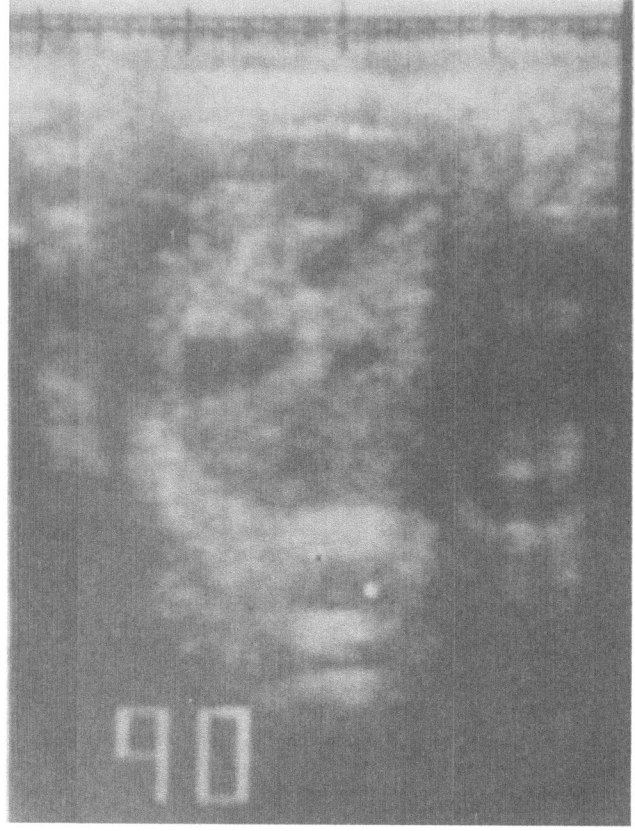

**Figure 14.13** Transverse scan of abdomen of fetus at 39 weeks gestation demonstrating multiple loops of dilated large and small gut secondary to meconium ileus

an unusual cause of ascites, but it may result from cardiac anomalies with subsequent intrauterine heart failure. Rh isoimmunization and non-immune hydrops are much more common causes of ascites.

## Genitourinary tract abnormalities

The kidney, bladder and external genitalia are well visualized using high scanners. Fetal kidneys may be visualized from as early as 16 weeks gestation and are best seen in the transverse plane (Figure 14.15). Once fetal kidneys are identified they may be examined in detail by longitudinal, oblique and transverse scan planes. The presence of normal kidneys rules out renal agenesis syndrome and B.O.R. syndrome. Renal agenesis is a sporadic defect whereas B.O.R. syndrome is autosomal dominant. Renal agenesis diagnosed prior to 20 weeks gestation is an indication for elective termination in this centre. The fetal bladder is observed in most fetuses and, when observed,

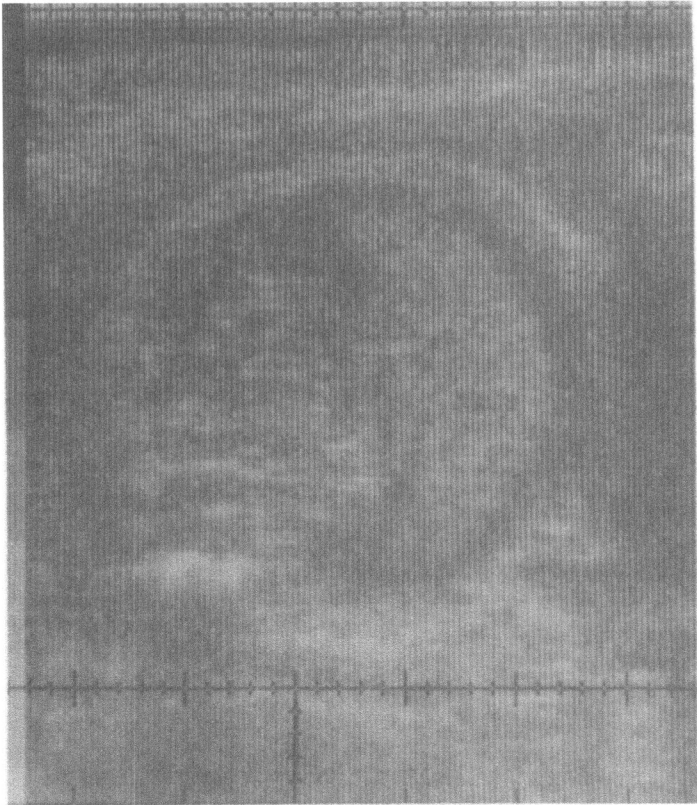

**Figure 14.14** Transverse scan of abdomen in a fetus with early ascites secondary to Rh isoimmunization. Note the well defined right lobe of liver

confirms the presence of functioning renal tissue and a patent collection system. Gross enlargement of the bladder suggests outflow obstruction, for example posterior urethral valve syndrome. Again, determination of fetal sex can be helpful in these instances. Gross distension of the bladder and ureters can present as a cystic mass in the abdomen with distension of the anterior abdominal wall ('prune belly syndrome'). Polycystic (unilateral or bilateral) kidneys may also present as abdominal cystic disease or as an enlarged renal mass.

Fetal sex determination by examination of external genitalia is a very reliable method beyond 32 weeks gestation. With the early real time scanners it was only possible to identify the male fetus with certainty; with the newer dynamic focus high resolution scanner it is now possible to identify female genitalia with the same degree of accuracy (Figure 14.16). Examination involves longitudinal and transverse scanning of the region. The fetus is normally in a position of flexion and visualization of the genitalia can be difficult. However, identification of the general area of the scanning during

**Figure 14.15** Transverse scan of a normal fetus demonstrating kidneys and spine. Note the more echogenic renal medulla and echolucent cortex. The fetal spine and shadowing are evident

fetal movement almost always provides a satisfactory scan. Gently rocking the male fetus while scanning produces a characteristic undulating motion of the scrotum ('jello sign'), which helps differentiate the structure from a folded fetal foot or loop of cord. Identification of the scrotum then permits determination of testicular descent. Recently intrauterine hydrocoele has been reported[11].

The need for fetal sex determination in late pregnancy arises infrequently. However, we have followed two fetuses at risk for X-linked coagulation disorders (haemophiliac) and knowledge that both of these fetuses were

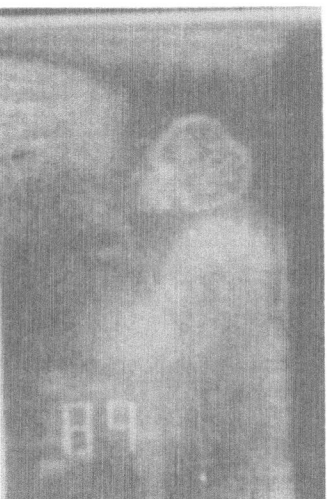

A

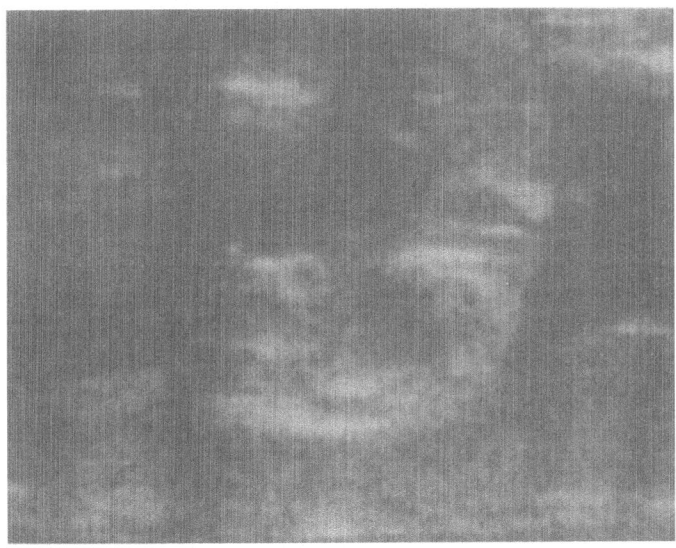

B

**Figure 14.16** A, Scrotum and penis in a normal male at 38 weeks. Note bilateral testes descended. B, Normal female genitalia. The labia majorum and apposed labia minorum are noted

female was reassuring to the parents and physician. In the future, combination of fetal sex determination with specific system observation (e.g. musculoskeletal) may be helpful in determining the risk and prognosis of fetus at risk for sex-chromosome linked disorders (e.g. Duchenne muscular dystrophy).

## Musculoskeletal systems

It is now possible to determine accurately the structural and morphometric characteristics of the fetal skeletal system. Short limb dystrophies can be identified in early gestation, and termination of pregnancy offered. Serial estimates of bone length (e.g. femur) can be useful in identifying fetuses with disorders such as achondroplasia, Ellis–Van Creveld syndrome, osteogenesis imperfecta and thanatophoric dwarfism. We have recently followed a patient whose first child had chondroplasia punctata, an autosomal recessive disorder with associated severe mental retardation. By determining normal femur length and growth in a fetus in her subsequent pregnancy we were able to reassure the mother that this infant did not suffer from the same disorder (Figure 14.17). This field of ultrasound measurement is just beginning to be explored by the innovative work of J. Hobbins and co-workers at Yale University[11]. No doubt as experience and resolving power increase, detection of other disorders such as talipes, abnormal facies, and minor short limb disorders will become possible and practical. In this regard we are investigating calvarium curvatures as a measure of later gestation identification of Down syndrome and other major chromosomal anomalies.

Functional assessment of the fetal musculoskeletal system is also a developing area. Fetal movements are easily visualized using dynamic scanners and attempts are under way to categorize these movements. For example, with a sudden fetal stimulus either arising from fetal self-stimulation, such as striking the face with its hand, or external stimuli, such as a loud noise or sudden palpation, one frequently observes upper limb extension with return to flexion and lower limb sudden extension which may

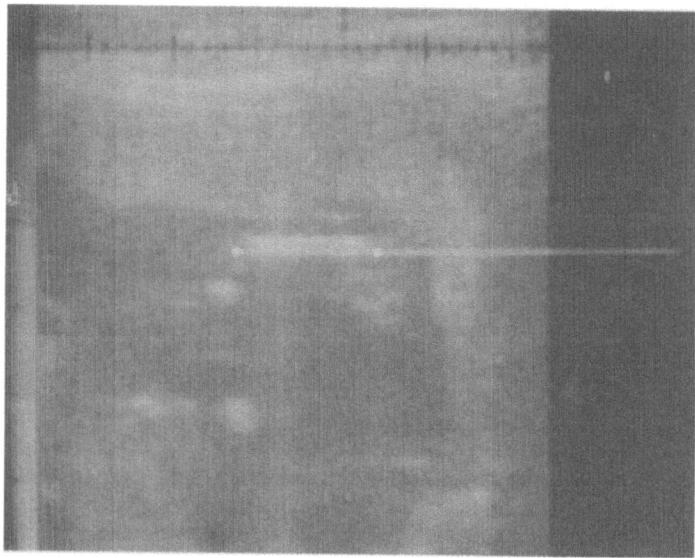

**Figure 14.17** Normal femur in longitudinal scan at 16 weeks in an infant at risk for chondrodysplasia punctata. The femur length is normal as was the fetus at delivery

be sustained for several seconds before return to flexion. These movements almost certainly reflect a variant of an intrauterine Moro reflex. Often reflex activity is likely to be described, allowing the observer to perform a passive neurological examination *in utero*. Such observations may further define the functional integrity of the neurological and musculoskeletal systems. At present, application of these principles is in an early state; however, gross examples of abnormal musculoskeletal function such as intrauterine seizure disorder have been observed[12]. We have followed fetuses at risk for myotonia congenita and Wernig–Hoffman disease and by observing normal fetal movements, and in particular normal fetal tone, were able to offer the parents a high degree of probability that their fetuses were normal.

## ASSOCIATED ULTRASOUND FINDINGS

In addition to structural and functional assessment of the fetus, observation of its environment can also provide clues to the presence of an anomaly. Abnormalities of amniotic fluid volume are frequently associated with congenital disorders. Oligohydramnios (largest pocket < 1 cm in two perpendicular planes) is always observed in fetuses with renal agenesis (Potter's syndrome, B.O.R. syndrome) and may be seen with severe cardiac defects (atrial myxoma) and omphalocoele. Hydramnios (>8 cm pocket of fluid) is associated with major anomalies in about 20 % of instances[2]. Defects in fetal swallowing or obstruction of the upper gastrointestinal tract (e.g. oesophageal atresia) are almost always associated with hydramnios. A complete search for anomalies is indicated whenever abnormalities of amniotic fluid volume are noted.

The incidence of congenital anomalies is double or trebled in infants with single umbilical arteries. In late gestation (32 weeks and beyond) the umbilical cord is well visualized and arterial position can be recognized. Determination of the number of umbilical arteries is a routine part of our screening examination for anomalies (Figure 14.18).

## CLINICAL SIGNIFICANCE OF ULTRASOUND
## CONGENITAL ANOMALY SCREENING

Early detection of major congenital anomalies (<20 weeks) allows the parents to consider termination of pregnancy. Ultrasound is helpful in detecting anomalies such as anencephaly, skeletal dystrophies, renal agenesis, meningomyelocoele and major gut disorders (gastroschisis) in the first trimester. Unfortunately, most major anomalies are not detected until beyond 20 weeks gestation and therefore termination is not a consideration. Nonetheless, it is extremely important to continue to search for these anomalies since knowledge of their presence can have a profound effect on the subsequent management of the pregnancy. In the case of some disorders, recognition and prompt treatment in the neonatal period can be lifesaving. For example, identification of diaphragmatic hernia or pleural effusion undoubtedly adds to neonatal survival. Other disorders, for example congenital heart block, are compatible with normal intrauterine life and

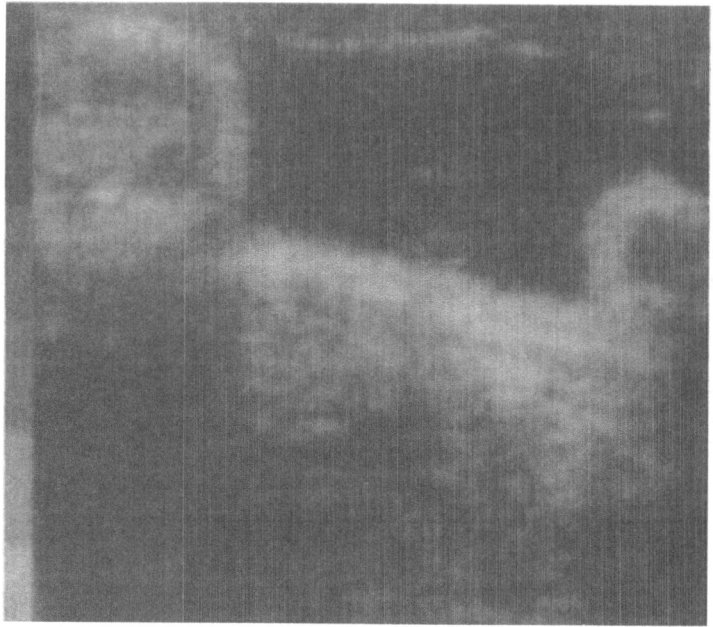

**Figure 14.18** Transverse scan of umbilical cord demonstrating two arteries and a single vein

recognition of the aetiology of the associated bradycardia can prevent needless emergency delivery for 'fetal distress'. At the other extreme, in fetuses with anomalies incompatible with extrauterine life (e.g. anencephaly, hypoplastic left ventricle syndrome) unnecessary caesarean section can be avoided. The latter is particularly important since the incidence of abnormal antepartum tests is high in abnormal fetuses. In our experience a major congenital anomaly is observed in one third of all abnormal contraction stress tests. Without ultrasound scanning in these patients many would be delivered by caesarean section of a fetus with a hopeless prognosis. In the past year at one major teaching hospital (Health Sciences Centre) at the University of Manitoba, at least three fetuses were delivered by caesarean section with gross anomalies. In two of these fetuses unrecognized anencephaly was present and the third fetus had gastroschisis and encephalocoele. All could have been easily detected by prior ultrasound scanning. As a result of this experience we recommend an ultrasound scan prior to delivery in the following circumstances:

(1) Premature labour and premature rupture of the membranes.
(2) Breech presentation at any gestation.
(3) Abnormalities of fetal growth.
(4) Suspect oligohydramnios and hydramnios.
(5) Multiple gestation.
(6) Diabetics, advanced maternal age (< 35), previous abnormal infants.

## SHOULD EVERY PREGNANT PATIENT HAVE AN ULTRASOUND SCAN?

The value of ultrasound in identifying structural and functional anomalies of the developing fetus is established. At the time of writing there are no known fetal complications associated with diagnostic ultrasound exposure. As systems for identifying fetal compromise in the antepartum and intrapartum periods improve and as neonatal intensive care improves, the relative contribution of major anomalies to perinatal mortality is rising. An aggressive approach to management of the fetus with distress has led to higher caesarean section rates and as a result to an increase in the number of abnormal fetuses delivered by caesarean section. When this circumstance arises it is a tragedy, since unnecessary maternal morbidity and even mortality and a compromised reproductive career are caused for a fetus with a hopeless prognosis. An urgent question in modern obstetrics is whether the benefits of routine ultrasound scanning in all pregnancies outweigh the economic and psychological costs as well as the unknown risk of ultrasound exposure *in utero*. In an attempt to address this issue we have reviewed our experience in the fetal assessment units at the University of Manitoba. In a 6 month period as part of a fetal assessment programme we obtained ultrasound scans using a linear array real time ultrasound method on 1400 consecutive referred high risk patients. All were 26 weeks or greater at the time of first examination and in all the primary indication for referral was for fetal biophysical profile scoring to detect fetal compromise[13]. Overall, 25 of these 1400 fetuses had major congenital anomalies at delivery (1.79%) (Table 14.2). Eighteen of these anomalies were recognized during fetal biophysical profile assessment (72%) (Table 14.2). In 12 of these examples an anomaly incompatible with fetal life was recognized, preventing intervention (48%). Three of these 12 (25%) fetuses developed overt evidence of fetal distress in the ante- or intrapartum period and death occurred. Although it is difficult to prove, we feel confident that if the obstetricians were not forewarned in these cases intervention would have occurred. In comparison, three of the nine undetected anomalies were delivered by caesarean section for fetal distress (33.3%). In all of the detected anomalies considered compatible with extrauterine life the neonatologists were forewarned and in at least two of these fetuses (both with pleural effusion) detection was likely to have been lifesaving[13].

Thus, observations suggest that, at least in an obstetrical high population, routine ultrasound scanning in late gestation for detecting of congenital anomalies may be justified. Controlled studies in earlier pregnancy are under way in several centres to further define the value of routine scanning and the appropriate time for the examinations. It seems likely that dynamic ultrasound will play a dual role in most pregnancies, that is to detect major congenital anomalies as well as to estimate fetal risk for asphyxia.

## CONCLUSION

The tremendous and continuing development of ultrasound imaging systems have and will continue to have a profound effect on antepartum detection of

**Table 14.2** Incidence of congenital anomalies and antenatal detection rate* in 1400 patients referred for antepartum fetal risk assessment

| Type of anomaly | No. at delivery | No. detected | % detection rate |
|---|---|---|---|
| 1 *Central nervous system anomalies* | | | |
| Anencephaly | 4 | 4 | 100 |
| Hydrocephalus | 4 | 3 | 75 |
| Meningomyelocoele | 1 | 1 | 100 |
| 2 *Cardiothoracic anomalies* | | | |
| VSD | 2 | 1 | 50 |
| Cardiomegaly (pulmonary stenosis) | 1 | 1 | 100 |
| Pleural effusion | 2 | 2 | 100 |
| 3 *Gastrointestinal anomalies* | | | |
| Oesophageal atresia | 2 | 1 | 50 |
| Duodenal atresia | 1 | 1 | 100 |
| Omphalocoele | 1 | 0 | 0 |
| 4 *Genitourinary anomalies* | | | |
| Renal agenesis | 2 | 2 | 100 |
| Renal cyst (unilateral) | 1 | 1 | 100 |
| Posterior urethral valve | 1 | 1 | 100 |
| 5 *Others* | | | |
| Down's syndrome | 1 | 0 | 0 |
| E-trisomy | 1 | 0 | 0 |
| Congenital ichthyosis | 1 | 0 | 0 |
| *Total* | 25 (1.79%) | 18 | 72 |

\* *en passant* detection rate

major congenital anomalies. Modalities such as dynamic ultrasound imaging systems now permit assessment of both structural and/or functional anomalies. This list of detectable anomalies is increasing rapidly and the result of these observations is improved obstetrical care from both maternal and fetal viewpoints. At the present time, delivery of a structurally abnormal fetus by caesarean section is an all too frequent occurrence. In the future it is likely that expansion of ultrasound scanning facilities will make the tragic event a rarity. It is also probable that more fetuses with correctable defects will be salvaged as a result of intrauterine detection. Newer ultrasound systems permit the equivalent of an intrauterine physical examination of the fetus. It seems likely that in the not too distant future such an examination will become an integral part of the management of all high risk pregnancies. Finally, ultrasound scanning for detection of congenital anomalies is not a field for the novice ultrasonographer. In this centre, we have established a team approach and a concentrated experience in anomaly screening. We include specialists in areas of specific organ system (e.g. neurosurgery for hydrocephalus) in all decisions concerning estimation of fetal prognosis and management. We would caution against irrevocable decisions (e.g. abortion) in instances of suspected anomaly without consultation with an experienced ultrasonographer.

# References

1 Morrison, I. (1976–79). *Annual Perinatal Mortality Report*. (Winnipeg: College of Physicians and Surgeons of Manitoba, Canada)
2 Kramer, E. E. (1966). Hydramnios, oligohydramnios and fetal malformations. *Clin. Obstet. Gynecol.*, **9**, 508
3 Kucera, M. (1971). Rate and type of congenital anomalies among offspring of diabetic women. *J. Reprod. Med.*, **7**, 61
4 Donald, I., MacVicar, J. and Brown, T. G. (1958). Investigation of abdominal masses by pulsed ultrasound. *Lancet*, **1**, 1188
5 Little, D. J. and Campbell, S. (1980). The diagnosis of spina bifida and intracranial anomalies. In Sanders, R. and James, A. F. (eds.) *The Principles and Practice of Ultrasonography in Obstetrics and Gynaecology*, pp. 179 ff. (New York: Appleton-Century-Croft)
6 Platt, L. D., Manning, F. A., Gray, C., Guttenburg, M. and Turkel, S. B. (1979). Antenatal detection of fetal A-V dissociation utilizing real time B-mode ultrasound. *Obstet. Gynecol.*, **53**, 59s
7 Platt, L. D., Manning, F. A. and Sipos, L. (1978). Human fetal breathing movements in high risk pregnancies. *Am. J. Obstet. Gynecol.*, **132**, 514
8 Lange, I., Manning, F. A., Baskett, T. F. and Morrison, I. (1980). Antepartum diagnosis of fetal pleural effusion. *Am. J. Obstet. Gynecol.* (In press)
9 Kearney, K., Vigneron, N., Frischman, P. and Johnson, J. W. C. (1978). Fetal weight estimation by ultrasonic measurement of abdominal circumference. *Obstet. Gynecol.*, **51**, 156
10 Cooperberg, P. L. (1980). Abnormalities of the fetal genitourinary tract. In Sanders, R. and James, E. (eds.) *The Principles and Practice of Ultrasonography in Obstetrics and Gynaecology*, pp. 161 ff. (New York: Appleton-Century-Croft)
11 Hobbins, J. C., Granum, P. A. T., Berkowitz, R. L., Silverman, R. and Mahoney, M. J. (1979). Ultrasound in the diagnosis of congenital anomalies. *Am. J. Obstet. Gynecol.*, **134**, 331
12 Platt, L. D. and Hohler, C. (1980). Personal communication
13 Manning, F. A., Lange, I., Baskett, T. F. and Morrison, I. (1980). Antepartum fetal biophysical profile testing: a prospective study. *Am. J. Obstet. Gynecol.* (In press)

# Index